Edexcel GCSE

Geography A

Geographical Foundations

Student Book

Andy Palmer • Michael Witherick • Phil Wood • Nigel Yates
Series editor: Nigel Yates

A PEARSON COMPANY

Published by Pearson Education Limited, a company incorporated in England and Wales, having its registered office at Edinburgh Gate, Harlow, Essex, CM20 2JE. Registered company number: 872828

Edexcel is a registered trade mark of Edexcel Limited

First published 2009.

British Library Cataloguing in Publication Data
A catalogue record for this book is available from the British Library.

ISBN 978 1 846905 00 1

Edited by Graham Bradbury
Original illustrations © Oxford Designers and Illustrators 2009
Illustrated by Oxford Designers and Illustrators
Picture research by Louise Edgeworth
Printed and bound in Great Britain at Scotprint, Haddington

Acknowledgements
The authors and publisher would like to thank the following individuals and organisations for permission to reproduce photographs:

(Key: b-bottom; c-centre; l-left; r-right; t-top)

Alamy Images: 67photo 217r; Arco Images GmbH 255; Wendy Connett 250t; Ashley Cooper 227; dk 229; dmark 74r; Patrick Eden 4b, 149; Julio Etchart 79; Leslie Garland Picture Library 91r; Mark Goble 166; Melvin Green 68; Michael Hatfield 198tr; David Hoffman Photo Library 197; Scott Hortop 196; imagebroker 94, 251, 252r; Images&Stories 7t, 31; ImageState 219l; International Photobank 243; Jesper Jensen 147; Stan Kujawa 122; Barry Lewis 212; Mediacolor's 146; Paul Melling 247r; Gianni Muratore 231; Ian Nellist 93; OS Photography 167; Chris Pancewicz 75; Photofusion Picture Library 219r; Stan Pritchard 92; Robert Read 169; Mark Salter 160; Alistair Scott 145; Skyscan Photolibrary 61; Sheila Smart 7b, 236; The Photolibrary Wales 189; Tom Hanley 178b; Westend 61 GmbH 108; Janine Wiedel Photolibrary 178t; Wilmar Photography 125; Art Directors and TRIP photo Library: Helene Rogers 234; Michael Thornton 139; Bob Turner 245; Chris Wormald 185; Bournemouth Echo: 83; Corbis: 137l; Bill Barksdale 137r; Bettmann 247l; Fridmar Damm/zefa 228; Louise Gubb/CORBIS SABA 250b; Lindsay Hebberd 40br; Peter Hulme/Ecoscene 142; Catherine Karnow 254; David Lomax/Robert Harding World Imagery 194; Gideon Mendel 217l; Richard Morrell 159t; Andy Newman/epa 244t; Roger Ressmeyer 112; Molly Riley/Reuters 13; Skyscan 63l, 192; Hans Strand 55; Liba Taylor 144; Mary Evans Picture Library: 198tl; Geoslides/Geo Aerial Photography: 74l; Getty Images: DigitalGlobe 155; Peter Essick/Aurora 40tr, 138; Dan Kitwood 182; Haywood Magee 224; Majid 110; John Peters/Manchester United via Getty Images 233; Caroline Schiff/The Image Bank 50; Bob Thomas/Stone 244b; Gandee Vasan/Photographer's Choice 5t; Robert Harding World Imagery: JJ Travel 51; R H Productions 161b; Hazel Brow Farm: 180, 180 (inset); iStockphoto: 40bl, 96; Jeremy Edwards 91l; Thomas Shortell 159b; Michael Utech 97l; Jupiter Unlimited: Brand X Pictures 278bl, 281; Goodshoot 278tl; Stockxpert 165, 237, 278r, 288; Lonely Planet Images: Mark Newman 141; Ordnance Survey: Reproduced by permission of Ordnance Survey 2009. All rights reserved. Ordnance Survey Licence Number 17, 20; PA Photos: AP 150; AP Photo/ Saurabh Das 34; Wong Maye-E/AP 213; Photofusion Picture Library: Dorothy Burrows 5b, 259; Photolibrary.com: Digital Vision 4t; Mauritius 97r; John Warburton-Lee Photography 252l; Rex Features: Dean Houlding 161t; Paul Marnef 40tl; Alex Segre 47; SUNSET 248; Dan Tuffs 113; Science Photo Library Ltd: NASA 11; Sunderland arc : 198b; Thieler, E. R., Martin, D., and Ergul, A., 2003. The Digital Shoreline Analysis System, version 2.0: Shoreline change measurement software extension for ArcView. USGS Open-File Report 03-076. Available online at http://woodshole.er.usgs.gov/project-pages/dsas/. : 63r

All other images © Pearson Education
Also see page 304 and the backcover for figures, tables and text.

Every effort has been made to contact copyright holders of material reproduced in this book. Any omissions will be rectified in subsequent printings if notice is given to the publishers.

The websites used in this book were correct and up to date at the time of publication. It is essential for tutors to preview each website before using it in class so as to ensure that the URL is still accurate, relevant and appropriate. We suggest that tutors bookmark useful websites and consider enabling students to access them through the school/college intranet.

Contents: delivering the Edexcel GCSE Geography A Geographical Foundations specification

Welcome to Edexcel GCSE Geography A Geographical Foundations

Why should I choose GCSE Geography?

Because you will:

- learn about and understand the world that you live in
- develop skills that will help you in other subjects and your future career
- get to complete practical work away from the classroom
- learn how to work as a team
- learn by investigating, not just listening and reading

What will I learn?

You only have to switch on the news or pick up a newspaper to see that we live in a fast-pace, ever-changing world. GCSE Geography gives you the chance to learn about those changes: from those on your own doorstep to those of global proportions. There are four units:

Unit 1: Geographical skills and challenges

Have you ever wondered...

- Why we have maps and what they can tell us?
- How to use web mapping sites such as Google Earth?
- How and why our climate is changing?
- How we can protect our planet for a longer life?

In this unit you will get a chance to learn and use a variety of geographical skills, including cartographic (map), ICT and GIS. You will also investigate the major challenges faced by the planet today, a key topic for anyone living in the 21st century. Using all the information that you learn in this section, and throughout the book, you will be able to make your own decision on this subject. This is an introductory unit and the issues and skills raised here are revisited throughout the course.

Unit 2: The natural environment

Have you ever wondered...

- Why coastlines retreat, causing collapse of land and houses into the sea?
- What causes major flooding and what we can do to prevent it?
- How to prevent and survive an avalanche?
- Why people choose to live in volcanic areas?
- How we can reduce our carbon footprint?
- Why some countries have limited water supplies while others have more water than they need?

This unit builds an overall understanding of physical geography around us in the natural world. You will also learn about environmental issues.

Unit 3: The human environment

Have you ever wondered...

- How we use the planet to make a living for ourselves?
- What the appeal of the countryside is?
- Why there are so many derelict factories?
- Why there are so many elderly people in the UK?
- Why people move from one country to another?
- Whether there are any negative effects to tourism?

This unit develops an overall understanding of human geography and the issues affecting the diverse people living on our planet.

Unit 4: Investigating geography

This is the unit where you can really get stuck in! It will involve undertaking research, carrying out fieldwork and then writing it up. The research and fieldwork can be undertaken out of class, but the writing up will all be in class time. This means you have to spend less time at home doing your geography coursework!

How will I be assessed?

The great thing about the course is that the three 1-hour exams can be spread over your two year GCSE course. For unit 4, you will need to write up your fieldwork task in the classroom under controlled conditions while supervised by your teacher.

- Higher and Foundation examination papers are available.
- Units 1, 2 and 3 exams are resource based. You will have a booklet containing maps, photographs and diagrams to help you answer the questions.
- Units 1, 2 and 3 exam questions will range from short questions to larger extended-writing questions.
- Unit 4 is the controlled assessment unit. You will complete a fieldwork task, and analyse and write up your results in class.

About this book

Objectives provide a **clear overview** of what you will learn in the section. Objectives increase in difficulty from ● to ◉

Skills Builder exercises will **develop your geographical skills** and understanding in a specific topic.

ResultsPlus features combine real exam performance data with examiner insight to give **guidance on how to achieve better results**.

Clear and accessible diagrams **highlight key concepts and enable skills practice**.

Chapter 2 Challenges for the planet

Objectives

- ● Understand that the Earth's climate has changed a great deal.
- ◉ Describe the reasons for those changes, both human and natural.
- ◉ Explain the effects of climate change and how we might respond.

Skills Builder 1

Study Figure 1.

Describe the trend in global temperature:

(a) over the past 10,000 years

(b) over the past 1,000 years.

ResultsPlus

Build Better Answers

Describe how changes in the orbital geometry of the Earth might cause the climate to change. (4 marks)

■ **Basic answers** (0–1 marks)
State that the Earth varies in its path around the sun but do not relate to climate change.

● **Good answers** (2 marks)
Describe one reason why the orbital changes affect the amount of solar radiation received – usually variations in the path of the orbit and less solar radiation because the Earth is further away on average.

▲ **Excellent answers** (3–4 marks)
Offer two descriptions, orbit and axis perhaps, and/or have good detail about the cycles of these variations.

The causes, effects and responses to climate change

How and why climate has changed since the last ice age

There is a large amount of evidence that the climate of the Earth is very variable. For most of its history the Earth has been a good deal warmer than it is today (see Figure 1). The two large ice sheets that we have today – one in Antarctica and one in Greenland – have not always been there, and are therefore quite unusual.

Figure 1 shows the global temperature change over the last 10,000 years and, in more detail, the changes over the past 1,000 years, including a prediction into the future. It is clear that the climate has changed a great deal in the past 10,000 years. Human beings were in existence throughout this time, but most experts agree that their impact on the world around them was quite modest. As a species human beings might have been responsible for hunting a number of other species to extinction, such as the mammoths, and they cleared substantial areas of forest. But it wasn't until the nineteenth century that numbers really began to grow and, as they industrialised, they began to contribute to **greenhouse gas** emissions. So **climate change** in the past needs to be explained by different factors – probably relating to the Earth's orbit, the output of the sun, or catastrophic events in the past.

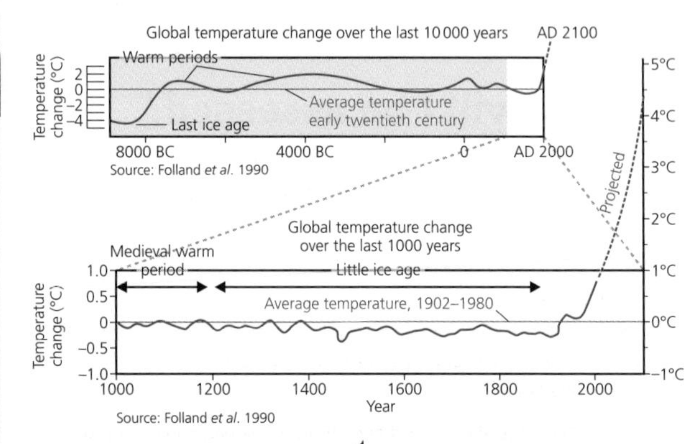

Figure 1: Global temperature changes. The top graph shows global temperature changes over the last 10,000 years. The bottom graph is a more detailed version of the white section in the top graph. It shows global temperature changes over the last 1,000 years.

Changes in the Earth's orbital geometry

The Earth orbits the sun on a slightly variable orbit. It also 'wobbles' on its axis a little, and the angle of tilt of the axis also varies. These three factors can affect the amount of energy received by the Earth by as much as plus or minus one per cent, which is enough to affect global temperatures. These variations are known as the **Milankovitch mechanism** and are thought to be the main reason for the series of **ice ages** experienced in the distant past.

Changes in solar output

We used to assume that the sun was 'constant' in its output, but we now understand that the energy it transmits does change although, again, by quite small amounts. However, although the changes aren't great, they have a large impact on global temperatures. These factors might act together to either reduce or increase global temperatures. Of course, they might also cancel each other out.

Catastrophic events

Volcanic activity and collisions between the Earth and large extraterrestrial objects such as meteors or even comets result in large quantities of material being ejected into the atmosphere. There is a lot of evidence linking dramatic events such as these with changes in the Earth's temperature. Some of these catastrophic events have been thought to be responsible for major extinctions such as that of the dinosaurs at the end of the Cretaceous period.

The Laki volcanic eruption in 1783

In 1783, millions of tonnes of poisonous gas and particles were sprayed over much of Iceland, as 27 km of volcanic vents poured out liquid rock for eight months. This 'fissure eruption', which split the Laki mountain, produced the world's largest lava flow for a thousand years –13 km^3 of material spread out over 500 km^2 of southern Iceland. The smaller particles combined with water vapour, producing a fog that was so dense that local people didn't see the sun for weeks. The eruption had two effects:

- In Iceland itself, it killed off the vegetation, which in turn led to the death of many of the island's animals, through starvation. Because they relied on meat to survive, a third of the island's population eventually died in the famine.
- The impact was also felt a long way from Iceland – in much of Europe and further afield – as the particles fired into the atmosphere by the eruption blocked incoming solar radiation and reduced global temperatures.

Sudden events such as these can change history. The bad harvests in western Europe that followed the Laki eruption led to widespread suffering and even starvation. This in turn led to an increase in civil unrest, which some historians think contributed to the French Revolution that began in 1789.

Activity 1

1. State two ways in which the amount of solar energy received by the Earth can vary.
2. Outline two ways in which global temperature is affected by sudden events beyond the control of humans.
3. Research two events that have been thought responsible for major changes in the Earth's climate.

Figure 2: Laki volcano cones after the eruption in 1783

Quick notes (Laki volcanic eruption):

- Climate changes can be triggered by sudden events.
- Sudden events are usually unexpecte
- The impac can spreac place at w event occu

Case Study: The 'best place' in the UK to retire to

The magazine, *Yours*, has ranked retirement destinations according to a range of factors – house prices, council tax levels, shopping facilities, crime rates, hospital waiting times, the availability of NHS dentists and the weather. A seaside resort famous for having the world's longest pier has been named as the best retirement place in the UK. Southend-on-Sea in Essex is described as a 'bargain' retirement location (Figure 19). It was ranked top because it is relatively flat, with a pedestrianised centre, a low violent crime rate and a council tax that is almost £100 lower than the UK average. It has 10 km of award-wining beaches, more than eighty parks and lots of activities for older people. It is also close to London. Southend lies within the 'Outer South East' and contributes to the big migration flow shown in Figure 2 (on page 223).

Poole in Dorset was placed second, because of its natural harbour and waiting times for hip operations there are lower than the national average. Whitehaven in Cumbria was third, with below-average waiting times for a hip replacement and house prices of nearly half the national average. Other places to make it into the top ten include Swansea in Wales, Clacton-on-Sea in Essex, Stirling in Scotland, Leamington Spa in Warwickshire, Skegness in Lincolnshire, Weymouth in Dorset, and Southport on Merseyside.

Figure 18: Southend-on-Sea – the top place to retire to

Case study quick notes:
This case study illustrates the factors that create the 'pull' of a retirement location.

Key terms are highlighted in the text, summarised at the end of each chapter, and are detailed in full in the glossary at the end of the book to enable you to **develop your geographical language**.

Activities provide extra **support** to ensure understanding and opportunities to **stretch** your knowledge.

Engaging photos **bring geography to life**.

Quick notes pull out the key information in examples and case studies for **quick revision** reference.

Real-life case studies to show the **theory in practice**! Each case study includes a set of quick notes that pull out the key points.

Throughout this student book you will find a dedicated suite of revision resources for **complete exam success.**

We've broken down the six stages of revision to ensure that you are prepared every step of the way.

Zone in: How to get into the perfect 'zone' for revision.

Planning zone: Tips and advice on how to effectively plan your revision.

Know zone: All the facts you need to know and exam-style practice at the end of every chapter.

Chapter overview: Provides a summary of **the key issue** that the chapter examines. Keep this issue in mind as you work through the Know Zone pages.

Key terms: A matching exercise to ensure that you can **understand and apply geographical terminology**.

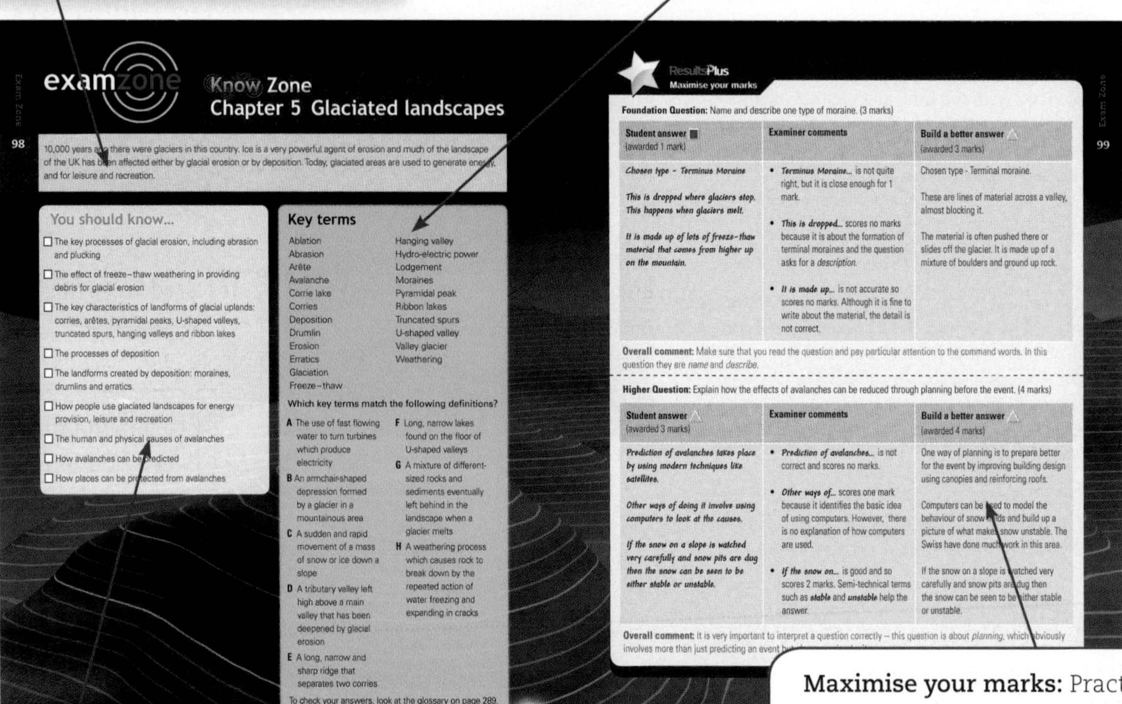

examzone Know Zone
Chapter 5 Glaciated landscapes

98

10,000 years ago there were glaciers in this country. Ice is a very powerful agent of erosion and much of the landscape of the UK has been affected either by glacial erosion or by deposition. Today, glaciated areas are used to generate energy, and for leisure and recreation.

You should know...

- ☐ The key processes of glacial erosion, including abrasion and plucking
- ☐ The effect of freeze–thaw weathering in providing debris for glacial erosion
- ☐ The key characteristics of landforms of glacial uplands: corries, arêtes, pyramidal peaks, U-shaped valleys, truncated spurs, hanging valleys and ribbon lakes
- ☐ The processes of deposition
- ☐ The landforms created by deposition: moraines, drumlins and erratics
- ☐ How people use glaciated landscapes for energy provision, leisure and recreation
- ☐ The human and physical causes of avalanches
- ☐ How avalanches can be predicted
- ☐ How places can be protected from avalanches

Key terms

Ablation, Abrasion, Arête, Avalanche, Corrie lake, Corries, Deposition, Drumlin, Erosion, Erratics, Glaciation, Freeze–thaw, Hanging valley, Hydro-electric power, Lodgement, Moraines, Pyramidal peak, Ribbon lakes, Truncated spurs, U-shaped valley, Valley glacier, Weathering

Which key terms match the following definitions?

A The use of fast flowing water to turn turbines which produce electricity

B An armchair-shaped depression formed by a glacier in a mountainous area

C A sudden and rapid movement of a mass of snow or ice down a slope

D A tributary valley left high above a main valley that has been deepened by glacial erosion

E A long, narrow and sharp ridge that separates two corries

F Long, narrow lakes found on the floor of U-shaped valleys

G A mixture of different-sized rocks and sediments eventually left behind in the landscape when a glacier melts

H A weathering process which causes rock to break down by the repeated action of water freezing and expanding in cracks

To check your answers, look at the glossary on page 289.

ResultsPlus
Maximise your marks

99

Foundation Question: Name and describe one type of moraine. (3 marks)

Student answer (awarded 1 mark)	Examiner comments	Build a better answer (awarded 3 marks)
Chosen type - Terminus Moraine *This is dropped where glaciers stop. This happens when glaciers melt.* *It is made up of lots of freeze-thaw material that comes from higher up on the mountain.*	• *Terminus Moraine...* is not quite right, but it is close enough for 1 mark. • *This is dropped...* scores no marks because it is about the formation of terminal moraines and the question asks for a *description*. • *It is made up...* is not accurate so scores no marks. Although it is fine to write about the material, the detail is not correct.	Chosen type - Terminal moraine. These are lines of material across a valley, almost blocking it. The material is often pushed there or slides off the glacier. It is made up of a mixture of boulders and ground up rock.

Overall comment: Make sure that you read the question and pay particular attention to the command words. In this question they are *name* and *describe*.

Higher Question: Explain how the effects of avalanches can be reduced through planning before the event. (4 marks)

Student answer (awarded 3 marks)	Examiner comments	Build a better answer (awarded 4 marks)
Prediction of avalanches takes place by using modern techniques like satellites. *Other ways of doing it involve using computers to look at the causes.* *If the snow on a slope is watched very carefully and snow pits are dug then the snow can be seen to be either stable or unstable.*	• *Prediction of avalanches...* is not correct and scores no marks. • *Other ways of...* scores one mark because it identifies the basic idea of using computers. However, there is no explanation of how computers are used. • *If the snow on...* is good and so scores 2 marks. Semi-technical terms such as ***stable*** and ***unstable*** help the answer.	One way of planning is to prepare better for the event by improving building design using canopies and reinforcing roofs. Computers can be used to model the behaviour of snow winds and build up a picture of what makes snow unstable. The Swiss have done much work in this area. If the snow on a slope is watched very carefully and snow pits are dug then the snow can be seen to be either stable or unstable.

Overall comment: It is very important to interpret a question correctly – this question is about *planning*, which obviously involves more than just predicting an event

Maximise your marks: Practice exam questions with student answer and examiner commentary to show you how to develop stronger answers (see next page).

You should know: A check-yourself list of the concepts and facts that you should know before you sit the exam. Use this list to **identify your strengths and weaknesses** so you can plan your revision wisely.

Don't panic zone: Last-minute revision tips for just before the exam.

Exam zone: Some exam-style questions for you to try, an explanation of the how you will be assessed, plus a chance to see what a real exam paper might look like.

Zone out: What do you do after your exam? This section contains information on how to get your results and answers to frequently asked questions on what to do next.

ResultsPlus

These features conmbine expert advice and guidance from examiners to show you **how to achieve better results**. Some are based on the actual marks that students have achieved in past exams.

Choose a stretch of coastline or coastal area you have studied where cliff recession is occurring or has occurred. Describe the effects of this cliff recession. (3 marks, June 2007)

How students answered

Most students answered this question poorly. They often only gave one effect when effects were required by the question. The question also required specific effects in the location studied rather than general effects of recession.

68% (0–1 marks)

Many students gave two effects, but still did not link these directly to the location that they had studied.

20% (2 marks)

Some students answered this question well. They gave at least two effects and referred to the studied location, using names of roads or buildings that had been affected by cliff recession.

12% (3 marks)

Exam question report: These show previous exam questions with details about how well students answered them.

- Red shows the number of students who scored low marks (less than 35% of the total marks)
- Orange shows the number of students who did okay (scoring between 35% and 70% of the total marks)
- Green shows the number of students who did well (scoring over 70% of the total marks).

They explain how students could have achieved the top marks so that you can make sure that you answer these questions correctly in future.

ResultsPlus
Build Better Answers

Suggest reasons why some countries are more successful than others in recycling waste. (4 marks)

■ **Basic answers** (0-1 marks)
Only give a description of differences in recycling, with no explanation.

● **Good answers** (2 marks)
Offer some explanatory statements relating to the role of government and/or level of economic development.

▲ **Excellent answers** (3-4 marks)
Provide full explanations of how some governments, such as Germany, encourage recycling by education, legislation and financial incentive.

Build better answers These give you an opportunity to answer some exam-style questions. They contain tips for what a basic ■, good ● and excellent ▲ answer will contain.

Exam tip: These provide examiner advice and guidance to help improve your results.

Watch out! These warn you about common mistakes and misconceptions that examiners frequently see students make. Make sure that you don't repeat them! The ■, ● and ▲ symbols highlight the severity of the error.

Maximise your marks These are featured in the Know Zone pages at the end of each chapter. They include an exam-style question with a student answer, examiner comments and an improved answer so that you can see how to build a better response.

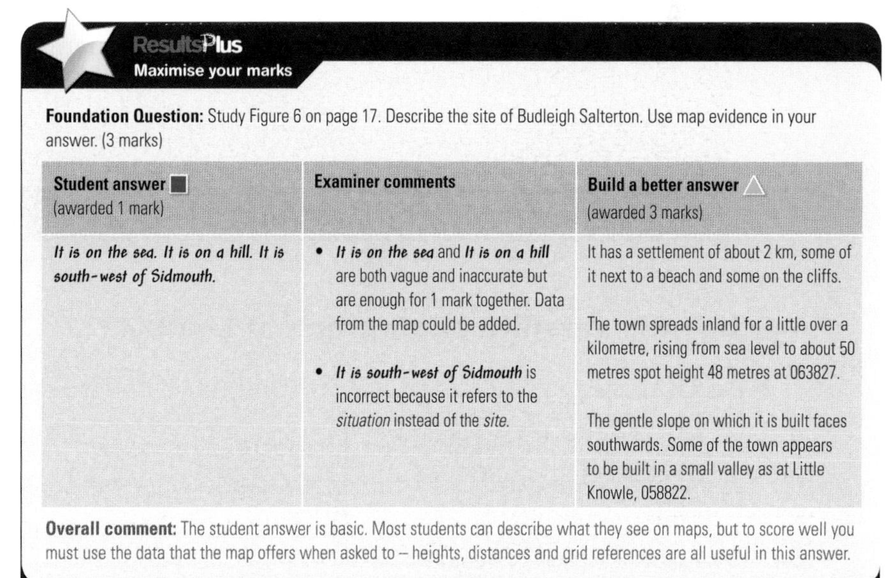

ResultsPlus
Maximise your marks

Foundation Question: Study Figure 6 on page 17. Describe the site of Budleigh Salterton. Use map evidence in your answer. (3 marks)

Student answer ■ (awarded 1 mark)	**Examiner comments**	**Build a better answer** △ (awarded 3 marks)
It is on the sea. It is on a hill. It is south-west of Sidmouth.	• *It is on the sea* and *It is on a hill* are both vague and inaccurate but are enough for 1 mark together. Data from the map could be added. • *It is south-west of Sidmouth* is incorrect because it refers to the *situation* instead of the *site*.	It has a settlement of about 2 km, some of it next to a beach and some on the cliffs. The town spreads inland for a little over a kilometre, rising from sea level to about 50 metres spot height 48 metres at 063827. The gentle slope on which it is built faces southwards. Some of the town appears to be built in a small valley as at Little Knowle, 058822.

Overall comment: The student answer is basic. Most students can describe what they see on maps, but to score well you must use the data that the map offers when asked to – heights, distances and grid references are all useful in this answer.

Unit 1 Geographical skills and challenges

Your course

This unit contains two sections:

Section A will show you how to apply geographical skills, such as map, ICT and GIS understanding, to your geographical investigations.

Section B investigates the challenges that our world is currently facing: climate change and sustainability. This section will inform you of all the issues, allowing you to make your own decision on the subject.

Your assessment

- You will sit a 1-hour written exam worth a total of 50 marks.
- There will be a variety of question types: short answer, cartographic, graphical and extended answer, all of which you will practice throughout this chapter and the rest of the book.
- **Section A** contains questions on geographical skills, such as cartography and graphics. You will have a resource booklet which contains the maps, diagrams and graphs that you will need to answer the questions.
- **Section B** contains questions on the two main challenges facing the planet: climate change and sustainable development. You might be asked to refer to material in the resource booklet to answer the questions.

Study the satellite photograph of London.

- Describe the growth of London north and south of the River Thames.
- Many of the 2012 Olympics sites are located in the Lea Valley, south of an area dominated by reservoirs. Where is this on the photograph?

Chapter 1 Geographical skills

Basic skills

Labelling and annotation

Labelling is simply indicating what something is – a feature, for example, as shown in Figure 5 in Chapter 3 (page 59). **Annotation**, on the other hand, involves adding some notes to explain something – perhaps the processes that brought about the feature – as shown in Figure 3 in Chapter 3 (page 58).

Maps are already 'labelled' because they use words and symbols to provide a whole range of information about an area. But they are seldom annotated, except when used in newspapers or other media. If your school subscribes to http://www.maps-direct.com/schools/ you can download and annotate maps of your local area to explain various features.

Labelling and annotating graphs and maps require two particular skills that geographers need to develop – describing **distributions** and describing **trends**.

Distributions

Take a look at Figure 1. Try to describe the **pattern** of the dots in such a way that someone who cannot see them can picture the pattern in their mind. The best way to do this is as follows:

1. Give an overview – is the pattern even or uneven? It is very unlikely that the distribution will be completely even, so you should offer a view as to how even it is – 'very', 'fairly' or 'quite' should be enough.
2. If it is generally quite even, are there any oddities, any gaps? These are called 'anomalies'. Identify where these are – 'there are fewer dots in the south-east corner', for example. Make sure that you use the appropriate geographical language – 'south-east' rather than 'bottom right'.
3. If the pattern is uneven, then you use helpful words to describe it. Are there 'clusters' of dots, for example, and, if so, where are they? You can break the 'map' down into sectors to do this more efficiently.

Trends

Trends are frequently shown by using graphs. Take a look at Figure 1 in Chapter 2 (page 30) and think about how you would describe a trend. As with your description of distribution, there is a useful method to follow here:

1. What happens overall? Compare the end with the start. Is it higher, lower or the same?
2. How does the trend vary? Is it a constant pathway or can it be divided into distinct 'periods'?
3. Have a close look for patterns – when certain trends seem to repeat.
4. Remember that when you are describing trends it is useful to understand the idea of rates of change.

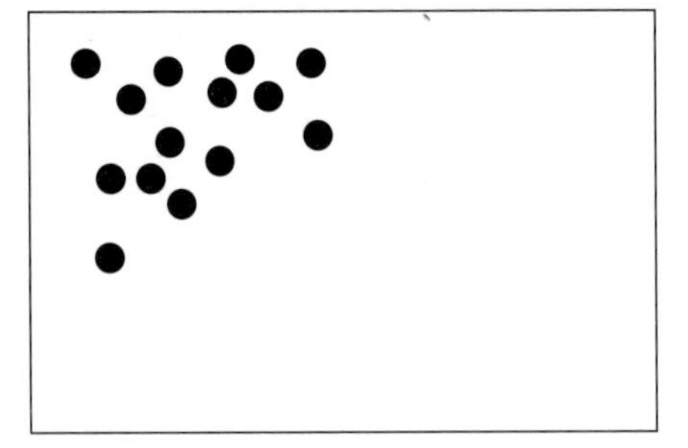

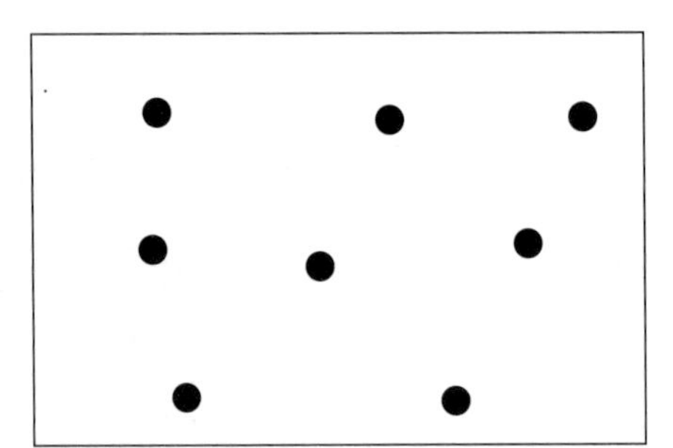

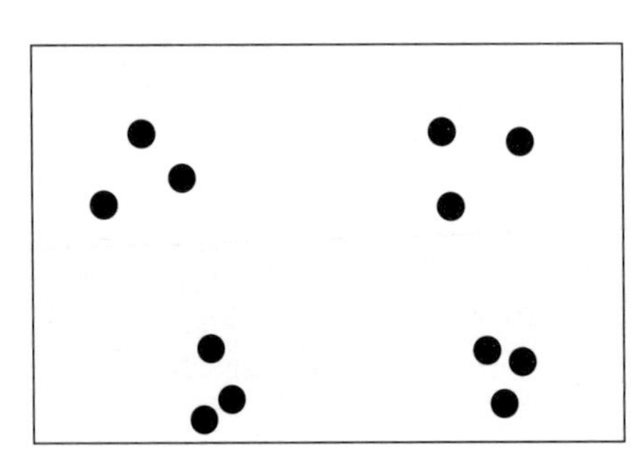

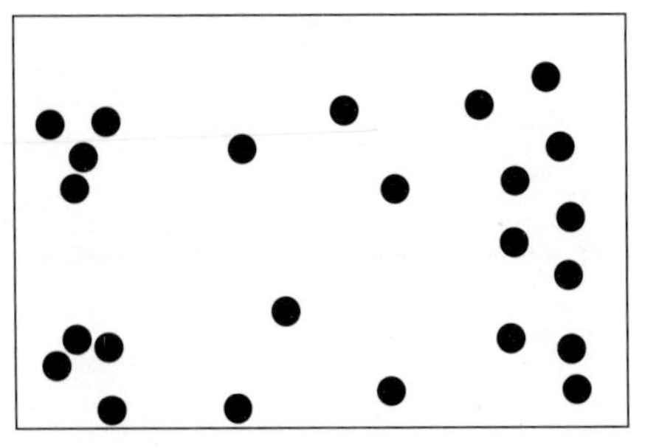

Figure 1: Four different distribution patterns (the dots could be rural settlements in a region or fast-food outlets in a town)

Drawing, interpreting and annotating field sketches

Field sketches (drawn in the field) or sketches drawn from photographs are useful when trying to identify particular features. Sketches are helpful because they simplify the view, by cutting out all the 'background noise' that clutters up the real world picture. They do not need to be works of art – in fact they should not be if they are to do their job properly. Figure 5 in Chapter 15 (page 263), for example, shows a sketch of the features shown in Figure 4 (Chapter 15 page 263). What you include on such a field sketch depends on what you are trying to show – what point you are trying to make. Of course such a sketch could also be annotated to explain the function of these sea-defences.

Aerial and satellite photographs

Using photographs to show an area or a landscape is not a hard task. Photographs can contain a great deal of relevant information – but some of it may not 'speak for itself'. So it is often helpful to add descriptive labels and explanatory annotations to photographs to make a point. This will help others to interpret the photographs better because good interpretation depends on a certain amount of knowledge and understanding of the place shown.

Take a look at Figure 15 in Chapter 2 (page 47). What is your immediate reaction? What features strike you as worth noting? Now consider the following question: 'What evidence is there that the **location** shown is in a High Income Country (HIC)?'

Not all photographs are taken in this way. Look at Figure 12 in Chapter 2 (page 40). This comprises two **satellite images** of Las Vegas. If you were asked to describe the characteristics of the city what would you note? Remember that 'characteristics' might be anything about the shape, the form and the distribution of key features. The aerial photograph of a golf course in Figure 12 on page 40 helps us to interpret what we are looking at and to conclude that:

1. There isn't much grass or vegetation either in it or around it.
2. In fact the surrounding area has no vegetation at all.
3. The few areas of vegetation in the city and on its margins include a number of rather strange squiggly lines.
4. The city is laid out on a grid pattern, with large square blocks of buildings.
5. The city has grown very rapidly in the 28 years between 1972 and 2000.
6. There are very few routes in or out of the city – in fact there is no obvious route at all in the west.

Figure 2 is a photo taken from the air too, but obviously not from a satellite, because it is not looking straight down from above. This is an **oblique aerial photograph**, taken from an aircraft. What are the key features to note in this photo?

Describe the pattern of the city landscape shown in Figure 2. (4 marks)

■ **Basic answers** (0–1 marks)
State that there are high buildings in the middle of the photograph, but little else is identified.

● **Good answers** (2 marks)
Identify the foreground as being a sports stadium and the higher buildings in the middle/background.

▲ **Excellent answers** (3–4 marks)
Also describe a pattern of rising building height towards the centre, and recognise the 'river' frontage and areas of open ground/brownfield sites in the foreground.

Figure 2: An oblique aerial photograph of a city

Writing coherently, showing the importance of good literacy skills in expressing geographical points

In order to make yourself understood in your exam answers, it is important to write coherently and use geographical terms to support your answers. Have a look at Figure 16 in Chapter 2 (page 48) and compare the answers on the left, to this question: *'Describe the distribution of tropical rainforest shown in Figure 16.'*

Answer A

There are lots of bits of forest in the middle. There are some big blobs too in various places. Altogether it looks like a really complicated pattern.

Answer B

The rainforest is distributed to the north and south of the equator. Overall the distribution is uneven and the largest regions of forest are found in South America and south-east Asia.

It is obvious that Answer B is stronger, but the gap between the two students is wide mainly because Answer B uses better terminology and has a clearer structure.

Examination questions can be broken down into 'bits' and you need to recognise these in order to answer questions effectively and coherently.
A typical question from Unit 1 might be: In many areas of the world tropical rainforests are under threat. With the use of examples, describe and explain the varied ways in which tropical rainforests are managed.

There are three tasks you need to complete to answer the question:

Task 1 – Identify the topic

Task 2 – Identify the focus

Task 3 – Identify the command word

- The topic is easily spotted – the management of tropical rainforests.
- The 'focus' is the varied ways in which this management is carried out.
- The command instruction is to describe and explain these methods – what are they and why are these methods chosen?

Keeping answers relevant is one of the hardest tasks for students. It is very tempting to write 'all I know' style answers which include information that is both relevant and irrelevant. Sticking to the three step formula above will keep your answers on track and make them coherent.

Cartographic skills

Atlas maps – distributions and patterns

Atlases contain vast amounts of information gathered together on maps of many different types. Describing the distributions and patterns shown on these maps is a vital skill. You should remember that maps are made by people and that 'rules' have to be made about how to draw them.

If you and some friends were each asked to make a map of your local area, you would have to make a number of immediate decisions: What **scale** to use? What to include? What not to include?

Remember that Atlas maps include a lot of data such as longitude and latitude. If you can, use the data that is included in your answers.

The same decisions have to be made by all mapmakers and so, unsurprisingly, no two different types of map for an area are going to look alike. Maps are 'models' of the real world, scaled down and showing some things but not others.

Atlas maps of the world face a problem that you do not when trying to draw a local map. The earth is a sphere and showing that on a flat piece of paper in two dimensions is very difficult.

In the Mercator projection (see Figure 16 in Chapter 21 on page 48 for an example) the shape of countries is kept much as you would see them from space but because that is done their sizes are not accurate. On the other hand, the Peters projection (Figure 3) shows countries according to their real size so Greenland (2 million km^2) is dwarfed by Africa (30 million km^2) but the shapes look unfamiliar. Fairly often the Mercator maps trim off Antarctica altogether and the equator is about two-thirds of the way 'down' the map. This puts Europe right in the middle and makes it appear larger and more important than it otherwise would.

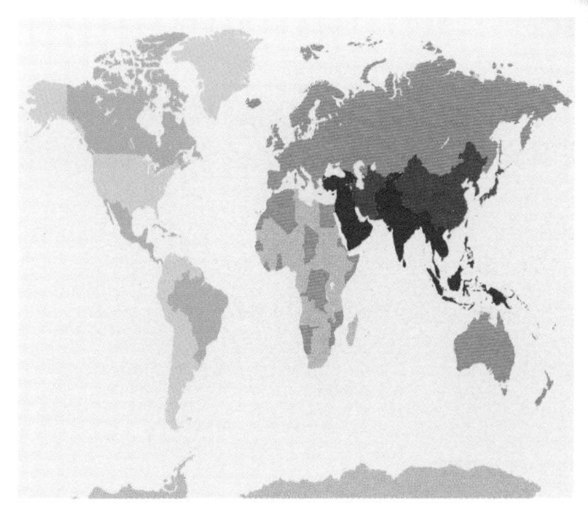

Figure 3: A map of the world, using the Peters projection

If you drew your own map of the local area it is likely that you would put your own home in the middle too, just as Mercator did.

Some atlas maps concentrate on only one or two features of a country. For example, a map showing the population distribution will not usually include anything else that just 'gets in the way' of that one idea. Take a look at Figures 2 and 3 in Chapter 12 (page 205) which show the global population distribution in rather different ways. Figure 3 is a choropleth map which uses shading to show densities, usually from lighter = lower to darker = higher. The **density** is not the same throughout any of these areas so differences are disguised. This is obvious when you look at Figure 2, which shows, for example, that the population distribution in the USA is very uneven. Figure 3 allows you to quote figures, whereas Figure 2 gives a better impression of where people actually live but does not allow you to quote data.

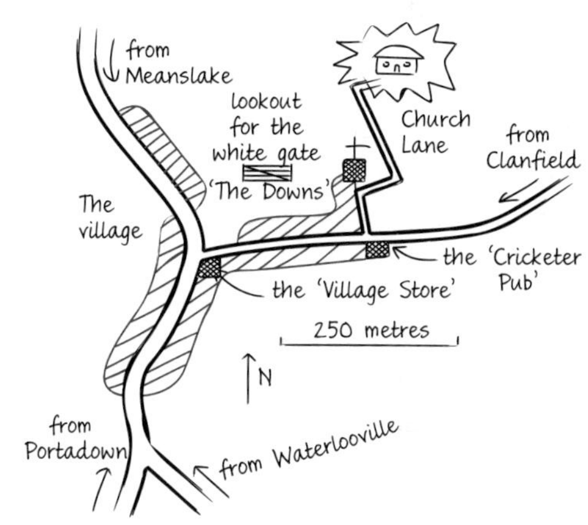

Figure 4: A sketch map

Sketch maps

The key to drawing a sketch map is to keep in mind what exactly it is 'for'. What are you drawing it to show? If it is to show how to get to a specific location, maybe as part of your fieldwork, then you might just want to show the key features of the route and any obvious problems that might occur along the way as in Figure 4. It isn't important to keep to an exact scale but some idea of the distances will help. Of course a photograph of the location would help.

If, as in Figure 5, the aim is to show a particular event or one aspect of an area then you might wish to add rather more detail and aim for greater accuracy in your sketch. This much more detailed sketch map could be further annotated both with text and photographs to 'explain' various features of this famous eruption event.

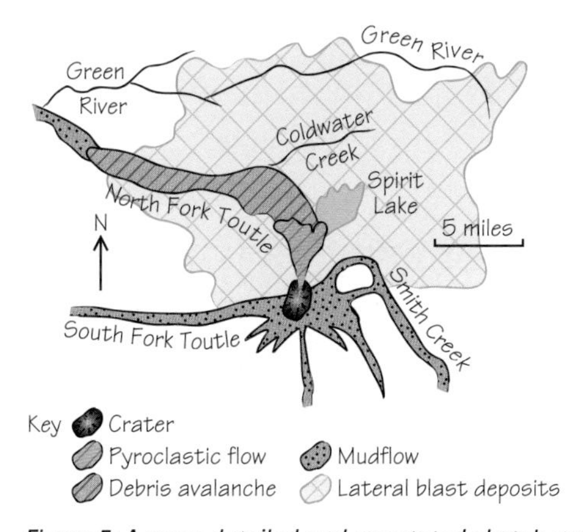

Figure 5: A more detailed and annotated sketch map with a key and showing an event

Ordnance Survey maps

Ordnance Survey maps are produced at a number of scales, but you need to be familiar with maps at the 1: 50,000 scale. This scale means that 1 centimetre on the map represents 50,000 centimetres on the ground – which is 500 metres. (You don't have to use centimetres. Whatever unit of length you use, 50,000 on the ground are represented by one on the map.) The table below shows a number of things that you can 'read' off a map.

Map information	Source	Can you put a number on it?	Comment
Altitude	Contours and spot heights	Yes – you can use the contours and estimate heights from these and the spot heights.	Avoid terms such as mountain unless certain. Use comparative comment.
Relief	Contours and distances	Yes – you can measure the gradient.	Often confused with altitude. Use terms like flat, steeply sloping, undulating.
Aspect – the direction that a slope faces	Contours and compass point	Yes – you could use the compass point. e.g. a south-west-facing slope.	Scale might obscure detail of local variations.
Surface drainage	Rivers, lakes, drainage ditches	Yes – you could work out the length of streams in a given area.	Not all rivers are marked on maps.
Land use (partial)	Use of map symbols – marsh, moor, bare rock, woodland, etc. Can only infer usage of 'white' areas. Same in urban areas – some uses are obvious (tourist information), but others are not.	Yes – partially. One could work out the percentage of coverage of a particular usage.	Deduction possible but conclusions will always be tentative. Beware of reading too much into place names.
Geology and soils	Can only infer these using other categories	No	As above – land use may be a key indicator here.
Settlement size	Area of settlements only – not their population	Yes – you can measure the area occupied by a town or city, or even the percentage of an area occupied by housing.	No clues about density because the height of buildings or how many people occupy them is unknown.
Settlement form	Shape from map	No	
Settlement distribution	From map	Yes, there are techniques to measure distribution and you can always use distances.	Use distances for measurement – one grid square = one square kilometre.
Settlement function (partial)	A few guesses possible using other information	No	Use tourist information, location, routeways as clues.
Transport and communication systems	Roads, tracks, waterways, railways	Partially – you can describe the shape of a network.	No clues about usage of these transport systems. Many forms of communication are not shown, e.g. the internet and telecommunications.

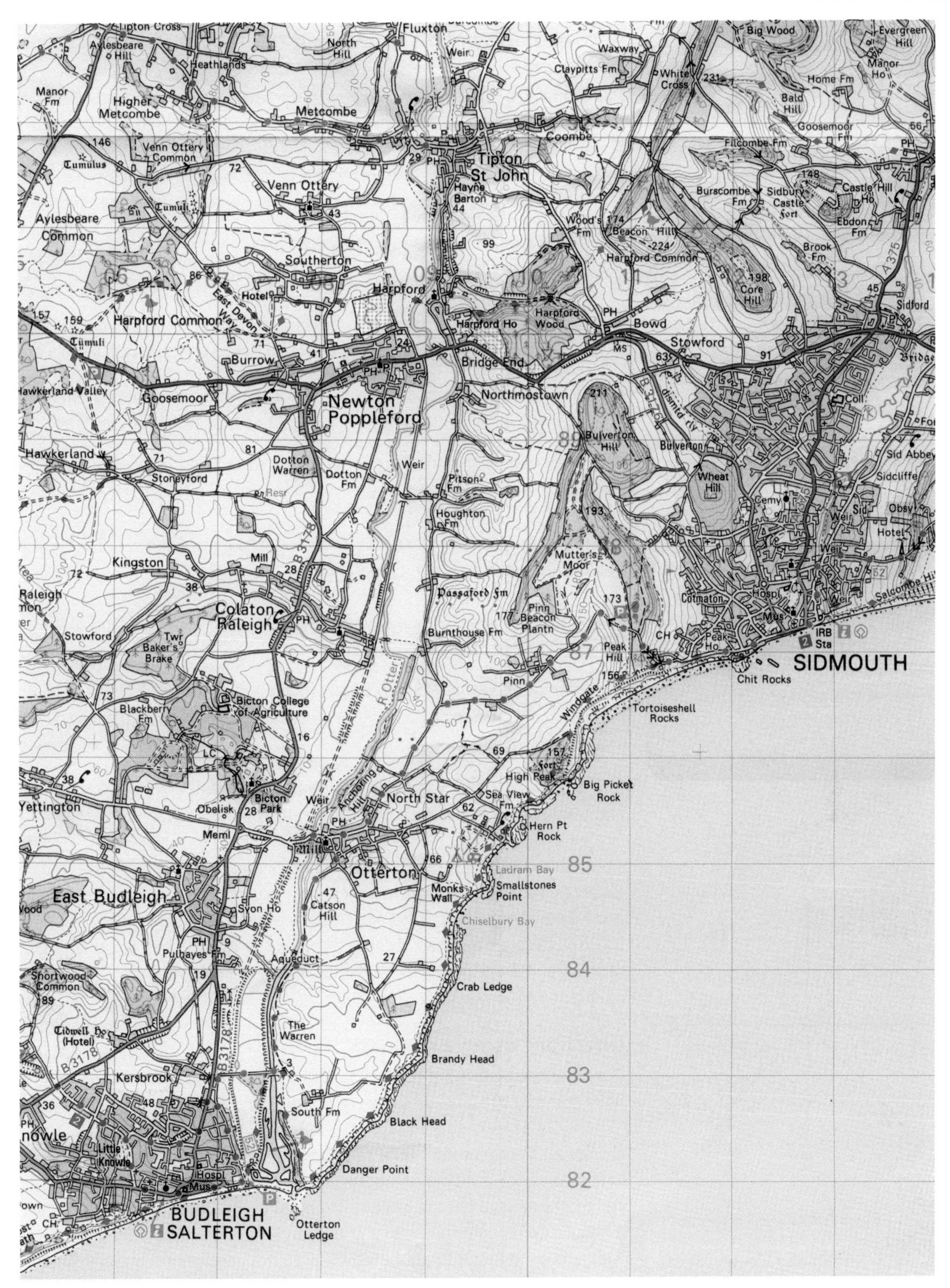

Figure 6: An extract from an Ordnance Survey 1: 50,000 map of Exeter and Sidmouth

Recognising symbols, using grid references and measuring distances

Like all maps, I: 50,000 OS maps show some features and not others. One can 'read' certain things from the map, as indicated in the table on page 16. A key is always included in examination questions, so recognising the symbols should cause few problems. However, it is best to learn at least the basic symbols so that map reading becomes easier. Some things are quite simple:

- Green bits mean woodland – of various types.
- Blue areas are either water, tourist information or motorways.
- Roads are colour coded from blue (motorways), red ('A' roads). orange/brown ('B' roads), yellow (local roads) and white (tracks).
- Contours are the thin brown lines that join places at the same height – they are at 10 metre intervals, that is to say 10, 20, 30 metres, etc. above sea-level.
- To help with heights there are spot-heights on a map – little black dots with figures against them.

Most symbols – such as those for churches, public houses and windmills – give some 'clue' to the feature represented.

Using the map extract (Figure 6) you can practise your understanding of the two types of grid reference:

- The village of East Budleigh is largely found in 0684 – a four-figure reference that identifies one of the blue-lined grid squares that each cover a square kilometre.
- The church with a tower in East Budleigh is located at 066849 – a six-figure reference that divides each grid square into a hundred little squares.

These grid squares can also be used as a quick method of measuring distances. For example, the 'crow flies' distance between East Budleigh (0684) and Newton Poppleford (0789 and 0889) is roughly five grid squares – so it is 5 km. To calculate the road distance between the two it is best to use a piece of string and divide the journey up into smaller sections with I cm of string representing 500 metres (1: 50,000).

Directions using a compass

Directions can be given in two ways. A broad indication is offered by an eight-point compass direction, as shown in Figure 7.

A more accurate method would be to give the ' bearing' – the direction, in degrees, from one point to another, as measured from North. The most common error is to get the whole thing back to front. So the direction of the triangulation point at the summit of Beacon Hill (112909) from the church with a tower in East Budleigh (066849) is 50° (or a little north of North East). If you took the bearing from the triangulation point to the church, then it would be 230° (or a little south of South West).

ResultsPlus
Build Better Answers

Describe the pattern of woodland shown in the OS map extract (Figure 6 on page 17). (4 marks)

■ **Basic answers** (0–1 marks)
State that woodland is found in most areas and that it is evenly distributed.

● **Good answers** (2 marks)
Also refer to the larger patches being on the eastern half of the map.

▲ **Excellent answers** (3–4 marks)
Also note that most of the woodland is on higher ground on steeper slopes.

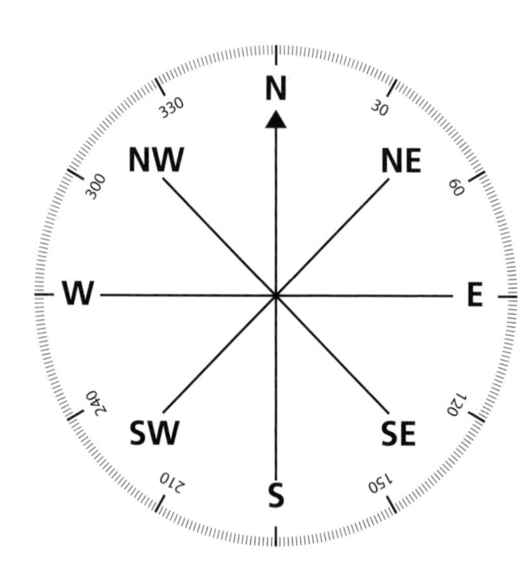

Figure 7: Eight points of the compass

Understanding the construction of cross-sections

You may be asked to outline the principles behind drawing a cross-section. A **cross-section** shows the **topography** (topography is another way of saying variations in **relief**). To draw a cross-section you decide what part of the landscape you wish to show, for example across a valley, and then draw a type of graph with the distance shown on the x-axis (horizontal) and the height above sea level as shown by the **contours** on the y-axis (vertical). You will need to choose your scale carefully for the y-axis or variations in height will be very hard to see.

Annotating cross-sections

Cross-sections show several aspects of the landscape and they can be annotated to show variations in slope angle, height and any other features that are of interest – such as the location of settlements, roads and rivers.

Heights can be read off the contours, but take care because contours are given at 10 metre intervals. So the top of a hill might, for example, be at 109 metres although the highest contour is the 100 metre contour. Spot heights may help you out here.

It is important to recognise different types of slope. The closer together the contours, the steeper the slope. So slopes that begin gradually and get steeper will look like the top one in Figure 8. These are known as concave slopes. When slopes begin steeply at the bottom and then become more gradual, they are known as convex slopes. Most slopes are made up of several different 'bits' – both concave and convex.

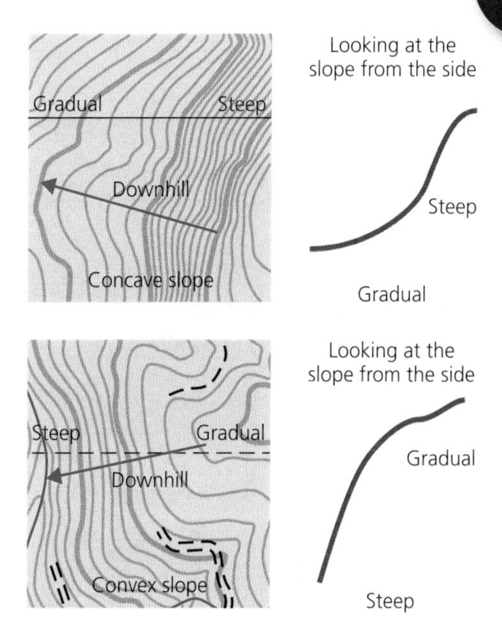

Figure 8: How contours show concave and convex slopes

Patterns of vegetation, land use and communications

I: 50,000 OS maps contain a certain amount of information about vegetation. Woods, orchards, parkland and marshland are shown. Where the map shows areas of bare rock, scree or cliffs you can assume that there is little or no vegetation present.

Much of a map, however, will be 'white' and you can only make a few educated guesses about what is going to be there. It is almost certainly farmland but you would need more evidence than the map provides to know what it looks like (Google Earth would help with this).

You will certainly need to answer questions about the pattern and distribution of different types of land-uses, including vegetation – and you might be asked to suggest reasons for these patterns. Describing patterns is covered on page 20 and the same techniques should be used here.

Now you should look at the pattern of communications on the map extract. Try to see this as a series of lines, as shown in Figure 9. The main roads **trend** east–west or north–south with the main settlements being the focus. Elsewhere there is a dense and complex pattern of local lanes. The larger settlements have a dense pattern of local roads.

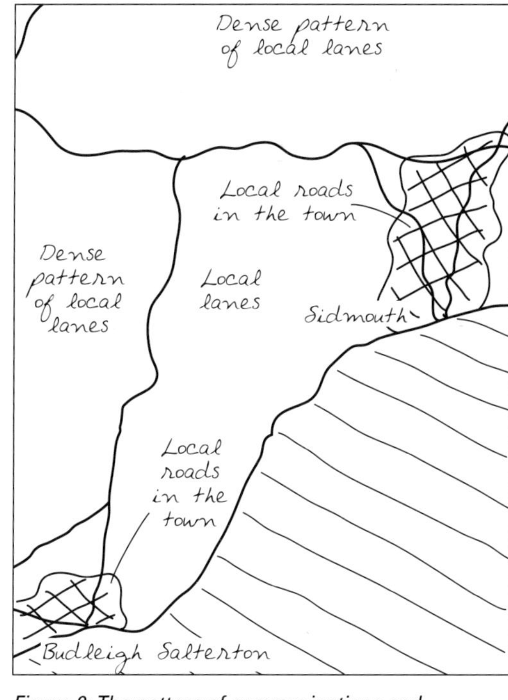

Figure 9: The pattern of communications and settlements on the Exeter and Sidmouth map extract

Nucleated

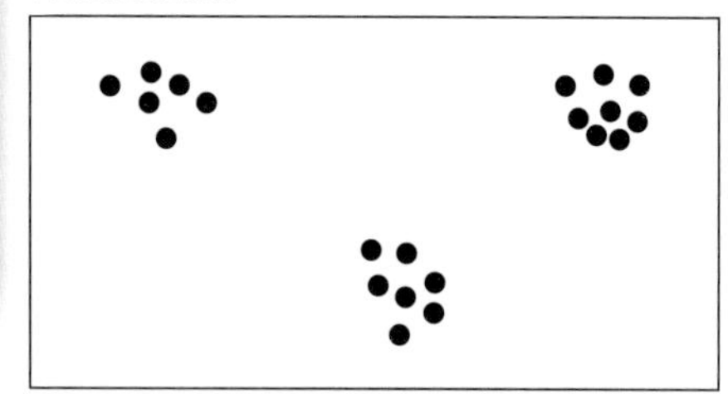

Linear

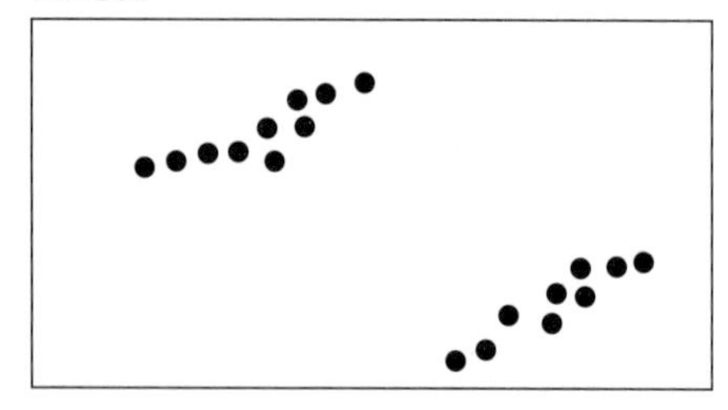

Dispersed

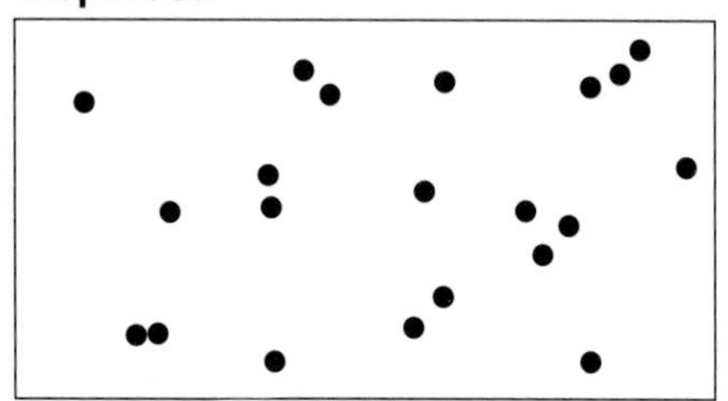

Figure 10: Three patterns of settlement

Figure 11: The distribution of nucleated settlements on the Exeter and Sidmouth map extract

Describing and identifying the site, situation and shape of settlements

The **site** of a settlement means the physical characteristics of the place (think of a building 'site'). The **situation** of a settlement is its location in relation to other places. Knowing its situation would help you find a place, and knowing its site might help you recognise it when you get there.

Describing a site from a map is easy enough, although you cannot 'see' everything. Remember 'SAGA' – Slope, Aspect, Ground conditions, Altitude. So for Tipton St John (0991): **Slope**: Moderate slope (see below); **Aspect**: West-facing; **Ground conditions**: Above the floodplain, so probably free-draining; **Altitude**: Between 25 and 50 metres above sea level.

The slope can be calculated, as follows. The land rises about 35 metres between the floodplain to the immediate west of the village (altitude about 10 metres), the spot height (44 metres) at 092912 and the 50 metre contour above it. This is a distance of about half a grid square so about 500 metres. So the slope angle can be calculated by dividing the distance 'along' (500 metres) by the distance 'up' (25 metres) to give a ratio – roughly 1 in 20, a moderate slope.

The situation of Tipton St John is:

1. 2 kilometres north of the main road (this is the A3052)
2. About 5 km north-west of Sidmouth
3. About 10 km north-north-east of Budleigh Salterton
4. On the eastern slopes of the Otter valley.

Settlements come in different shapes and sizes, as shown in Figure 10.

Nucleated settlements are villages and towns in which the bulidings are clustered. There may be a central feature – a village green, a market square or just a crossroads or a bridge. Sometimes villages and small towns are strung out along a road, in which case they are described as linear. In some areas there are very few villages at all – just single farms or small groups of buildings (hamlets). This is dispersed rural settlement.

Distributions and patterns of physical and human features

In examination questions you might be asked to describe a whole range of physical and human features on a map. The technique is exactly the same as the description of any other pattern. So if you were asked to describe the distribution of settlements on a map extract imagine them as simple dots. Take no notice of their size unless you are actually asked to describe the distribution of one type of settlement. Figure 11 shows the distribution of hamlets, villages and towns on the Exeter and Sidmouth map extract. Individual 'farmhouses' have not been included.

The pattern is obviously not even and the distribution is closely related to the physical features, both the coastline and the major river valley.

Human activity from map evidence, including tourism

Maps do not show much detail about human activity. It isn't even possible to make a guess about the population of places because you cannot tell how high buildings might be. Of course larger settlements are likely to have more people in them and, as a result, some of the buildings are likely to be shops, offices and other non-residential uses that cannot be read off the map.

The area has a sprinkling of tourist information. Most of it is within a kilometre or two of the coastline. The coast itself is not very developed because most of it is dominated by cliffs and not easily accessed. Between Sidmouth and Budleigh Salterton there are no settlements on the coast and very little access to the coast which partly explains the lack of tourist information. However, Sidmouth has changed a great deal in the last 50 years. Look closely at the 1950 map of the town and the local area (Figure 12) and now look at the modern map on page 17.

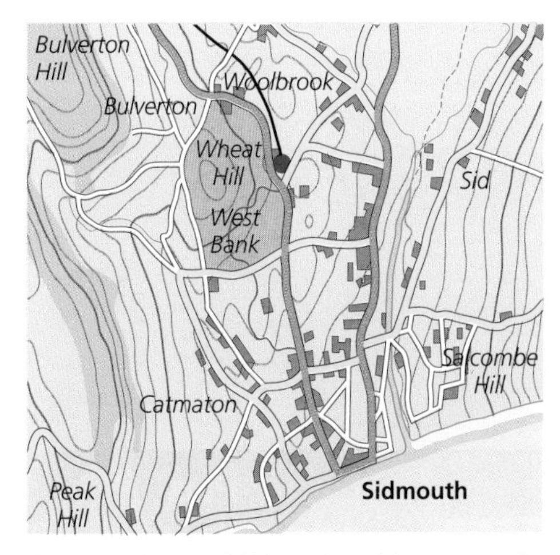

Figure 12: A map of Sidmouth and the surrounding area dating from 1950

It is obvious that Sidmouth has grown a great deal in recent years. The town may have very few older buildings and not as much history as some other places to attract visitors. The town's modern road layout also suggests this – it has many culs-de-sac (roads that are closed at one end) which are a feature of modern town planning.

Occasionally maps will show works and obvious evidence of some industrial activity, but not on this extract. Once again, do not dismiss negative evidence – it is important. You would be right be say that most of this map extract is dominated by rural land-use but the economic activity of most of the people who live here is unlikely to be related to agriculture. The city of Exeter which lies a few kilometres further west probably provides the jobs for many of the inhabitants of the small villages and hamlets. It is also probable that this is an area of retirement and a look at the population data would confirm this.

Using maps with photographs, sketches and directions

The value of maps in 'describing' an area can be helped a great deal by the use of photographs or sketches which can add the details that aren't shown on the map.

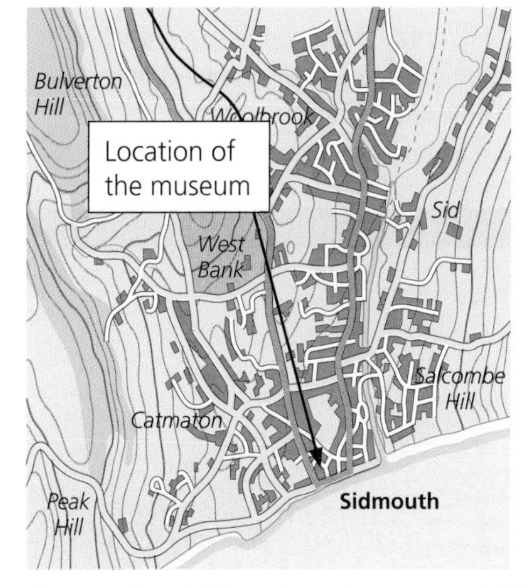

Figure 13: Part of Sidmouth, as shown on a 1: 25,000 scale map

A map showing the site of a settlement and its surrounding area would be more useful if the function of the buildings could be shown by photographs or annotations added to the map to explain certain features and directions.

This is especially the case when giving directions. Suppose you wish to find the museum in Sidmouth (1287). A label or photograph would be very useful.

Take a look at Figure 13, which shows a more detailed map of the location of the museum (this map is at a larger scale – 1: 25,000) and a label showing where the museum is. You could also insert a photo of the museum instead of the label so that there can be no mistake when you get there.

Exam Tip

Try to use examples and evidence in your answers, such as 'Sidmouth obviously caters for tourists, for example, there is a tourist information office located at 128873'.

Graphical skills

Graphs and charts

You will need to learn how to construct a series of graphs and charts when you come to do your controlled assessment. Quite often, an examination question for this Unit will leave out parts of a graph or a chart and give you the data to complete it. So you should get some practice in constructing them.

Line graphs

A line graph is a way to summarize how two pieces of information are related and how they vary depending on one another. Line graphs are often used to show how something varies over a period of time. The x-axis is used to show the time with the y-axis used to show the variable. Look at Figure 1 in Chapter 14 (page 241) which shows the growth of international tourism.

Pie-charts

Pie-charts are popular because they convey information in an easily 'readable' form – and they are quite simple to construct. The only real task here is to convert the pieces of data into percentages and then into proportions – the slices – of 360° (the whole pie). This technique has been used in constructing the pie-charts in Figure 5 in Chapter 2 (page 34) showing the various sources of greenhouse gases. A disadvantage of this technique is that we can get a view of the varying importance of different categories without ever knowing exact numbers.

Divided bar charts

Divided bar charts are also useful, especially when comparing data from one area or time with data from another, as in Figure 9 in Chapter 2 (page 36) which shows forecasts of the future threat of flooding according to different scenarios of sea-level rise.

ResultsPlus
Watch out!

- Do not confuse rate of change with total figures – if the population growth rate on a graph is shown to be falling, the population is still increasing, but more slowly.

Geographical enquiry skills

Much of this topic is preparation for the tasks that you will undertake when you complete your controlled assessment. In this section you will be introduced to the key parts of that task and given some hints about how to carry them out successfully.

Geographical questions, hypotheses and issues

We all know what questions are. Broadly speaking 'geographical' questions involve searching for answers to why some areas are different from others. Take a look at the questions posed in Figure 1 in Chapter 15 (page 260). These questions set the scene by allowing you to highlight the differences between areas.

These types of questions should start to make you think about the possible reasons to explain these geographic patterns. That is the process of developing **hypotheses**. Now some of these reasons are unlikely to have much to do with the geography of a country, although some will. Once you have developed a few ideas you have got to the stage of forming a number of hypotheses. A hypothesis is a testable statement.

Issues surrounding a chosen topic could also be explored. An issue is a debatable point where it is unlikely that the data will give a clear answer. The hypotheses above can all, with care, be established as either right or wrong. Issues are not so easy to resolve. The following is an issue posed in the form of a statement:

'Declining services pose serious problems for many rural communities'

Issues can often be a good focus for research and enquiries but need to be turned into a research programme by establishing questions and hypotheses. In this case it would be important to establish:

- What is meant by a 'problem'?
- Whether some people in a rural community might benefit (winners) whilst others might not (losers)

Sequences of investigation and enquiry approaches

One of the most common failings of fieldwork design and the eventual written account of that work is a lack of focus. Much depends on your original hypotheses. If these are clear then much else will be clear too. You will have some choice about this; the task questions are set by the examination board but you will have a role to play both in the choice of which question to pursue and how to focus that question. So if the chosen task question is 'How has service provision changed in your chosen rural area?' (see Chapter 15) you can approach this by developing a number of hypotheses that will help you focus. These could be posed as follows, for example:

1. Larger villages have more services than smaller villages
2. Villages closer to large towns have fewer services than those further away form large towns

To test the hypothesis we need to establish a sequence of investigation.

1. Identify the population of the rural communities to be studied
2. Investigate the services that exist in these communities
3. Develop a method of measuring these services – remember one 'large' shop might be more important than two smaller shops

Outline the sequence of tasks that you would carry out if you were to investigate the impact of congestion charging in cities. (4 marks)

■ **Basic answers** (0–1 marks)
Suggest that information about congestion charging can be researched using the internet.

● **Good answers** (2 marks)
Establish an order by outlining that the impacts might be positive or negative and suggest what some of these impacts might be.

▲ **Excellent answers** (3–4 marks)
Include a clear sequence of enquiry based on different interest groups (players) and establish a series of possible methods for assessing their views, for example questionnaires.

Extracting and interpreting information from sources

The skills that you need here are exactly the same as those that you have covered in earlier parts of this Unit. Remember that examination questions will ask you the methods that you would use to interpret a whole range of resources. It is worth bearing in mind that whatever the source of the evidence a number of common elements can be used when extracting information:

1. Say what you see – what is the overall impression?
2. Look for patterns, trends or groups in the information
3. Look for oddities (anomalies)

Describing, analysing and interpreting evidence

Making sense of evidence is about the hardest part of this exercise. What does all this data mean?

1. It is almost impossible to prove something to be true; evidence that suggests something is just that – it suggests something.
2. The 'fact' that 56 out 60 people tell you that a town has poor facilities is not the same thing as saying that the town does have poor facilities. Who are these people, where do they come from, what do they mean by 'poor' – poor compared to what? Poor weather for an inhabitant of Tahiti is not quite the same as 'poor' weather for an inhabitant of Scotland.
3. Not all sources of information are completely reliable. Both primary and secondary sources may well have 'vested' interests which influence their view. If you ask a motorist who commutes by car what he thinks about congestion charging, he/she may have a very different view from a retired and non-driving pensioner living by a main road into the city.

ResultsPlus Exam Tip

Avoid statements such as 'this is proved by…' because this is very unlikely to be the case. Use 'the evidence suggests…' instead.

Drawing and justifying conclusions from evidence

The conclusion is a return to your original hypothesis; however, this is not easy. Almost all conclusions are likely to be:

1. Partial – because it is subject to bias because of the evidence being selected. This is likely to happen even unconsciously.
2. Tentative – because of limited evidence. No piece of research can cover all the angles or gather all the data.

Have a look at the section in Chapter 15 about concluding your study (page 272).

Evaluation

Evaluating is a tough skill. It involves taking a very critical look at what you have done and how it might be flawed. Considering how you would do it better is a more positive way of tackling this.

Many students believe that collecting larger quantities of data will improve the quality of their data collection exercise. That is certainly not a foolish conclusion but it is not always the best approach. Often it is quality that matters more.

Remember that evaluation needs to step back and take a critical view of what you have done. This includes looking at the original hypothesis. Look at the relevant section in Chapter 15 (page 272) for more information.

ICT skills

Collecting and annotating photograph and satellite images

There a number of sources for free digital and satellite photographs: http://www.freedigitalphotos.net/; http://www.royalty-free-stock-clips.com/; and http://visibleearth.nasa.gov/. In this section the skill that is being tested is your ability to download photographs and other images and your understanding of how to annotate them.

Using databases such as census and population data

There is a great deal of information about populations on the web. The most useful national site is http://www.statistics.gov.uk/neighbourhood-statistics.asp. This can then be searched for local information by clicking on the appropriate country and this will bring up a screen for you to enter a postcode or the name of your chosen 'target'; you'll also need to enter the type of area that you need statistics for.

Information from the last census can be gathered on: http://www.statistics.gov.uk/census2001/get_facts.asp. This will allow you to explore a wide range of census information. One of the most useful is: http://www.statistics.gov.uk/census2001/quick_pictures.asp. From here you can select a local authority area. Information about local patterns of deprivation is available on: http://www.communities.gov.uk/communities/neighbourhoodrenewal/deprivation/deprivation07/. This site can be used to generate local maps of multiple deprivation if you click on 'interactive mapping'.

Using the internet

The internet is a very powerful tool and has an enormous amount of available information. To use it wisely it is important to establish the source of information. Typing 'volcanic eruption' into Google will generate about 1,500,000 'hits' so it is better to narrow your search a little.

1. Try to identify a particular research question or hypothesis. For example, you might be investigating the impact on people of volcanic eruptions, in which case 'impact volcanic eruption population' will narrow your search down to 190,000 hits and produce more focused results.
2. Check out the origin of the information. Is the authorship clear? If it has a bibliography then follow one or two of those links which might help establish where the article has come from.
3. There will be sites that are not well supported by links or any obvious authorship – it doesn't make them 'wrong' but it should make you suspicious.
4. Don't forget that some topics are very controversial indeed, with people having different points of view. You should check out the motives and the vested interests of different groups.

Describe the sequence of steps that you would take to download a satellite image to help your data presentation for your coursework. (4 marks)

■ **Basic answers** (0–1 marks)
Identify a source for satellite images, such as Google Earth.

● **Good answers** (2 marks)
Also include that the satellite image can be saved either as a picture or by using Microsoft publisher.

▲ **Excellent answers** (3–4 marks)
Also comment that these images should be inserted into the text at an appropriate place and, with a text box, annotations can be added to help link the image to the points being made in the text.

Video and television programmes

'I saw it on the News – so it must be true.' Sadly, this is not quite good enough today. As with the internet, it is best to be on your guard when taking information from videos, films and television. Filmmakers and documentary makers are not just in the business of informing people. They are frequently seeking to make money, improve their reputations and further their careers. (Video footage on YouTube is particularly hard to evaluate – some of it is useful but much of it is more about entertainment than conveying information accurately.)

ResultsPlus
Exam Tip

Remember that examination questions about enquiry and ICT skills will ask you only about the principles involved, not the details of your own work.

Data presentation and analysis techniques

As with all the ICT skills, an examiner can only ask you 'generic' questions about the methods that you would use and what ICT skills you would employ. The following is a typical question: 'Complete this table and then explain how you would use ICT to construct a graph or diagram to show the information.'

Percentage of workforce in different employment categories

	Country A	Country B
Primary	3	12
Secondary	14	24
Tertiary	83	—

You will need to go through the process step by step. To calculate the 'missing' percentage, add 12 and 24, and take that away from 100 = 64. This data could be shown on two pie-charts or on two divided bar charts. These can be accessed either through Insert/chart/pie or Insert/chart/column. Data can then be typed in to produce two graphs that can be compared.

Using spreadsheets and data handling software

The most familiar way of creating a spreadsheet is to use Excel. Excel can also generate charts and diagrams to help you analyse the data.

Researching and presenting investigative work

The research element here involves the ability to conduct internet research and how to integrate it into your own work. All such material should be sourced and references given. If, for example, you were doing fieldwork on deprivation in your local town and needed to follow it up, one route would be to research neighbourhood statistics online and then download a map which could be annotated and integrated into your text.

Presenting data from an official statistics site (for example levels of deprivation in areas in England) alongside photographs and primary data would allow you to draw conclusions about an area and compare it with others.

Presentation of investigative work using ICT employs a wide range of techniques, but the key is to keep a clear view of the enquiry question and the hypotheses so that each piece of research is put into the correct context.

Geographical Information Systems (GIS) skills

Capturing and representing geographical information

As with the ICT skills, you need to be able to explain how you do this and, sometimes, explain why you do it. So the examination questions are about the principles. Obviously it helps you answer these questions if you have had some experience.

The following exercise allows you to create a map in the Aegis system.

- The map is blank when you open the **folder** named **World** and find the file named **world map and data.aed**. Your first job is to activate it by clicking anywhere on the map. This lets you add data to it.

Click the chart wizard button

Click **Next** and then **area shading**. Then **choose from the list of available data** – for example, life expectancy.

Click **Next** and **Next** again – leaving the label field blank. Click **Next** and then choose **linear in range**. Click **Next** again and then choose how many classes for the data. You may want to change this from 4 to 6 to get a wider range of colours on the map. Now click **Next** to move on.

You can change the colour shading if you want. Click on the data class and then on the colour picker table. Obviously it's good to use just one colour shade, but you can also use certain colours to draw attention to the data (for example dark red for the country with the highest death rate). Click **Next** to see the finished graph. Don't forget to add a key.

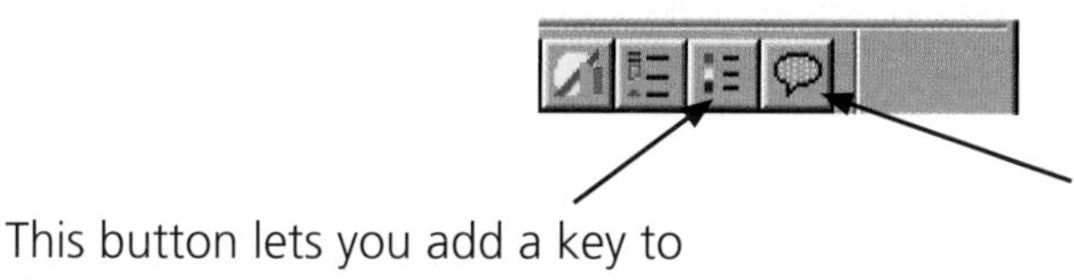

To identify different countries, click this button.

Finally, **copy the map and key into Word**. The best method is to make a screenshot of the finished map. Centre the map in the middle of the screen window and then press **Alt** and **Print Screen**. Paste the screenshot into Word, right click and choose **Format picture**. Select **Layout** and **In front of text**, then select **Picture** and crop the map until it is the correct size. Make sure you save the work in **My Documents**.

Using web mapping sites

Google Earth allows you to capture images and add information to them that can complement map and photographic evidence. Have a look at Figures 2, 3 and 6 in Chapter 15 (pages 261 and 265).

The Google Earth software also allows you to print out directions.

- Remember that GIS is useful in showing data in many different ways. It does not make the data any more or less reliable than it would be in a different form.

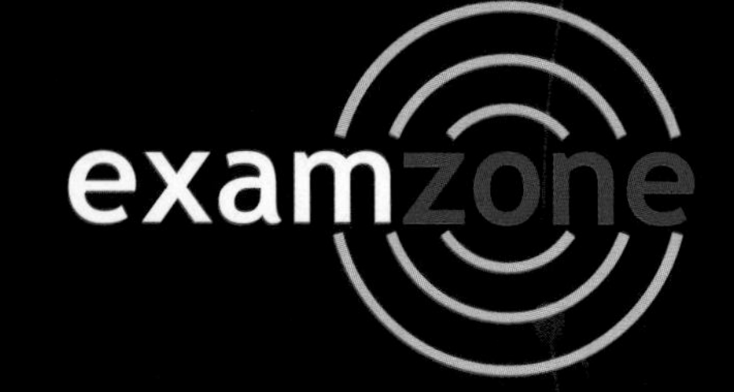

Know Zone
Chapter 1 Geographical skills

Geographical skills are an essential part of modern life. You need them to find your way around and to understand and interpret the world around you. They have a practical value that will not just make you more knowledgeable about the world, but also make you move around it more effectively and efficiently.

You should know...

- ☐ How to label and annotate diagrams, maps, graphs and sketches
- ☐ How to draw sketches and annotate them
- ☐ How to interpret different types of photographs
- ☐ How to write clearly
- ☐ How to describe distributions and patterns from atlas maps and OS maps
- ☐ The main symbols on OS maps
- ☐ How to describe direction using compass points
- ☐ How to draw cross-sections
- ☐ How to recognise and describe patterns of vegetation, land use and communications
- ☐ How to describe the site, situation and shape of settlements
- ☐ What can be inferred form maps about the human activity in an area, including tourism
- ☐ How to use maps with photographs, sketches and written directions
- ☐ How to complete and interpret a range of graphs and charts
- ☐ How to identify and evaluate geographical questions and hypotheses
- ☐ The appropriate sequence of an investigation
- ☐ How to extract information from a range of sources
- ☐ How to analyse and interpret evidence gathered for an investigation
- ☐ How to draw conclusions
- ☐ How to evaluate the methods of data collection, their presentation and the analysis
- ☐ How to use ICT skills to gather data such as photographs and satellite images
- ☐ What databases are and how to use them
- ☐ How to use the internet to investigate case studies
- ☐ How to extract information from television and video
- ☐ How to present data using ICT
- ☐ How to use spreadsheets using ICT
- ☐ How to capture and show geographical information using Aegis or similar systems
- ☐ How to use Google and other mapping sites

Key terms

Aerial photograph
Altitude
Annotate
Aspect
Contours
Cross-section
Density
Distribution
Evaluation
GIS
Ground condition
Hypothesis
Label
Location
Oblique aerial photograph
Pattern
Relief
Satellite image
Scale
Site
Situation
Sketch map
Topography
Trend

Which key terms match the following definitions?

A The ground occupied by a settlement

B A testable statement

C The direction in which a slope faces

D The number of people (or things) per unit area

E Where things are found

F A reflection upon a piece of work, outlining its strengths and weaknesses and suggesting possible improvements

G A line on a map connecting places at the same height above sea level

H The shape of the land, especially in terms of its altitude

To check your answers, look at the glossary on page 289.

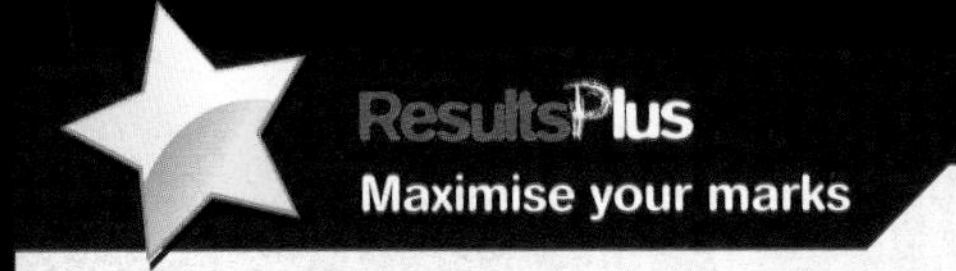

Foundation Question: Study Figure 6 on page 17. Describe the site of Budleigh Salterton. Use map evidence in your answer. (3 marks)

Student answer ■ (awarded 1 mark)	**Examiner comments**	**Build a better answer** △ (awarded 3 marks)
It is on the sea. It is on a hill. It is south-west of Sidmouth.	• ***It is on the sea*** and ***It is on a hill*** are both vague and inaccurate but are enough for 1 mark together. Data from the map could be added. • ***It is south-west of Sidmouth*** is incorrect because it refers to the *situation* instead of the *site*.	It has a settlement of about 2 km, some of it next to a beach and some on the cliffs. The town spreads inland for a little over a kilometre, rising from sea level to about 50 metres spot height 48 metres at 063827. The gentle slope on which it is built faces southwards. Some of the town appears to be built in a small valley as at Little Knowle, 058822.

Overall comment: The student answer is basic. Most students can describe what they see on maps, but to score well you must use the data that the map offers when asked to – heights, distances and grid references are all useful in this answer.

Higher Question: Study Figure 6 on page 17. Describe the pattern of land use in the area of the map to the west of the River Otter. Use map evidence in your answer. (4 marks)

Student answer ■ (awarded 1 mark)	**Examiner comments**	**Build a better answer** △ (awarded 3 marks)
This is a pretty empty part of the map. *The biggest place is Budleigh Salterton but not much else is around except a few hamlets.* *There is a quite a bit of woodland, for example near Colyton Raleigh.* *Elsewhere there is nothing at all.*	• ***This is a pretty empty...*** is not awarded any marks because it does not use geographical language. • ***The biggest place is...*** is awarded a mark because Budleigh Salterton is identified, but the student, who obviously spots a few other places, fails to address pattern. • ***There is a quite...*** is awarded 1 mark because there is woodland on the map, but once again the student has not addressed its pattern. • ***Elsewhere there is...*** would not get a mark because this student considers 'white' land to be empty, which is incorrect.	This is a lightly populated area of the map with a few villages. The exception to this is Budleigh Salterton on the coast. Many of the other settlements, for example East Budleigh (0684) and Newton Poppleford (0889), are located just to the west of the river, above the floodplain. The woodland is distributed quite evenly in small patches that are rarely bigger than 0.5 km^2. The land use elsewhere is largely farming, although there is park land around Bicton College (071865).

Overall comment: This type of question often tests students because they are unclear what pattern means. Make sure that you get a lot of practice using terms such as pattern, distribution and trend.

Chapter 2 Challenges for the planet

Objectives

- Understand that the Earth's climate has changed a great deal.
- Describe the reasons for those changes, both human and natural.
- Explain the effects of climate change and how we might respond.

Skills Builder 1

Study Figure 1.

Describe the trend in global temperature:

(a) over the past 10,000 years

(b) over the past 1,000 years.

ResultsPlus
Build Better Answers

Describe how changes in the orbital geometry of the Earth might cause the climate to change. (4 marks)

Basic answers (0–1 marks)
State that the Earth varies in its path around the sun but do not relate to climate change.

Good answers (2 marks)
Describe one reason why the orbital changes affect the amount of solar radiation received – usually variations in the path of the orbit and less solar radiation because the Earth is further away on average.

Excellent answers (3–4 marks)
Offer two descriptions, orbit and axis perhaps, and/or have good detail about the cycles of these variations.

The causes, effects and responses to climate change

How and why climate has changed since the last ice age

There is a large amount of evidence that the climate of the Earth is very variable. For most of its history the Earth has been a good deal warmer than it is today (see Figure 1). The two large ice sheets that we have today – one in Antarctica and one in Greenland – have not always been there, and are therefore quite unusual.

Figure 1 shows the global temperature change over the last 10,000 years and, in more detail, the changes over the past 1,000 years, including a prediction into the future. It is clear that the climate has changed a great deal in the past 10,000 years. Human beings were in existence throughout this time, but most experts agree that their impact on the world around them was quite modest. As a species human beings might have been responsible for hunting a number of other species to extinction, such as the mammoths, and they cleared substantial areas of forest. But it wasn't until the nineteenth century that numbers really began to grow and, as they industrialised, they began to contribute to **greenhouse gas** emissions. So **climate change** in the past needs to be explained by different factors – probably relating to the Earth's orbit, the output of the sun, or catastrophic events in the past.

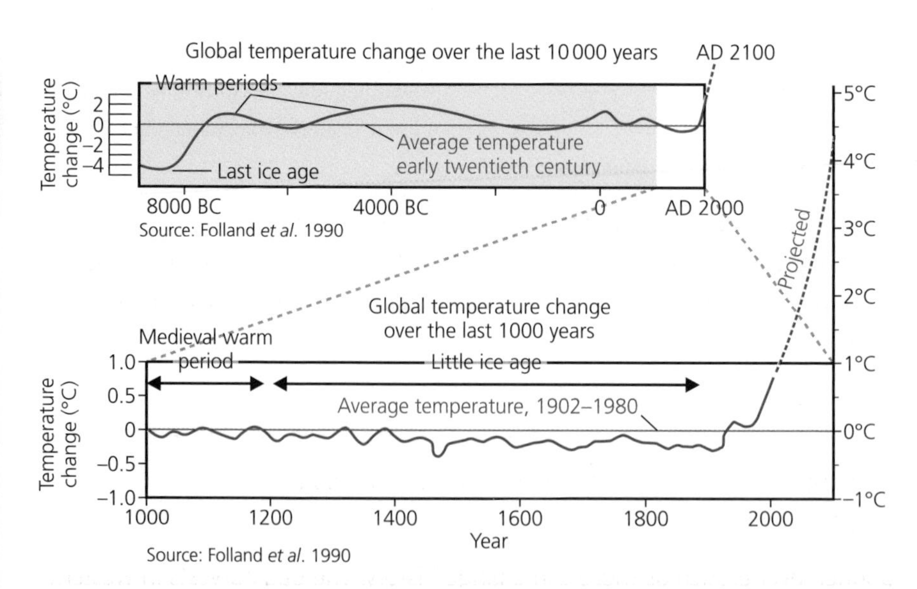

Figure 1: Global temperature changes. The top graph shows global temperature changes over the last 10,000 years. The bottom graph is a more detailed version of the white section in the top graph. It shows global temperature changes over the last 1,000 years.

Changes in the Earth's orbital geometry

The Earth orbits the sun on a slightly variable orbit. It also 'wobbles' on its axis a little, and the angle of tilt of the axis also varies. These three factors can affect the amount of energy received by the Earth by as much as plus or minus one per cent, which is enough to affect global temperatures. These variations are known as the **Milankovitch mechanism** and are thought to be the main reason for the series of **ice ages** experienced in the distant past.

Changes in solar output

We used to assume that the sun was 'constant' in its output, but we now understand that the energy it transmits does change although, again, by quite small amounts. However, although the changes aren't great, they have a large impact on global temperatures. These factors might act together to either reduce or increase global temperatures. Of course, they might also cancel each other out.

Catastrophic events

Volcanic activity and collisions between the Earth and large extraterrestrial objects such as meteors or even comets result in large quantities of material being ejected into the atmosphere. There is a lot of evidence linking dramatic events such as these with changes in the Earth's temperature. Some of these catastrophic events have been thought to be responsible for major extinctions such as that of the dinosaurs at the end of the Cretaceous period.

The Laki volcanic eruption in 1783

In 1783, millions of tonnes of poisonous gas and particles were sprayed over much of Iceland, as 27 km of volcanic vents poured out liquid rock for eight months. This 'fissure eruption', which split the Laki mountain, produced the world's largest lava flow for a thousand years –13 km^3 of material spread out over 500 km^2 of southern Iceland. The smaller particles combined with water vapour, producing a fog that was so dense that local people didn't see the sun for weeks. The eruption had two effects:

- In Iceland itself, it killed off the vegetation, which in turn led to the death of many of the island's animals, through starvation. Because they relied on meat to survive, a third of the island's population eventually died in the famine.
- The impact was also felt a long way from Iceland – in much of Europe and further afield – as the particles fired into the atmosphere by the eruption blocked incoming solar radiation and reduced global temperatures.

Sudden events such as these can change history. The bad harvests in western Europe that followed the Laki eruption led to widespread suffering and even starvation. This in turn led to an increase in civil unrest, which some historians think contributed to the French Revolution that began in 1789.

Activity 1

1. State two ways in which the amount of solar energy received by the Earth can vary.
2. Outline two ways in which global temperature is affected by sudden events beyond the control of humans.
3. Research two events that have been thought responsible for major changes in the Earth's climate.

Figure 2: Laki volcano cones after the eruption in 1783

Quick notes
(Laki volcanic eruption):

- Climate changes can be triggered by sudden events.
- Sudden events are usually unexpected.
- The impact of these changes can spread far beyond the place at which the initial event occurs.

Just forty years ago a decline in temperatures globally led to a debate about the possibility of a new 'ice age'. But since that time the rapid rise in global temperatures has alerted scientists to a quite different idea – that of **global warming** caused mainly by the human contribution to greenhouse gases.

ResultsPlus
Build Better Answers

Explain why higher levels of greenhouse gases cause a rise in global temperatures. (4 marks)

■ **Basic answers** (0–1 marks)
Identify at least one major greenhouse gas and add something about how they have increased in recent years.

● **Good answers** (2 marks)
Explain how this rise of greenhouse gases traps more solar radiation causing temperatures to rise.

▲ **Excellent answers** (3–4 marks)
Also understand that the solar radiation is trapped by greenhouse gases as it is radiated back from the ground surface, heating the atmosphere from below and/or that rising temperatures may cause other effects, further raising temperatures (methane released from permafrost for example).

The causes of current climate change on a local and global scale

In the past forty years concerns have been growing about the contribution made by humans to global warming. There are two parts to this discussion:

- The impact on global temperatures of increasing 'greenhouse gases' (Figure 3)
- The role of human beings in contributing to these increases.

The atmosphere is largely made up of nitrogen and oxygen. Neither gas is affected by solar energy which passes through them, both on the way 'in' and, when it is radiated back from the Earth's surface, on the way 'out'. It is a few other gases in the atmosphere that make life on Earth possible, because they absorb the radiation going back out from the Earth. They 'trap' the radiation – a bit like a greenhouse does – and make the atmosphere warmer. Without these 'greenhouse gases', life on Earth would not be possible – it would be too cold. But too high a level of these gases would make temperatures too high for life. The range within which life is possible is quite narrow, so the level of greenhouse gases needs to stay relatively stable. The most important greenhouse gases are:

- Water vapour
- **Methane**
- Carbon dioxide
- Nitrous oxide
- CFCs.

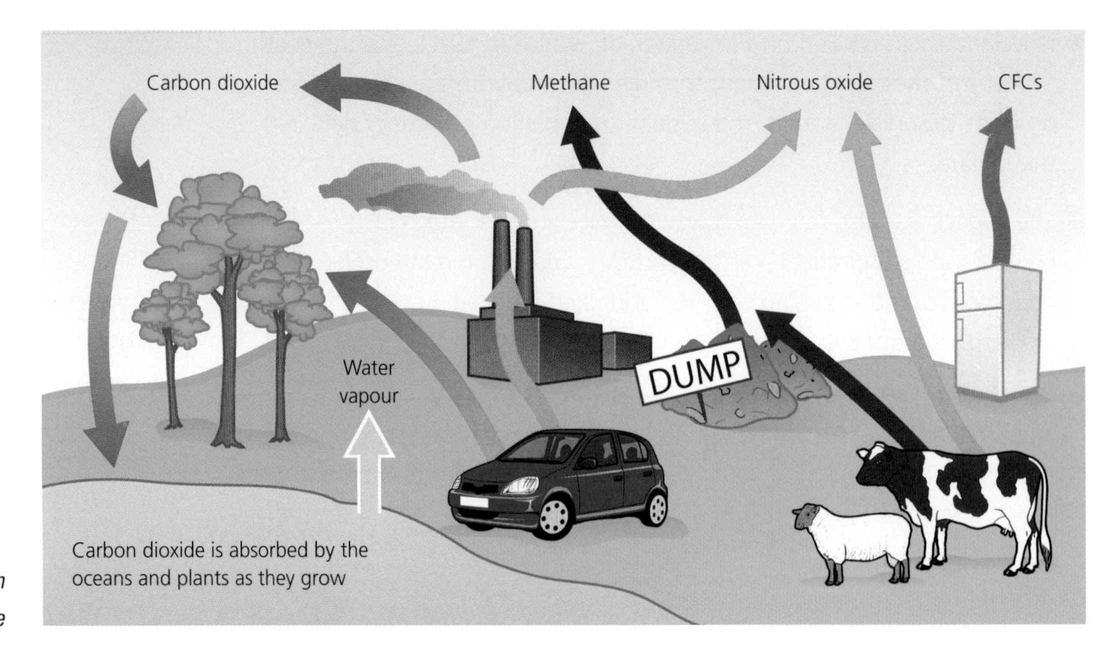

Figure 3: Greenhouse gases in the atmosphere

There are natural sources of all five gases, and in the past their level has varied according to the amount of vegetation and water on the Earth's surface and the amount of volcanic activity. By far the most important greenhouse gas is water vapour, which is almost entirely 'natural'. The oceans and the biosphere act as 'carbon sinks' – absorbing carbon dioxide (CO_2) by natural processes such as solution and photosynthesis. The various processes have long worked together, keeping the level of greenhouse gases fairly even.

In recent years, however, almost all climate experts have agreed that global temperatures are increasing, and most – but not all – blame human activity for most of this increase. The main focus of attention has been the rising levels of CO_2 and methane caused by human activity. Most of this comes about through the burning of **fossil fuels** such as coal, oil and natural gas.

Methane

Methane is twenty-four times more potent a greenhouse gas than carbon dioxide – the culprit normally at the centre of global warming discussions. Methane levels have risen as the number of livestock animals has increased. Rising incomes throughout the world have increased the demand for meat, meaning that the numbers of animals reared – especially cattle and sheep – have soared in the last few decades.

Livestock animals produce methane as part of their digestive process, belching it out while chewing cud and excreting it in their waste. According to the Worldwatch Institute, about 15 to 20% of global methane emissions come from livestock.

Carbon dioxide

Figure 4 shows a rapidly rising level of CO_2, with the rate of that increase getting faster. It is this increasing rate that leads some experts to identify this as a major challenge to the future of humanity.

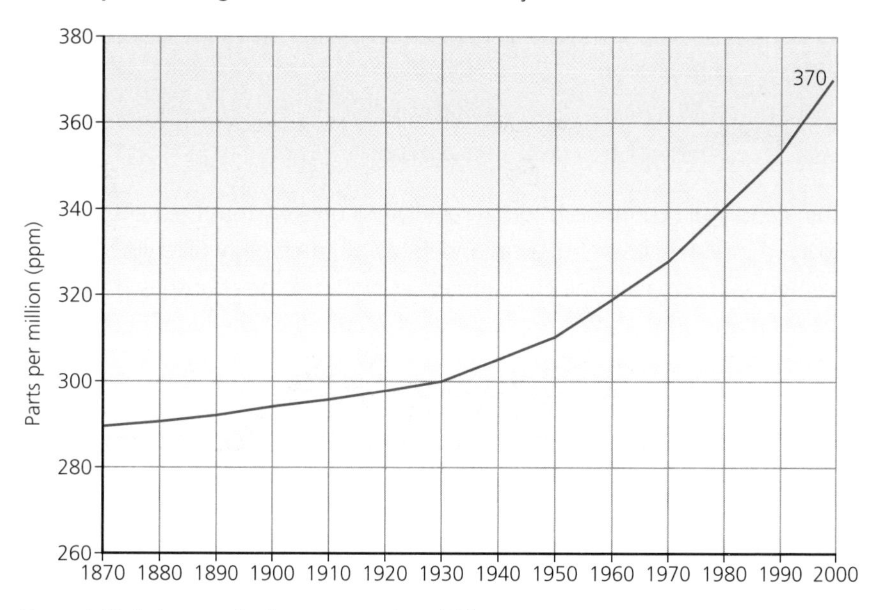

Figure 4: Global atmospheric concentration of CO_2

Skills Builder 2

Study Figure 3.

(a) Identify three greenhouse gases.

(b) Describe the process by which a greenhouse gas affects the temperature of the atmosphere.

(c) Why does eating more meat make the atmosphere warmer?

Skills Builder 3

Study Figure 4.

(a) Describe the trend in CO_2 emissions between 1870 and 2000.

(b) Suggest three ways in which human action may have contributed to the changes that you have described.

Figure 5 shows that the contribution of human activities to greenhouse gases is quite varied. The vast majority comes from increases in wealth around the world which have led to increased demand for:

- energy
- food
- consumer goods
- transport.

Skills Builder 4

Study Figure 5.

(a) Describe the main contribution to the emission of carbon dioxide.

(b) Describe the main contribution to the emission of methane.

(c) Identify the main differences between the ways in which humans contribute to CO_2 and methane emissions.

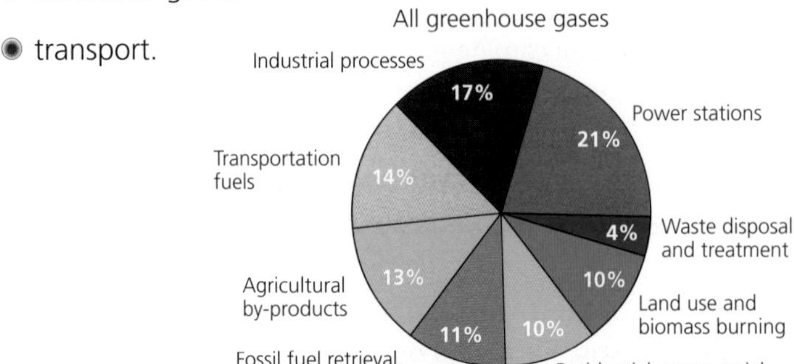

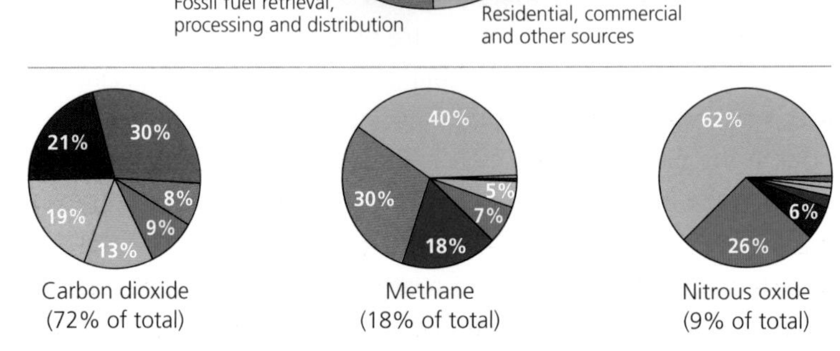

Figure 5: Greenhouse gas emissions by sector of human activity

The figures for car ownership in 2008 (below) illustrate the way in which economic growth causes increases in the use of resources – which in turn increases the emission of greenhouse gases.

	Car ownership	Population	Number of cars for every 100 people
USA	260 million	300 million	87
China	59 million	1,300 million	4
India	12 million	1,100 million	1

Figure 6: The Tata Nano car, a small car which is priced around £1,500

As China and India continue to grow rapidly it is unlikely that their citizens will give up dreams of owning a car and enjoying the type of lifestyle that is accepted as 'normal' in the USA and other high-income countries. The fast developing countries will not want to be lectured by the high-income countries about the need to slow down their rate of growth because of global warming. The high-income countries have already reached high levels of consumption, raising the world's greenhouse gases to dangerous levels on the way. Now the developing countries want to catch up, and their businesses are rapidly expanding their markets in these new lifestyle products. One of the most striking examples is that of Tata Motors, a branch of the huge Indian corporation, which has started production of a car priced at around £1,500 (Figure 6). It will sell in vast numbers – and every car bought and every bicycle and rickshaw replaced by a car will add to CO_2 emissions.

The negative effects that climate change is having on the environment and people

The future impacts of climate change on the environment and the population are not entirely clear – and the experts are still arguing about them.

Climate change will affect the weather and longer-term climatic conditions in many parts of the world. It will be drier and hotter in some regions but colder and wetter in others. Some experts suggest that hazardous weather events such as hurricanes and winter storms will become more frequent, but this is highly controversial.

Changing global temperatures and rainfall patterns will have an impact on the natural environment, as animals and plant species adapt to the new conditions. In recent years, for example, much has been written about the threat to polar bears, as the Arctic ice has been melting and their natural habitat has shrunk. But not everyone agrees that all the impacts of global warming are negative.

The impact on food production

Just as the natural environment will be affected, so will the global pattern of **crop yield**. The world's most important foods are rice and wheat, and these are grown in distinct regions. As Figure 7 shows, a small number of countries are responsible for growing a large proportion of the wheat crop that is sold on world markets. The wheat 'belts' in countries such as the USA, Canada and the Russian Federation are in the interior of these countries and it is likely that these regions will experience changes in their climate. What is much less clear is whether these impacts will be negative or not. However, it is worth remembering that any changes will need a fairly speedy response by governments, farmers and the organisations concerned. Almost every country depends on food imports and at any moment the world has about 40 days' worth of food supply in store. If climate change reduces harvests significantly then food prices will rise and some populations will be very badly affected, especially in sub-Saharan Africa.

Skills Builder 5

Study Figure 7.

(a) Identify the world's largest exporter of wheat.

(b) Identify the world's largest importer of wheat.

(c) Suggest reasons why Africa exports no wheat at all.

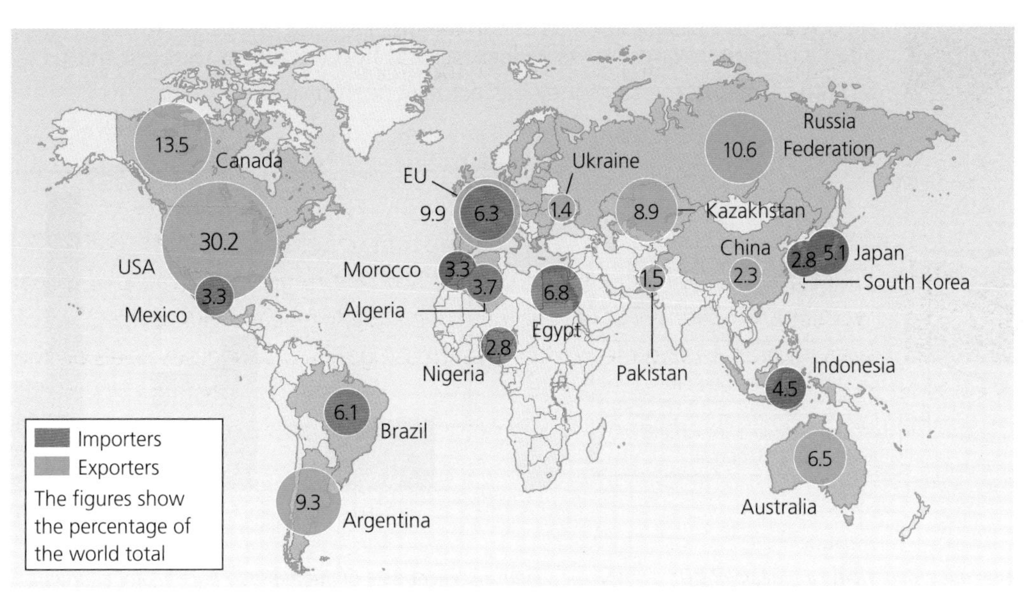

Figure 7: The main exporters of wheat and the main importers, 2007

Reductions in food supply are not the only risk to human health. Changing climate might also lead to changing patterns of disease. An example of this would be malaria. Having finally got rid of the disease in 1970, the Italian government is now having to deal with its return, as temperatures increase in the south of the country. Tick-borne encephalitis, a virus which attacks the nerve system, is also on the way back in Italy. While only eighteen cases had been reported before 1993, 100 have been since, mostly around Venice.

The impact on sea level

The best known impact of climate change is rising sea level, caused by two processes – the addition of water from the melting of land-based ice sheets and the expansion of the sea water itself as it gets warmer. Rising sea levels threaten the survival of small low-lying islands and coral reefs, and will have an impact on all coastal regions. In addition, low-lying deltas and flood plains may be contaminated by sea water.

The predictions about sea-level change depend on the various 'models' used to predict the rates of global warming, as shown in Figure 8. These vary from the 'worst case scenario' – with no change in the current rapid rate of increase in CO_2 – to models which assume that we will reduce our emissions. What will actually happen will depend on how we tackle climate change and, in particular, on our contribution to greenhouse gases. The most pessimistic forecast shows sea levels rising by nearly a metre by the end of this century, while the most optimistic forecast calculates the rise at less than 20 cm. But even that relatively small change poses serious challenges for low-lying regions and for islands. The impacts relating to the three forecasts are shown in Figure 9. This shows that the impact of flooding will be very severe if we do nothing about climate change. Even if we make the effort to restrict the rise in CO_2 levels to 750 ppm there will be at least another 10 million people flooded by the 2080s.

The affluent countries' record on helping the poor is extremely disappointing. In 2002, at the Johannesburg 'Earth Summit', they pledged more than $1bn to enable poor and vulnerable countries to predict and plan ahead for the effects of global warming, as well as to fund flood defences. But less than $180m of the promised money has actually been delivered.

Skills Builder 6

Study Figure 9.

(a) Identify the number of people likely to be flooded every year by 2020 if we do nothing about greenhouse gas emissions.

(b) Estimate the number of people affected by flooding by the 2080s if we limit emissions to 750 ppm.

(c) Other than sea-level rise, identify another reason why more people might be threatened by flooding in the future.

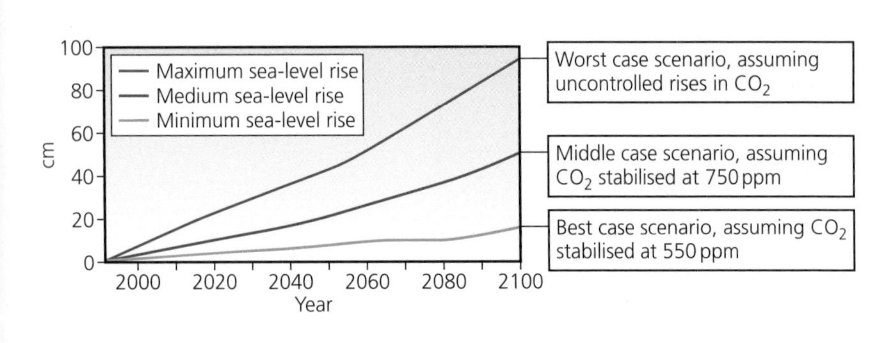

Figure 8: Three predictions for future sea-level rises, depending on levels of CO_2

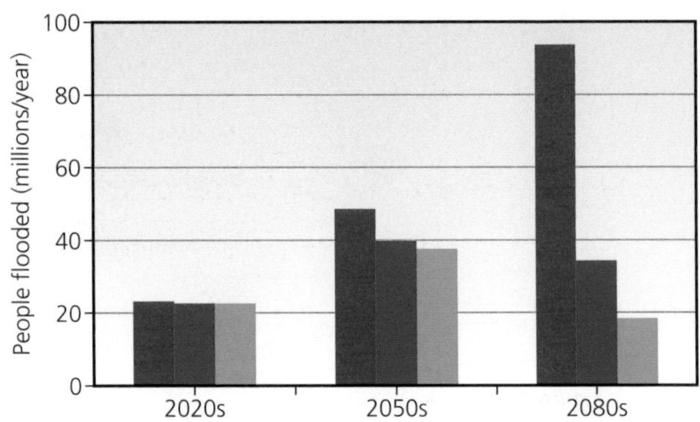

Figure 9: Forecasts of the average annual global number of people flooded under the three scenarios in Figure 8

The threat to Bangladesh

Bangladesh is one of the most vulnerable countries in the world (Figure 10). It is a hazard 'hotspot' with tropical cyclones and river flooding posing a threat to its population, many of whom live on the fertile floodplains and deltas of the Ganges and Brahmaputra rivers. Like many others they are threatened by climate change without contributing to it. The GDP per capita (person) is only $1,900 and often much lower among the rural population, 20% of whom live in extreme poverty. These chronically poor people own no land or assets, are often illiterate and may also be sick.

Whether or not tropical cyclones are more frequent because of global warming is unclear but their impact, combined with rising sea levels, poses a real threat to Bangladesh. In November 2007, Super-cyclone Sidr devastated the southern part of the country with wind speeds of 223 km/hr, killing more than 3,000 people and affecting another 7 million. The losses would have been far higher if it had not been for shelters built along the coast and good early warning by the Bangladeshi Met Office. This event offers a clear warning about the future, not least the secondary impact of such events when agricultural areas are ruined as salty seawater is driven inland by the storm surge, getting into groundwater, and sewage systems are disrupted.

Bangladesh needs $3.5bn over the next five years to strengthen and maintain its defences. Its government is bitterly disappointed with the lack of help from high-income countries. Having played a leading part at the Bali conference in negotiating an investment fund to deal with the impacts of global warming, it has not yet received any funds.

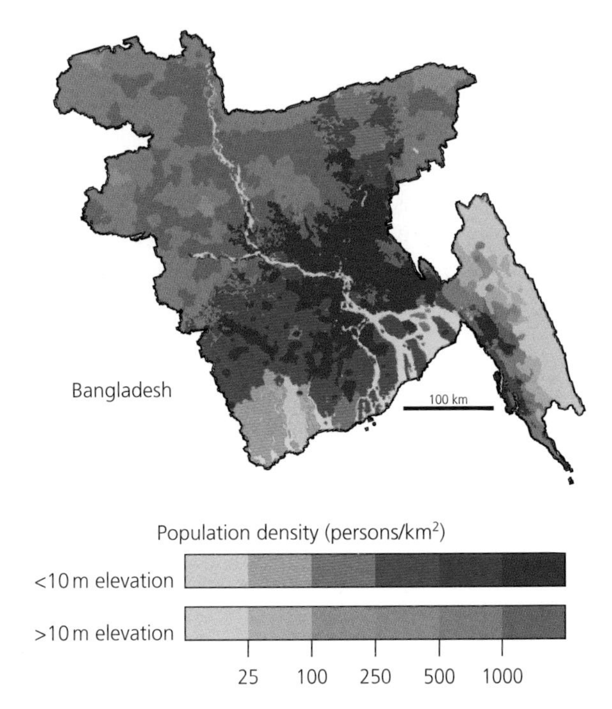

Figure 10: Population density in areas below 10 metres elevation in Bangladesh, and above 10 metres

Activity 2

1. Outline the reasons why Bangladesh is so vulnerable to sea-level rise.
2. Suggest how the Bangladeshi government can prevent damage to communities at risk from flooding.
3. Suggest reasons why richer countries have been unhelpful in assisting countries such as Bangladesh.

Quick notes (Bangladesh):

- The threat from global warming and sea-level rise varies from place to place.
- The poor and disadvantaged are more at risk form sea-level rise than the more affluent and powerful.
- Countries most at risk from climate change impacts are often those that contribute least to enhanced greenhouse gas emissions.

The responses to climate change, from a local to a global scale

There any many different approaches to tackling the threat of climate change. These operate on a number of scales, from individual action up to **international co-operation**. Some people, however, strongly disapprove of taking any action at all. These people have a number of 'reasons':

However these views are not very common, and most people – and governments – believe that action is necessary. Even those who doubt some of the more frightening claims about our future generally believe that it is better to play safe and, whenever and wherever possible, it is better to save resources rather than waste them. Most people are quite keen to reduce their impact by reducing their own 'carbon footprint'.

Individual action

'**Live sustainably**' campaigns stress the need to 'live simply' and change our own habits. To live sustainably means to live in a manner that preserves our environment, and take personal responsibility for the future of the planet. The path to living sustainably has many different routes. To live sustainably one would first need to reduce the impact on the Earth's resources. Of course there are a number of ways in which we could all achieve this (see the table on page 39).

National campaigns

National governments are very important players in dealing with climate change. However, they have three sets of responsibilities which are sometimes in conflict – economic growth, social equity (a fair society), and protection of the environment.

Economic growth inevitably involves the use of more resources – more energy, more minerals, more land and water. But it is very hard, maybe impossible, for governments to be elected if they suggest to people that they should become poorer to save the environment. Governments also need to take care that societies are 'fair' (by maintaining a level of social equity). Even governments that are made up of people who don't believe in social equity have to pay attention to it or they risk unpopularity and, if things get really bad, social unrest.

Are you living simply and sustainably?	Benefits	Costs and problems
Use local farmers' markets – do you really need imported food?	Supports local farmers and reduces the costs of transport and the carbon footprint.	Buying locally produced greenhouse-grown tomatoes in the spring is more damaging to the environment than buying imported tomatoes grown outside in Spain, for example. It costs more energy to heat the greenhouses than to transport Spanish tomatoes.
Reduce the unnecessary luxuries in your life – how many pairs of shoes do you have?	Reduces the waste of resources in their manufacture and their transport, quite apart from the packaging.	What about the foreign workers who make these 'unnecessary' goods? Will they still have jobs if we all cut back on these luxuries?
Get on your bike	Cars are great polluters. The average journey by car is just 8.7 miles, and with 33 million cars on the UK roads there is enormous scope here to reduce CO_2 emissions.	The old and the infirm cannot use bikes. It is also hard to imagine how the economy would continue to operate as effectively if we cut back on road freight.
Recycling and conservation	There are huge savings here, as materials are reused and resources saved. It is surely better to switch a light off than leave it on.	It isn't obvious how the rural peasants of Bangladesh benefit from our careful recycling. The impact on global warming is quite small and certainly doesn't improve the life of the poorest, especially if trips to the recycling bins are made in polluting vehicles.

For some governments, these pressures can outweigh their other responsibility – protection of the environment. The government of the USA, for example, refused to sign the Kyoto treaty to limit the emission of greenhouse gases. This was despite the increasingly concerned opinions of the population.

In the USA this conflict between protection of the environment and the hopes and desires of the population is really well illustrated by the growth and development of American cities. As Figure 11 shows, many of these cities are among the worst in the world as far as energy usage is concerned.

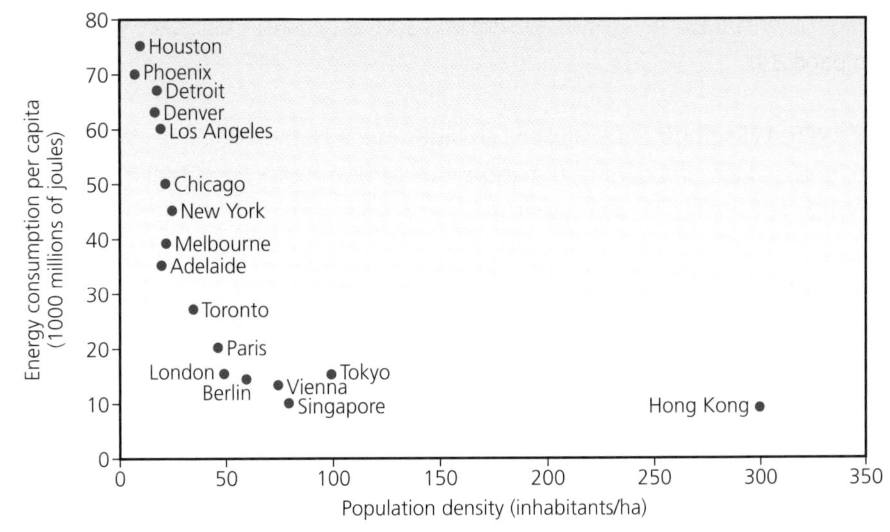

Source: Adapted from Newman, P. and Kenworthy, J., 1990

Figure 11: Population density and energy consumption, selected world cities

Using examples describe how local action might help tackle climate change. (4 marks)

■ **Basic answers** (0–1 marks)
Make a general point about attempts to tackle climate change and name one or more schemes.

● **Good answers** (2 marks)
Describe at least one 'local' attempt with some detail of the scheme/policy.

▲ **Excellent answers** (3–4 marks)
Describe local schemes/policies and show how they help tackle climate change by either reducing emissions or increasing carbon sink.

Urban sprawl has created cities that are completely dependent on motor cars. This debate has become a lot more lively since oil prices have risen. Nowhere is the conflict more obvious than in a city such as Las Vegas (Figure 12).

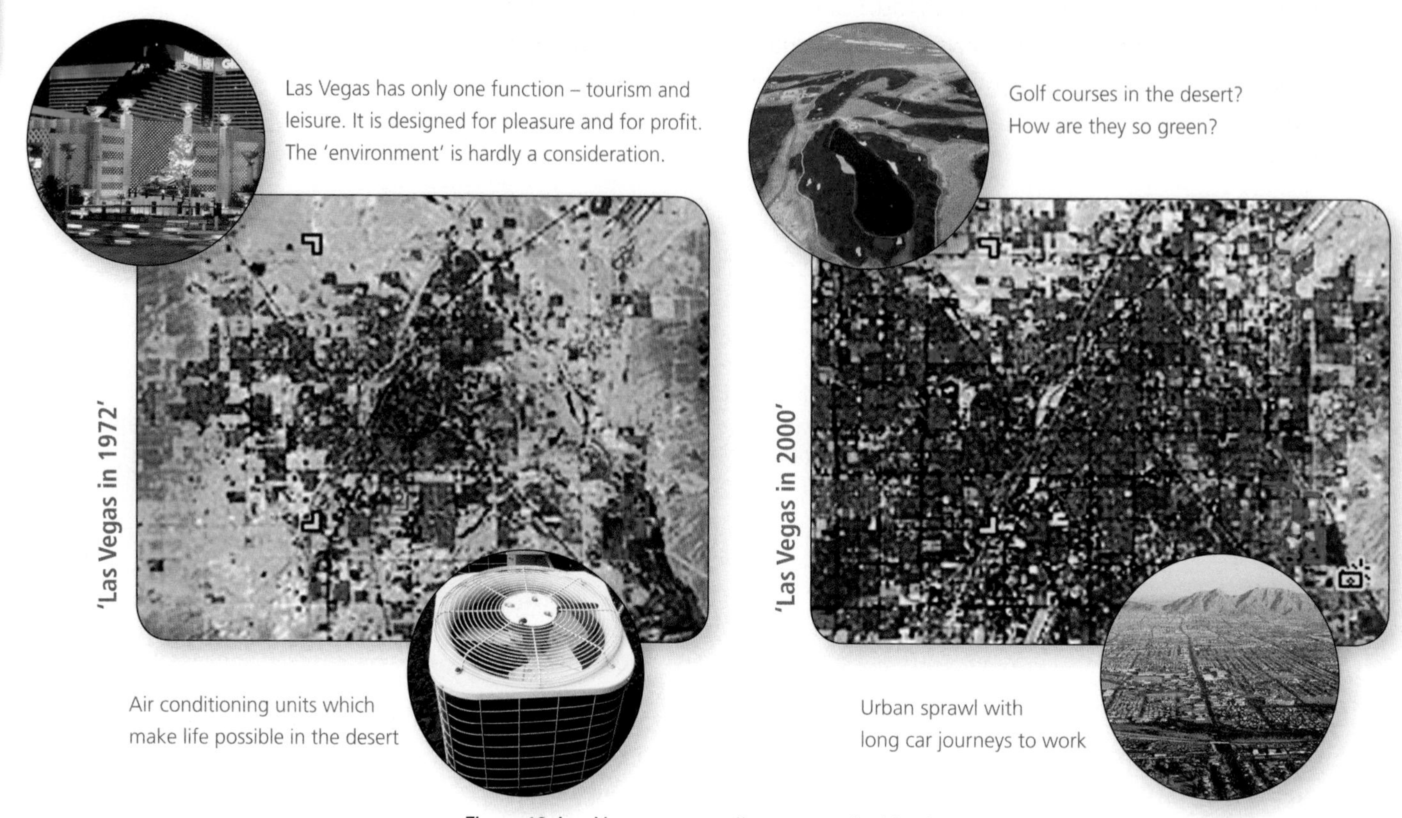

Figure 12: Las Vegas: a sprawling, unsustainable city

Cities do not grow by chance. Even ancient Rome and Athens had governments that controlled their growth in certain ways. In the USA many cities in the south-west have grown up in a period when cheap petrol and cheap land have encouraged urban sprawl. Some American cities have population densities so low they are less than those of English rural areas (Atlanta in Georgia, for example, has a lower population density than Surrey). Governments have the power to stop this sort of growth – but have chosen not to do so.

All the recycling and other lifestyle changes made by citizens of Las Vegas will not make much of a dent in the enormous carbon footprint of their city. Unsustainable cities illustrate that:

- Not all places are planned with the environment in mind.
- Some forms of modern technology, such as air conditioning, are bound to increase energy consumption.
- Tourism involving long-distance travel, especially air travel, poses serious challenges for the environment.

Activity 3

1. List three reasons why Las Vegas is a good example of an unsustainable city.
2. Using the weblinks below – and your own research – investigate the history of Las Vegas's growth, to show how it was (and is) dependent on modern technology.

http://www.lasvegasnevada.gov/

http://www.1st100.com/

http://en.wikipedia.org/wiki/Las_Vegas,_Nevada

International action

Governments find it difficult to be consistent at home – encouraging recycling whilst, at the same time, allowing the building of new runways and airport terminals – and they don't do much better in the international arena. International action is slow because many countries find it difficult to balance a commitment to economic growth with cutting emissions. For example, according to a statement from the US government at the Bali conference in 2007:

'the United States believes that any arrangement must also take into account the legitimate right of the major developing economies and indeed all countries to grow their economies…'

The following table shows a timeline of the international climate change debate.

YEAR	MEETING	OUTCOME
1988	The United Nations sets up the Intergovernmental Panel on Climate Change (IPCC).	The first report of the IPCC shows that the Earth has warmed by 0.5° C in the past century – it recommends 'strong measures'.
1992	The Earth Summit takes place in Rio de Janeiro.	The United Nations Framework Convention on Climate Change (UNFCC) is signed by 154 nations agreeing to prevent 'dangerous' warming from greenhouse gases – it sets voluntary targets for the reduction of these emissions.
1997	The Kyoto protocol is agreed.	The UNFCC's voluntary targets are replaced by legally binding controls on emissions. 178 countries signed a treaty that came into force in 2005.
2007	The Bali Conference.	The IPCC confirms that there is a greater than 90% chance that global warming in the past 50 years is caused by human activity. All nations agreed to negotiate a deal to tackle climate change but were unable to agree the details at Bali – they promised to finalise the details by 2009 but this has not yet happened.

Objectives

- Define sustainability in at least one way.
- Explain how urban transport can be managed to make it more sustainable.
- Explain and illustrate how the management of resources in tropical rainforests can have a variety of effects.

Sustainable development for the planet

Definitions and interpretations of 'sustainable development'

Concern about the environment began in the 1960s, when writers such as Rachel Carson pointed out that the uncontrolled use of pesticides was destroying bird life and possibly affecting human health. Carson also accused the chemical industry of lying about the effects of its products. A few years later, in 1972, a group of scientists known as the 'Club of Rome' published *Limits to Growth*, a book that suggested it would not be possible for 'industrial output' and 'population' to just keep on growing whilst maintaining the levels of 'natural resources' and 'food production'. They revised their pessimistic forecast in 1999 (see Figure 13).

Since the 1980s the idea of **'sustainable development'** has become very well known. In fact, the word 'sustainable' is now used by almost every global organisation – from governments to corporations, from schools to local authorities, from charities to churches. The word is used – and misused – so much that it is easy to lose track of what it really means.

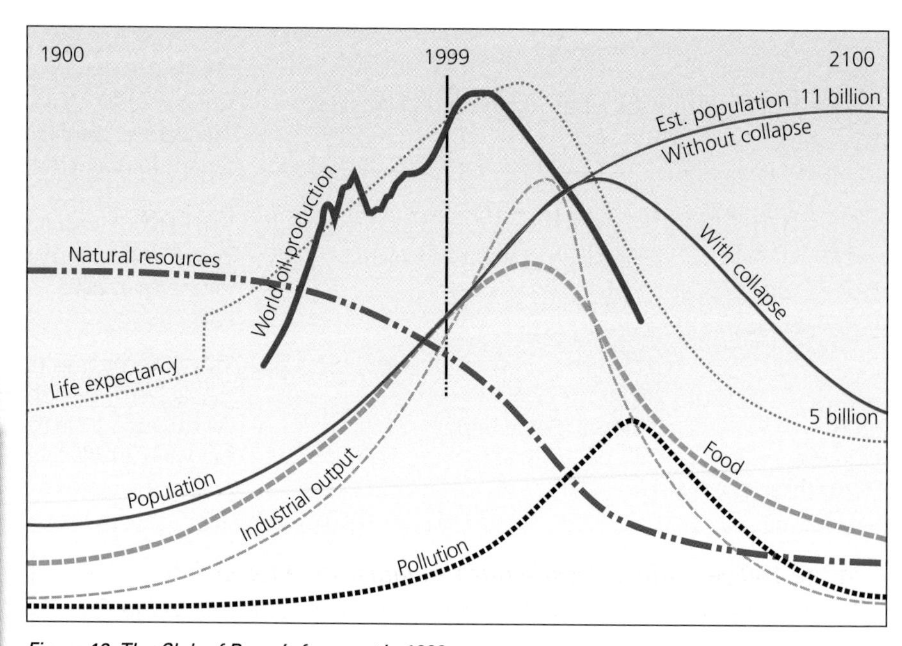

Figure 13: The Club of Rome's forecast in 1999

Skills Builder 7

Look at Figure 13.

(a) Identify one trend that is expected to go on rising until 2100.

(b) Describe the changes in pollution that are forecast.

(c) Explain the changes in pollution that are forecast.

The United Nations definition

The 'official' definition of sustainable development is the one in *Our Common Future*, the report of the 1987 United Nations World Commission on Environment and Development. (This is sometimes called the 'Brundtland report', after the chairperson, former Norwegian Prime Minister, Gro Harlem Brundtland.) According to this definition:

> *Sustainable development is development that meets the needs of the present without compromising [limiting] the ability of future generations to meet their own needs.*

Since the publication of the report, the focus of sustainable development has frequently been about 'future generations', with much less said about the 'present' generation. As the global gap between rich and poor has increased, some critics have suggested that the original meaning of the report, with its emphasis on poverty today, has been lost. The idea that sustainable development should close the gap between rich and poor *today* does not feature as much as the original authors intended in the many discussions that focus on recycling, the reduction of carbon footprints and the sustainable management of cities. Certainly after more than twenty years – and much effort and some success in promoting sustainable development within high income countries and elsewhere – there is no evidence that the poor of the world have benefited greatly.

Interpretations of 'sustainable' by large organisations

Most large global organisations (also known as transnational companies – TNCs) claim to have adopted the idea of sustainable development, but their interpretations may differ from the UN definition. For some, 'sustainable' seems to mean 'responsible':

> *A sustainable business is by definition a responsible business. Responsibility isn't an add-on or luxury; it's an integral part of the way we run our business to achieve profitable, consistent and sustainable performance.*
>
> BP Caspian description of sustainability

For some other organisations, 'sustainable' simply seems to mean 'long-term' – something that can be kept going:

> *In the 140 year life of Nestlé our fundamental approach to business has been the creation of long-term sustainable value for our consumers, customers, employees, shareholders and society as a whole.*
>
> Nestlé

Explain why there are different interpretations of sustainable development. (3 marks)

Basic answers (0–1 marks)
Offer a basic definition of sustainable development as 'development that helps people in the future'.

Good answers (2 marks)
Include present and future needs in the definition and 'explain' differences by suggesting that the topic is complex.

Excellent answers (3 marks)
Offer a good definition and make the point that either sustainable has become a useful word for large organisations to use or there may be some tension between satisfying current needs and looking after future generations.

Activity 4

Read the article about Nike on the following link: http://www.mcdonough.com/writings/inspiration_innovation.htm

1. Identify three ways in which Nike is trying to become more sustainable.
2. Summarise the methods that Nike have used to make the company more ecologically responsible.

It isn't always obvious what 'sustainable' means in company statements. If the word is removed then not a lot of meaning is lost. Large organisations spend a great deal of money on their public image and they want to be seen as 'caring'. No doubt, many of the people who work for them do care about the environment, but their first priority is to make a profit, whilst trying to keep their customers, shareholders and workers all happy.

Large organisations and their policies for sustainability

As we have seen, large organisations seem to use the word 'sustainable' rather freely. Like 'organic' and 'natural' it has become a useful word to help sales. There are, however, companies that have taken the message about environmental damage more seriously, and have successfully embraced a comprehensive view of sustainability.

Interface Inc – the first sustainable corporation?

Interface Inc is the largest manufacturer of commercial carpets in the USA and its mission statement is very different from those of many other large organisations:

> *Why is striving for sustainability so important?*
>
> *Here's the problem in a nutshell. Industrialism developed in a different world from the one we live in today – fewer people, less material well-being, plentiful natural resources. What emerged was a highly productive, take–make–waste system that assumed infinite resources and infinite sinks for industrial wastes. Industry moves, mines, extracts, shovels, burns, wastes, pumps and disposes of four million pounds [1,800 tonnes] of material in order to provide one average, middle-class American family with their needs for a year. Today, the rate of material throughput is endangering our prosperity. At Interface, we recognise that we are part of the problem. What's the solution? We're not sure, but we have some ideas. We believe that there's a cure for resource waste that is profitable, creative and practical. Our vision is to lead the way to the next industrial revolution by becoming the first sustainable corporation.*
>
> Adapted from: http://www.interfaceinc.com/goals/sustainability_overview.html. New website address: interfaceglobal.com

The company has identified seven goals to improve its environmental performance as shown in the table on page 45.

Activity 5

1. Outline the 'problem' identified in Interface's mission statement.
2. What do you think Interface means by the 'next industrial revolution'?
3. Develop a fact file about Interface and its policies by using the following links:

http://www.interfaceinc.com/

http://uk.youtube.com/watch?v=RcRDUIbT4gw

http://www.bigpicture.tv/videos/watch/e00da03b6

Eliminating waste	in every area of business
Benign emissions	eliminating toxic substances from products, vehicles and facilities
Using renewable energy sources	solar, wind, biomass, geothermal, tidal, etc.
Closing the loop	redesigning processes and products to close the technical loop using recovered and bio-based materials
Using resource-efficient transport	to reduce waste and emissions
Culture	Creating a culture amongst stakeholders that integrates sustainability principles and improves their lives
Business model	Creating a new business model that demonstrates the value of sustainability-based commerce

Adapted from: http://www.interfacesustainability.com/seven.html. New website address: interfaceglobal.com

Unilever

Another large organisation that has embraced sustainable development is Unilever. The company promotes sustainable development within its very large and wide-ranging operations. Its website – http://www.unilever.com/ourvalues – describes many of these policies in detail with some useful case studies.

A good example of Unilever's policies in action comes from Kenya, where a branch of the company Unilever Tea Kenya (once known as Brooke Bond) is looking again at how to use wood from eucalyptus trees as fuel to dry the picked tea-leaves. An increase in tea production is likely to lead to a reduction in fuel wood. In a twin-track policy, the company has:

- Changed its tree planting policy, increasing the density and also introducing coppicing which cuts back the tree to stimulate growth – this will increase wood supply by 15%
- Improved wood burning techniques by drying the wood out for longer, so cutting back the moisture in the wood to make it more efficient as a fuel – with more efficient boilers, this will reduce wood consumption by about 25%.

There are many examples of these changes in production methods. The company suggests that becoming 'more sustainable' may also allow it to become more profitable as well, at least in some areas of their operations.

With the use of examples describe how large organisations have made their operations more sustainable. (6 marks)

Basic answers (Level 1)
Describe in general terms how companies and other organisations try to make savings in their use of resources.

Good answers (Level 2)
Name at least one organisation/company and offer some details about one of its policies – perhaps using recycled materials or reducing packaging.

Excellent answers (Level 3)
Use at least two examples and identify more than one area of operation in which sustainable solutions have been sought/implemented – perhaps energy use as well as production for example.

Urban transport – the 'public versus private' debate

Making cities more sustainable is not easy. They are, by their nature, concentrations of many people in relatively small areas and they have to drain resources from outside their boundaries – food and fuel being two obvious examples. (This is in complete contrast to poor rural areas in low-income countries, which have very limited impact on the resources outside their area.)

Transport solutions and the density of cities

Making cities more efficient in their use of resources will not make them self-sufficient – but it might make them less wasteful. This is the usual goal of the transport policies that have been introduced in many cities. Public transport, properly organised, will always be more efficient and use fewer resources than private transport. But cities come in all different shapes and sizes, and pose different problems for planners. Transport solutions are obviously not the same in low-density cities as they are in high-density cities. As Figure 14 shows, in low-density cities (where everything is spaced out and there is limited public transport), the residents' fuel consumption is much higher than in denser, more compact cities – where public transport is usually better developed and used. Planning and implementing sustainable transport solutions is always going to be easier in high-density cities. Amsterdam, for example, has 700,000 bicycles and an efficient tram network.

Activity 6

1. Give two reasons why it is easier to develop public transport in some cities than in others.

2. Why might some American cities be regarded as unsustainable?

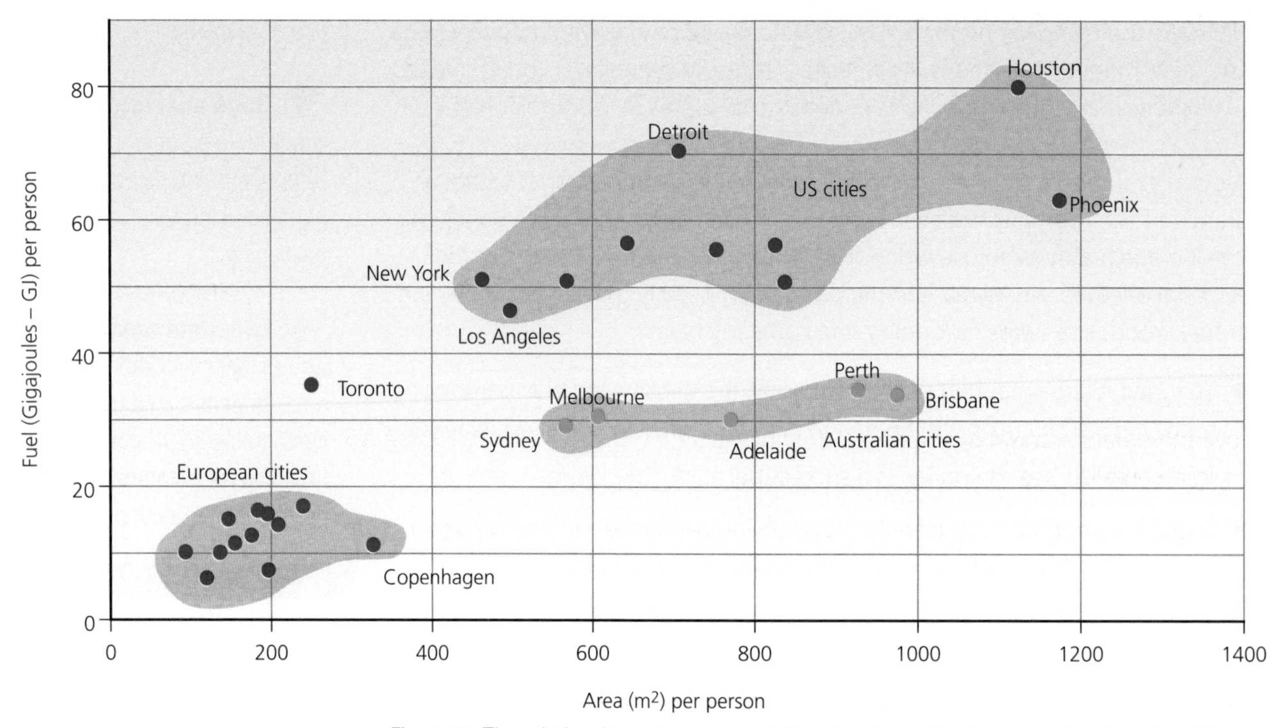

Figure 14: The relationship between population density and fuel consumption in major cities

Congestion charging

London was one of the first cities to introduce **congestion charging** (Figure 15). This is a simple idea – if drivers enter certain areas of the city, usually the centre, they have to pay a daily charge. In London, a central zone was set up in 2003, and a western extension was added in 2007. According to the Fifth Annual Impacts Monitoring Report:

- Before charging began, some 334,000 vehicles entered the central zone each day
- In 2006, around 70,000 fewer vehicles entered it each day – a drop of 21%
- Traffic levels in the western extension have dropped by 25%.

Figure 15: Cameras 'read' the registration plates of cars entering the Congestion Charge Zone, and their owners have to pay a daily fee.

It remains a controversial policy, and opinions are sharply divided between environmental organisations (such as Greenpeace) and motoring organisations. Environmentalists see the policy as a success because, they claim:

- Greenhouse gas emissions in the zone have been reduced by almost 20%
- Pollutants that adversely affect Londoners' air quality and health have fallen by 12%.

Others suggest that it is not curing the problem of congestion and pollution but actually making it worse. Often sponsored by motoring organisations, there are many websites explaining these views, such as http://www.abd.org.uk/london_congestion_charge_report2007.htm.

Activity 7

1. Using the congestion charge weblink given, and your own research, identify the main groups in favour of congestion charging, and opposed to congestion charging.
2. Suggest reasons for their differing views.

ResultsPlus
Exam Question Report

Choose a fragile environment you have studied which has been damaged by resource exploitation. Explain the effect of the damage on the fragile environment. (5 marks, June 2008)

How students answered

Most students did not choose a fragile environment, and so wrote answers that were completely irrelevant, or identified fragile environments in general but did not identify any specific location or case-study

45% (0–1marks)

Many students could describe the chosen environment, usually Ecuador's Oriente region, and outline the reasons why damage was inflicted, but did not explain the link between the exploitation and the damage.

43% (2–3 marks)

Some students answered this question well, concentrating on the effects and using data to help explain how oil spills caused damage to the rainforest and to its peoples. Explanation sometimes involved the 'need' to exploit the resource for the country recognising that some benefits occurred as well as some costs.

12% (4–5 marks)

Resource extraction from tropical rainforests

The sustainable extraction of non-renewable resources is not possible, given that they will eventually run out. What is more, the extraction can cause significant damage to the environment by affecting other systems, such as the vegetation. Tropical rainforests are lightly populated regions, often inhabited by groups who have little power in the capital city. These peoples have often lived sustainably in the rainforest for many thousands of years, living off its resources without plundering them.

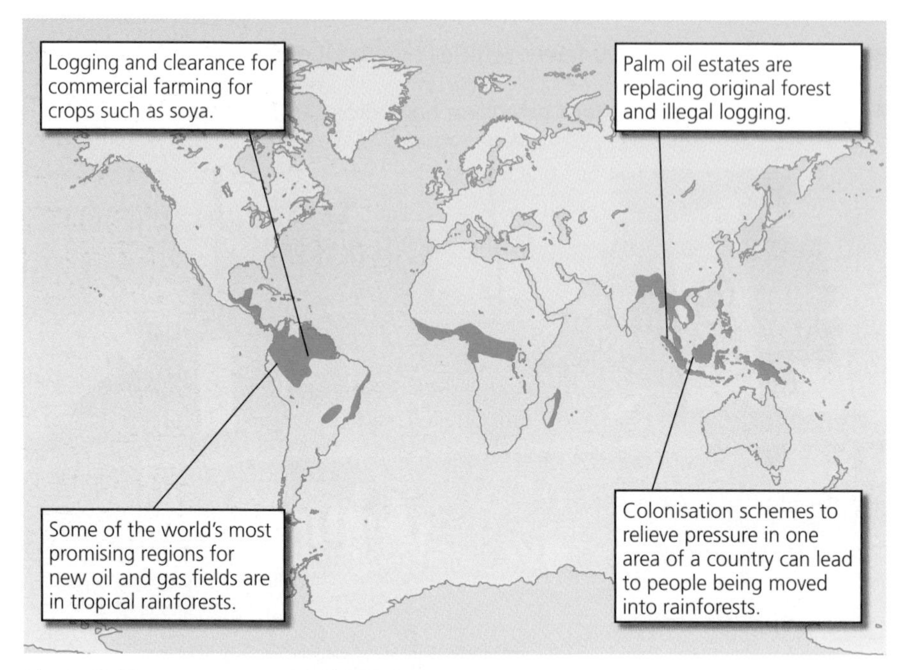

Figure 16: The pressures on the world's rainforests

Rainforests are 'fragile' because the vegetation grows very quickly but also breaks down very quickly. In the wet and hot conditions, leaves rot in days rather than weeks. As soon as they rot, the nutrients released are taken up by plant growth. If you remove the vegetation, you remove the source of the nutrients – and the soil rapidly become infertile and useless. Other vegetation, such as agricultural crops, can temporarily 'replace' the rainforest, but the whole point of growing them is to harvest them, not to let them rot back into the ground.

In such 'fragile' environments the impacts of human activities are more obvious and the damage done is hard to reverse (Figure 16). In the tropical rainforests, the impacts of **resource extraction** include the factors shown in the table on page 49.

Deforestation	To set up their operations, companies open roads through forests. These bring settlers who have access to timber and new land, and who may engage in slash-and-burn activities and logging.
Local conflicts	Indigenous and local peoples often gain the least from natural resources extraction, but stand to lose the most. Compensation from energy firms and the government, where it is awarded, is often very small. In addition, local communities are not always informed of extraction projects.
Biodiversity loss	Fragmentation of natural habitats caused by the installation of pipelines leads to species having smaller populations that are not viable in the long term. Companies operating close to (or even inside) protected areas don't always follow the rules laid down to maintain biodiversity.
Soil and aquatic pollution	Many things can go wrong as oil is brought to the surface and processed. Spills and toxic by-products are sometimes dumped near the site or stored in open waste pits, polluting the surrounding lands and water.
Air pollution	Some of the by-products of natural gas are burned in the open air. The flames pollute the atmosphere and can cause fires, threatening the lives of local inhabitants. Unnecessary 'flaring' is also a waste of gas which could provide energy to local people and reduce deforestation.

Oil from Ecuador

Poor countries have a long history of being exploited. Many are former colonies – and the whole point of colonies was to supply the 'home' country (the invading foreign power) with raw materials. When these countries gained their political independence they often found that they had little choice but to continue exporting their raw materials, often with the 'help' of transnational companies.

The Oriente ('east') is the half of Ecuador east of the Andes. It consists of 13 million hectares of tropical rainforest, lying at the headwaters of the Amazon river network (Figure 17). The region contains one of the most diverse collections of plant and animal life in the world, including many endangered species. The Oriente is also home to 95,000 indigenous people and 250,000 immigrants, who have followed the oil roads east in search of land and work.

The development of oil extraction in Ecuador has followed a pattern that is sadly familiar to most developing countries with natural resources. Since the first barrels were extracted in 1972, the industry has been dominated by multinational corporations, led by Texaco until its 1992 departure. There has been very little obvious government oversight and almost no attention paid to non-economic concerns. As a result, the oil industry has taken a predictable toll on the Oriente's environment and on the welfare of its inhabitants:

- Oil operations discharged 4.3 million gallons of toxic waste into the Oriente's environment every day.
- Toxic contaminants in drinking water have been reported at levels reaching 1,000 times the safety standards recommended by the US Environmental Protection Agency.
- Local health workers reported increased gastrointestinal problems, skin rashes, and birth defects and cancers – ailments that they believe to be related to this contamination

Despite the continued benefits of oil, at least for some, the country has suffered increasing levels of poverty, inequality and unemployment. Basic social indicators (maternal and infant mortality, child nutrition, literacy) have stagnated or got worse. At the same time, oil has opened the door to powerful foreign interests – including the multinational oil companies (many with revenues dwarfing Ecuador's GNP). Social inequalities in the country are more obvious, with a powerful class of higher-paid oil employees alongside an impoverished and desperate public – who had been brought up on promises of oil wealth.

Ecuador has an estimated 10 to 20 years of oil left in the Amazon. But at best the income from it could only pay off a fifth of the current national debt. The government's most optimistic projections for the future leave the country in much the same place as before – in poverty and in debt.

Quick notes
(Oil from Ecuador):

- The 'winners' from projects such as this one include powerful minorities in the country.
- Natural resource extraction can be very damaging to the environment.
- Remote places are more difficult to record and monitor.
- Transnational corporations have enormous resources and are difficult to prosecute successfully when they are accused of wrong-doing.

Activity 8

1. Define the term 'fragile environment'.
2. Identify the features of tropical rainforests that make them 'fragile'.
3. Draw up a simple matrix (grid) to show 'winners' and 'losers' from:
   (a) the extraction of oil from the Oriente
   (b) the development of oil palm estates in Papua New Guinea.

Figure 17: El Oriente in Ecuador

Palm oil production in Papua New Guinea

Palm oil is a reddish vegetable oil used mostly for cooking, but also as an ingredient in margarines, soaps, and cosmetics. It is extracted from the fruit of the oil palm.

There are four major oil palm projects in Papua New Guinea, and palm oil has now become the country's largest agricultural foreign exchange earner, overtaking coffee (Figure 18). Most of the projects consist of a 'nucleus estate' and several local smallholders who grow the oil palms. The smallholders supply the oil palm fruits to the nucleus estate company, which owns the mills needed to produce the oil. The nucleus estate companies, which are mostly foreign owned, supply the seedlings, expertise, tools, fertilisers, and so on, which are later paid for by the growers. The growers provide labour and bear all the cost of land clearing and all stages of the palm plantation establishment, including regular maintenance and harvesting. So much of the risk is on the growers because their return is dependent on the productivity of their plots – and on the world palm oil price level. This model is more profitable for the oil companies than owning conventional large-scale plantations.

Figure 18: Palm oil production

The government gives tax breaks for companies interested in developing this industry, and there are plans for expansion of oil palm plantations in nearly every province of Papua New Guinea. But the change that comes with this kind of externally imposed project is often disruptive and can undermine communities' existing social systems. Some of the problems include:

- A move away from the local customary processes of land use and allocation
- Waterway pollution during construction and operation of the industry
- Smallholders being totally dependent on the oil palm company – and price fluctuations
- Less biodiversity, as plantations reduce the habitats of endangered species.

Small landowners are concerned about this process. In 2003, for example, a group put an advertisement in a local paper, saying:

> We, the landowners are developing and will continue to develop our land on our own terms. We therefore sternly warn all those parties involved in wanting to use **our land** for oil palm to stay out!
> Any attempt to bring oil palm on **our land** will be strongly resisted.

Quick notes
(Palm oil production):

- The growth of plantation crops frequently involves the removal of rainforest.
- **Biofuels** have high environmental costs that may outweigh their benefits.
- Economic benefits from agricultural changes such as this one fall unevenly on the local population.
- There are considerable social costs involved when export crops such as this one are introduced.

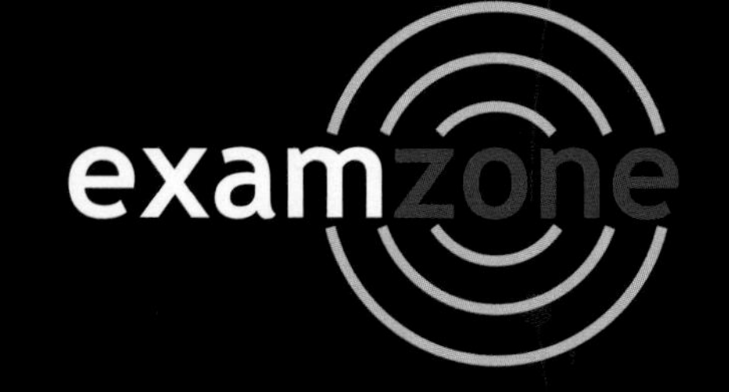

Know Zone Chapter 2 Challenges for the planet

Climate change is controversial. We know that the global climate is changing and many believe that human activity is the main cause. Some, but not all, scientists think the results will be disastrous. What we can do about it and what we choose to do about it are big issues for all of us.

'Sustainable' is a word that you see a lot, but its meaning changes a good deal and you need to take care when interpreting it. In this section, you need to know your examples and case studies to score well.

You should know...

- ☐ How the climate has changed over the last 10,000 years
- ☐ What caused these changes, including volcanic action, orbital geometry and solar output variations
- ☐ The causes of current climate change at both the local and global scales
- ☐ What causes these changes, including rising car ownership
- ☐ What the negative impact of climate change is on people and the environment
- ☐ How changes in climate might affect sea levels, food production and fragile environments such as small islands
- ☐ How we can respond to climate change, including both local and global responses
- ☐ The detail of some of these schemes, including 'live simply' campaigns and global agreements
- ☐ How to define sustainable development
- ☐ Why there are a number of different interpretations of sustainable development
- ☐ How some large companies have tried hard to implement policies to make them more sustainable businesses
- ☐ Why some cities are easier to make more sustainable than others
- ☐ What can be done in practice to reduce dependence on motor cars
- ☐ Why fragile environments are at risk from certain types of development
- ☐ Why resource extraction from rainforests poses real problems for the management of these areas
- ☐ Two contrasting examples of resource extraction from tropical rainforests

Key terms

Air pollution
Biodiversity
Biofuels
Climate change
Congestion charging
Corporation
Crop yield
Deforestation
Fossil fuels
Global warming
Greenhouse gases
Ice age
International co-operation
Live sustainably/simply
Methane
Milankovitch mechanism
Orbital geometry
Resource extraction
Solar output
Sustainable development

Which key terms match the following definitions?

A A system in which, if drivers enter certain areas of a city, usually the centre, they have to pay a daily charge

B Campaigns that stress the need to change our own habits to live in a manner that preserves our environment

C Fuel derived from biological material, such as palm oil

D The number and variety of living species found in a specific area

E A period in the history of the Earth when temperatures are low, large ice sheets develop and glaciers advance

F The addition of harmful chemicals, particles or biological material to the atmosphere

G Non-renewable resources that can be burned – such as coal, oil or natural gas – that have been formed in the Earth's crust

H The removal of the Earth's resources, usually by mining

To check your answers, look at the glossary on page 289.

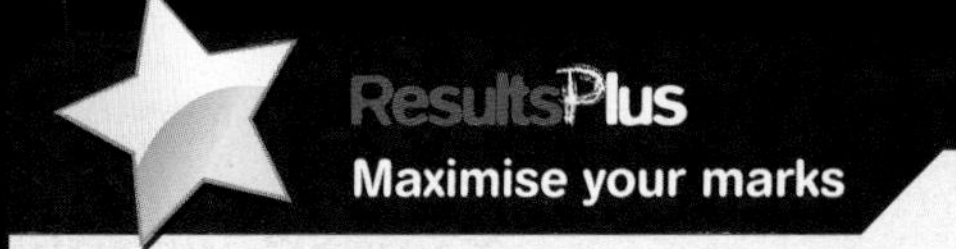

Foundation Question: What is 'sustainable management'? (3 marks)

Student answer ■ (awarded 1 mark)	**Examiner comments**	**Build a better answer** △ (awarded 3 marks)
Sustainable means protecting things so they last better. *This is what you have to do if you manage things and want to be here in the future to enjoy it rather than destroying it all now.*	• ***Sustainable means...*** is awarded no marks. The student's definition of ***sustainable*** and ***protecting things better*** is not precise enough. • ***This is what you...*** is awarded 1 mark because the idea of enjoying it in the future is a better definition of 'sustainable' than the previous sentence. However, the student does not address 'management'.	Sustainable development is development that tries to meet the needs of the present without stopping future generations from meeting their needs. People who manage companies, local areas and governments should help organise their 'businesses' to ensure that this can be achieved. This is sustainable management.

Overall comment: Many students offer *examples* when they are asked for *definitions*. Examples might help your definition, and it is useful to add them even when not specifically asked to do so, but you must remember to include the definition.

Higher Question: Explain how volcanic activity has an effect on the climate. (4 marks)

Student answer ■ (awarded 1 mark)	**Examiner comments**	**Build a better answer** △ (awarded 4 marks)
Volcanoes like the one in Iceland can be very damaging. *This eruption killed loads of people not just because of the explosion but because people couldn't find any food.* *The eruption blotted out the sun for days which made it colder for a while.*	• ***Volcanoes like the...*** This is true but not relevant to this question because it does not explain *how it* is ***damaging***. • ***This eruption killed...*** contains nothing relevant to the question. The student is focusing on the impacts on people rather than the effect on the climate. • ***The eruption blotted out...*** is worth 1 mark because it is correct. However, the command word in the question is *explain*. The student does not explain ***blotted out the sun***.	Many volcanic eruptions, such as the Laki eruption, produce large quantities of gas and small particles. The Laki volcano erupted for 8 months. The material entered the atmosphere and circulated in the northern hemisphere. These particles combined with water vapour to produce a fog that reduced the amount of sunlight reaching the surface. Particles in the upper atmosphere reflected back sunlight. Temperatures fell as a result of these processes, both in Iceland and beyond.

Overall comment: Do not be tempted to write all you know about a topic. The command word is *explain* so candidates who can offer some reasons will score higher.

Unit 2 The natural environment

Your course

This unit investigates the physical geography of the natural world and the issues relating to the environment. There are two sections:

Section A will cover the physical world and you will study **one** topic:

- Topic 1 (Chapter 3): Coastal landscapes
- Topic 2 (Chapter 4): River landscapes
- Topic 3 (Chapter 5): Glaciated landscapes
- Topic 4 (Chapter 6): Tectonic landscapes

Section B will cover environmental issues and you will study one topic:

- Topic 5 (Chapter 7): A wasteful world
- Topic 6 (Chapter 8): A watery world

Your assessment

- You will sit a 1-hour written exam worth a total of 50 marks.
- There will be a variety of question types: short answer, graphical and extended answer, which you will practice throughout the chapter that you study. You will answer **one** question from Section A and **one** question from Section B.
- **Section A** contains four questions, one on each of the physical world topics. You will have a resource booklet which contains any material that you will need to answer the exam questions.
- **Section B** contains two questions, one on each of the environmental issues topics. You might be asked to refer to material in the resource booklet to answer the questions.

Remember to answer the questions for the topics that you have studied in class!

Study the aerial photograph of Iceland, which is located just south of the Arctic Circle.

- At what time of year was this photograph taken?
- Justify your answer.
- Describe the pattern of rivers.

Chapter 3 Coastal landscapes

Objectives

- Know that there are two types of wave.
- Be able to describe the landforms of the coastline.
- Understand how processes impact on the coastal landforms.

Coastal processes produce distinctive landforms

The two types of wave – destructive and constructive

There are different types of wave that move towards the coast. You may have noticed this if you have stood on a beach and watched the waves coming in. Waves are produced by wind blowing across the surface of the water. The stronger the wind and the longer it blows, the bigger the waves tend to be.

When waves break, water rushes up the beach due to the energy from the wave. This is called the **swash**. When the water has lost its energy further up the beach, it runs back down again, under gravity. This is the **backwash**. Big waves break with lots of force and energy and this means they have the power to carry out **erosion** of beaches or rocks on the coast. These are called **'destructive waves'** and you can see their main features in Figure 1. As well as being tall and breaking with lots of power, they usually arrive quickly and have a high frequency – a lot of them come in a short period of time. Their backwash is greater than their swash and so sediment is taken away from the beach, back into the sea.

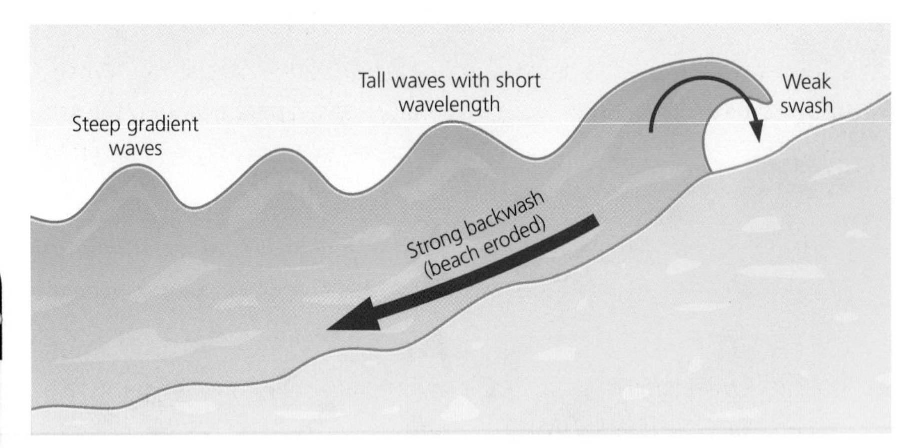

Figure 1: Destructive waves

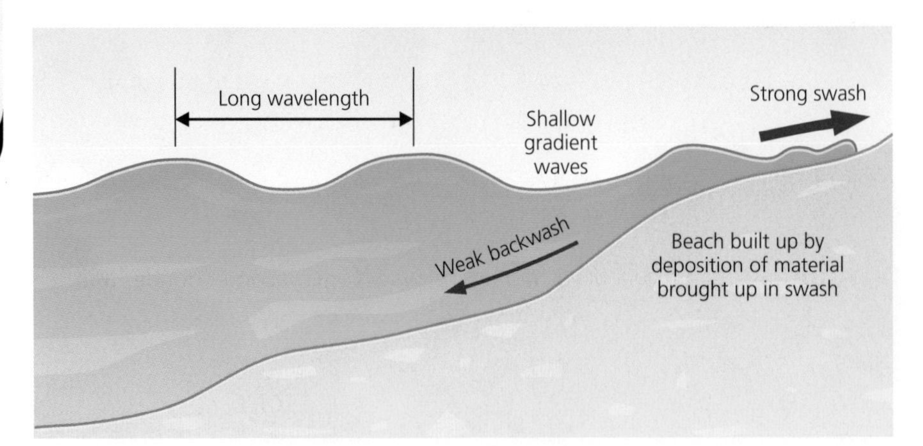

Figure 2: Constructive waves

ResultsPlus
Watch out!

Some students confuse the words 'destructive' and 'constructive'. Remember the difference by thinking of building up and adding material as being 'construction' – like on a building site.

'Constructive waves' are very different. In calm conditions – without much wind – the waves are usually small, weak and with a low frequency. They don't break with much force and they tend to add sand and other sediment to the coastline by **deposition**. The swash is greater than the backwash and so sediment is pushed up the beach, helping to build it up. You can see this in Figure 2. The differences between the two types of wave are summarised in the table.

	Destructive wave	Constructive wave
Wave height	High (more than 1 metre)	Low (less than 1 metre)
Wave energy	High	Low
Wave frequency	High (over 10 per minute)	Low (fewer than 10 per minute)
Swash:Backwash	Backwash > Swash	Swash > Backwash
Main process	Erosion	Deposition

The differences between destructive waves and constructive waves

Activity 1

1. Which parts of the British Isles do you think have the most destructive waves? Why is this the case?

2. Does a particular beach always have the same type of wave, or can it vary? Why might this happen?

The impact of processes on the coast

Coasts are subjected to a wide range of natural processes that can combine to produce some very distinctive landforms. These processes include **weathering**, erosion and **mass movement.**

Weathering can happen in many ways. Some common ones in coastal areas are:

- Salt crystal growth – this happens because sea water contains salt. When spray from waves lands on rocks, the water can evaporate leaving the salt behind. The salt crystals grow and create stresses in the rock, causing it to break down into small fragments.
- **Acid rain** – all rain is slightly acidic. If the air is polluted, it can be very acidic. When rain falls on rocks, the acid in it can react with weak minerals causing them to dissolve, and the rock to decay.
- Biological weathering – the roots of **vegetation** can grow into cracks in a rock and split the rock apart.

Wave erosion involves several different methods, including:

- **Hydraulic action** – this results from the force of the water hitting the cliffs, often forcing pockets of air into cracks and crevices in a cliff face.
- **Abrasion** – this is caused by the waves picking up stones and hurling them at the cliffs and so wearing the cliff away.
- **Attrition** – any material carried by the waves will become rounder and smaller over time as it collides with other particles and all the sharp edges get knocked off.
- **Corrosion** – the dissolving of rocks and minerals by sea water.

ResultsPlus
Exam Tip

Good quality examination answers usually show a clear understanding of processes. Don't just name the process – explain how it works.

There are several different processes of mass movement. In coastal areas, the main forms of mass movement are:

- **Rock fall** – which is one of the most sudden forms of mass movement. Rock fall occurs when fragments of rock weathered from a cliff face fall under gravity and collect at the base.
- Slumping – which often happens when the bottom of a cliff is eroded by waves. This makes the slope steeper and the cliff can slide downwards in a rotational manner, often triggered by saturation due to rain, which both 'lubricates' the rock and makes it much heavier. You can see how the weight added by rainwater and the erosion by waves combine to cause a rotational slump in Figure 3.

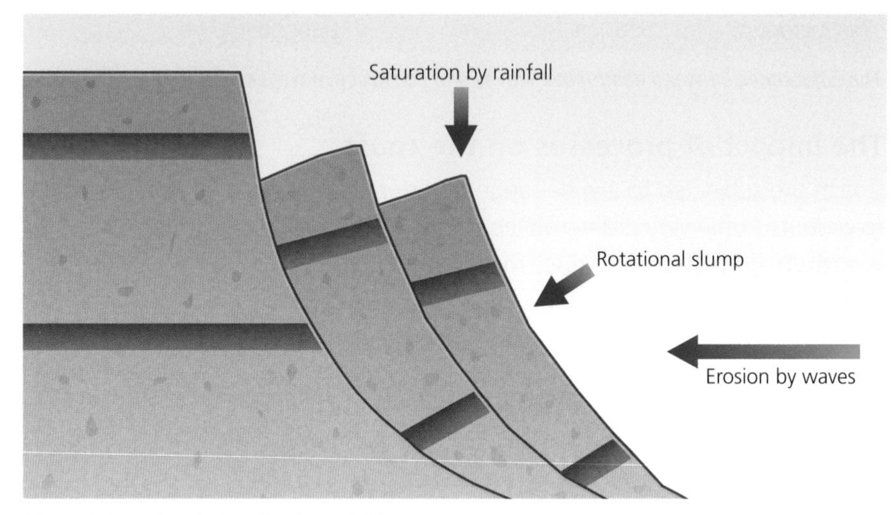

Figure 3: Rotational slumping in a cliff face

The landforms that are produced by these processes can be seen in many coastal areas – you may have seen many of them yourself, when on holiday.

It can be helpful to think of the formation of cliffs and wave-cut platforms as a sequence: undercutting – collapse – retreat – repeat.

Cliffs and wave-cut platforms are usually found together. The wave-cut platform is a fairly flat rocky area at the base of a cliff. Waves erode the coastline causing undercutting at the base of the slope. This forms a notch that gradually gets bigger. The rock above eventually loses its support and then it collapses. The debris is gradually washed away by the waves. The process is repeated and so the cliff slowly retreats backwards and becomes steeper. At Seven Sisters in Kent the cliffs are up to 160 metres high, whilst beneath them the wave-cut platforms extend up to 540 metres out to sea.

Headlands and bays are another pair of landforms usually found together, with headlands separating the bays along a coastline. The key to their formation is differences in **geology**. Where a coastline is made of different types of rock, the areas of weak, easily eroded rock are eroded back by the waves, creating bays. The areas of more resistant rock do not get eroded so rapidly. So they jut out into the sea, as headlands, separating one bay from another. The result is an 'indented' coastline, as you can see in Figure 4.

In Dorset, for example, Swanage Bay has been eroded in weak clay, whilst the headlands either side are made of more resistant chalk – Ballard Point – and limestone – Peveril Point.

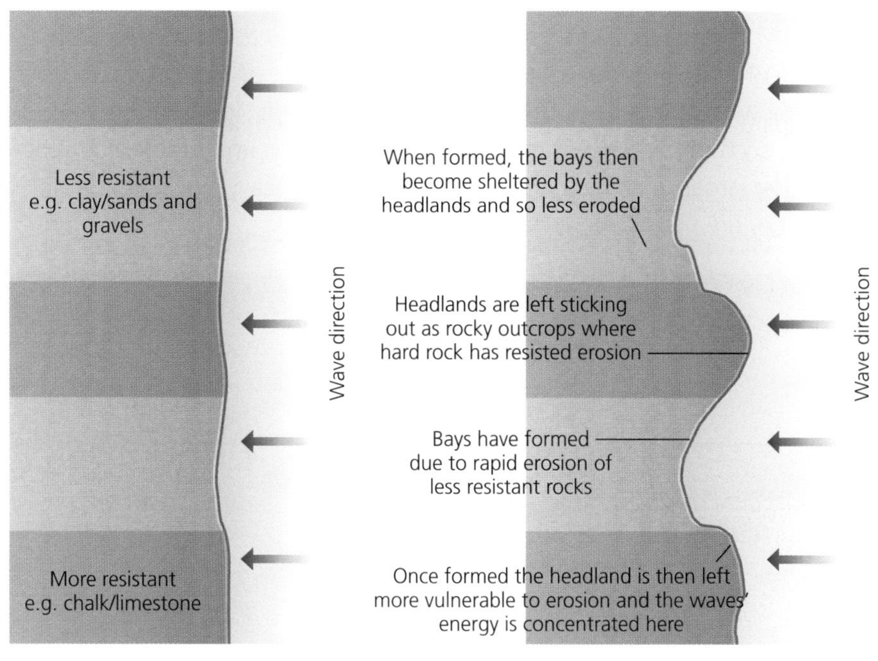

Figure 4: The formation of headlands and bays

Although headlands are made of resistant rock, they also gradually become eroded. Waves approaching the coastline bend around the headland and erode it from both sides. Any points of weakness, such as cracks or joints in the rock, become eroded more quickly than the surrounding rock. This forms small caves on either side. With more erosion taking place over time, the two caves may join to form an arch. More erosion, and weathering of the arch roof, leads to the roof collapsing. This leaves an isolated **stack**. The stack itself gradually becomes undercut by wave erosion, eventually collapsing to leave a **stump**, which may only be visible at low tide. All of these landforms are shown in Figure 5, and they can also be seen in Dyrholaey on the south coast of Iceland, as shown in Figure 6.

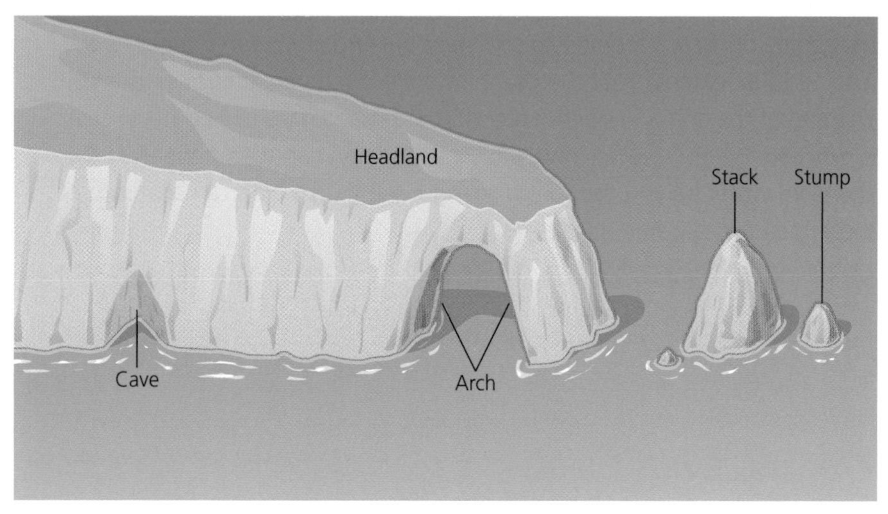

Figure 5: Landforms produced by the erosion of a headland

Activity 2

1. Find a photograph of a bay and headland coastline from the internet.
2. Label the bays and headlands, using drawing tools.
3. Annotate to suggest how the bays and headlands have been formed, using text boxes.

Skills Builder 1

Look at Figure 6.

(a) Name the landforms labelled A and B.

(b) Describe how these landforms might change over time.

(c) Explain how natural processes might cause these changes.

Figure 6: Landforms of erosion at Dyrholaey, southern Iceland

ResultsPlus
Build Better Answers

Describe and explain the process of longshore drift. (6 marks)

■ **Basic answers** (Level 1)
Offer ideas only about waves, rather than the movement of sand and sediment along a beach.

● **Good answers** (Level 2)
Accurately describe the movement of sand and sediment along a beach in a zig-zag motion.

▲ **Excellent answers** (Level 3)
Not only describe the zig-zag movement accurately, but also explain that it happens because of the wind and wave direction.

The process of longshore drift and its impact on the coastline

Where waves approach the coastline at an angle, when they break, their swash pushes beach material up the beach at the same angle. The backwash then drags the material down the beach at a 90° angle to the coast, due to the force of gravity. This produces a zig-zag movement of sediment along the beach known as **longshore drift**, which you can see in Figure 7. The smallest material, such as fine sand, is easily moved and so ends up furthest along the beach. The largest materials – pebbles perhaps – are heavy and so harder to move. They are not moved as far.

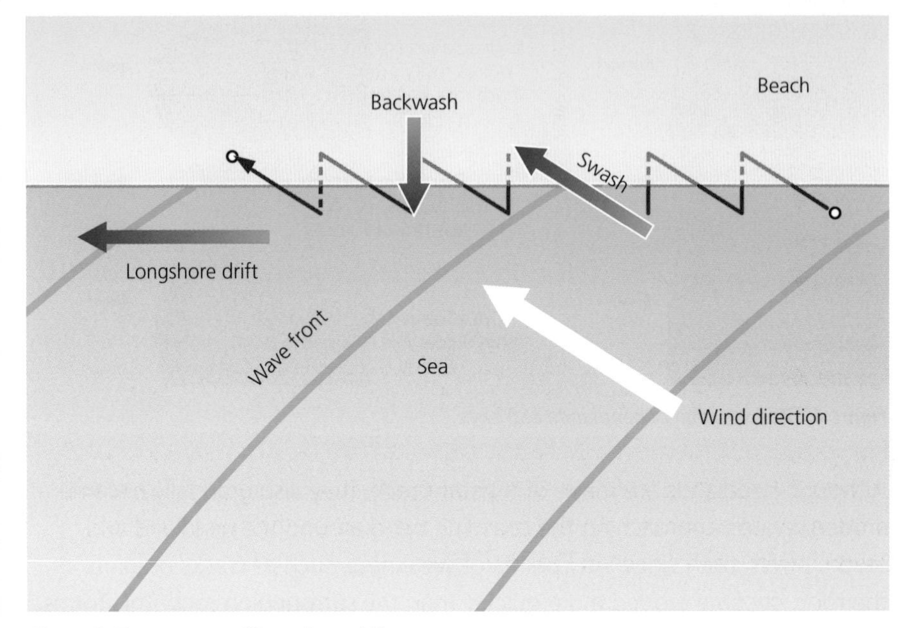

Figure 7: The process of longshore drift

The formation of beaches, spits and bars

Constructive waves add sediment to the coastline and lead to the formation of a beach. This is particularly the case on low energy coastlines that are sheltered from strong winds and waves, such as in a bay between two headlands. A good example is Swanage Bay in Dorset.

Longshore drift can carry beach sediment beyond a bend in the coastline leading to an extension of the beach into the open water, known as a **spit**. The end of the spit is in open water and often becomes curved because it is exposed to strong winds and waves. Deposition can then happen in the sheltered water behind the spit, forming a salt marsh. You can see both of these features of a spit in Figure 8.

Figure 8: Aerial view of Dawlish Warren Nature Reserve in Exmouth, Devon

If longshore drift continues to extend the length of the spit, it may join back up with the coastline the other side of an opening such as a bay. This results in the creation of a **bar** with a lagoon behind, such as the one at Slapton, Devon, which you can see in Figure 9.

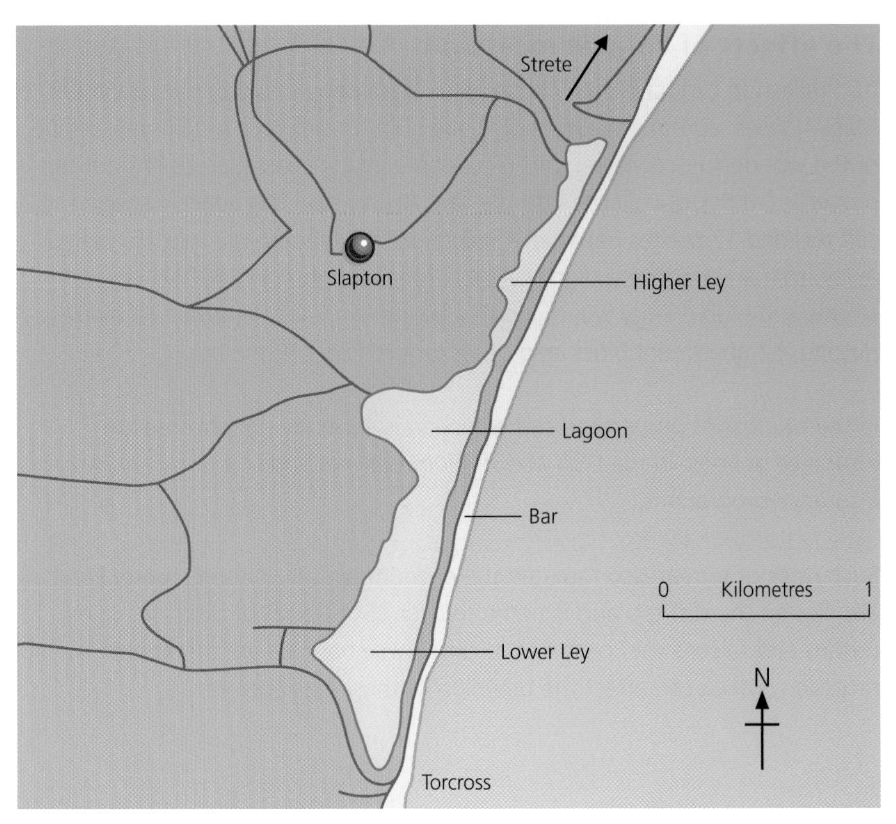

Figure 9: The bar and lagoon at Slapton in Devon

Skills Builder 2

(a) Draw a sketch of the spit in Figure 8.

(b) Label the spit, the curved end and the salt marsh.

(c) Annotate your sketch to explain how the spit has formed and developed.

ResultsPlus
Exam Tip

Examination questions sometimes ask about how landforms develop over time. With a spit, you could mention that if it continues to grow across a bay or other indentation in the coastline, then it could develop into a bar.

Objectives

- Know the factors affecting rates of cliff recession.
- Be able to explain how people and the environment are affected by coastal recession.
- Understand why methods of engineering used to protect the coast have both advantages and disadvantages.

ResultsPlus
Exam Question Report

Choose a stretch of coastline or coastal area you have studied where cliff recession is occurring or has occurred. Describe the effects of this cliff recession. (3 marks, June 2007)

How students answered

Most students answered this question poorly. They often only gave one effect when effects were required by the question. The question also required specific effects in the location studied rather than general effects of recession.

68% (0–1 marks)

Many students gave two effects, but still did not link these directly to the location that they had studied.

20% (2 marks)

Some students answered this question well. They gave at least two effects and referred to the studied location, using names of roads or buildings that had been affected by cliff recession.

12% (3 marks)

Coastal landforms are subject to change

Differential rates of cliff recession

Erosion and retreat of coastal cliffs can happen at very different rates. Some cliffs erode at rates of over 1.8 metres per year, such as those at Holderness in Yorkshire, whilst others are barely eroded at all. The rate of erosion and recession is due to the influence of factors such as the **fetch** (the distance of sea over which winds blow and waves move towards the coastline), the geology and coastal management strategies.

Coasts that face a major ocean, such as the south-west coast of England facing the Atlantic Ocean, have a very long fetch and the winds are strong and persistent. This produces destructive waves with high energy that can erode cliffs at rapid rates.

However, this is not the only factor that affects rates of erosion and recession. Indeed, the geology of south-west England is mainly granite, which is a very resistant rock. The actual rates of erosion are, therefore, very slow – typically only a few millimetres per year. One of the reasons for the high rates of erosion and recession at Holderness is that the geology is very weak clay.

The final factor that affects these rates is that of coastal management. If coastal defences such as concrete sea walls protect weak geology then rates of erosion will be much slower. Granite boulders are often placed in front of weak cliffs to protect them. The boulders erode only slowly and cliff recession is almost stopped.

The effects of coastal recession

Cliff recession can have many impacts on both people and the environment. Many houses, apartments and hotels are built on cliff tops to take advantage of the wonderful sea views. Cliff recession in many coastal areas has put these properties at risk of collapse into the sea. At Durlston Bay near Swanage the cliff receded 12 metres between 1968 and 1988, ending up only 25 metres away from an apartment block called Purbeck Heights. In 2000/01 severe storms and high energy waves led to a further retreat of another 12 metres, putting the apartment block and other properties at severe risk.

In the UK, loss of property to cliff recession is currently not covered by insurance policies. In the USA about $80m per year is paid out by a national insurance programme.

Such rates of retreat also threaten the environment. Durlston Country Park also lies on the cliff top and is home to over 250 species of birds, including puffins and falcons that nest on the cliff. These habitats are threatened by cliff recession, which can affect the breeding of these rare species.

Prediction and prevention of the effects of coastal flooding

It is widely predicted by scientists that sea levels will rise by about 1 metre by the year 2100, mainly due to global climate change. This is going to put many low-lying coastal areas around the world at risk of flooding. These places include Bangladesh, the Netherlands, the Maldives and even parts of eastern England, such as the Thames estuary. Insurance companies in the UK predict that flooding will be eight to twelve times more frequent here by 2100. These predictions have led governments to take action. In London, although the Thames Barrier is already in place (see Figure 10) to hold back very high tides, an Environment Agency project called Thames Estuary 2010 (TE2010) will install a series of new flood walls along the river. There are also plans to leave areas of open space, on to which flood water can spread without damaging buildings. Both of these measures should help prevent the potentially serious effects of flooding.

Figure 10: The Thames Barrier

Coastal flooding is also caused by strong winds and storms, which can increase the height of waves and tides. The Environment Agency monitors sea conditions 24 hours a day, 365 days a year. The Storm Tide Forecasting Service provides the Environment Agency with forecasts of coastal flooding, and surge and wave activity, together with warnings when hazardous situations are seen to be developing. If people are concerned about flooding from the sea, they can contact the Environment Agency's 24-hour Flood Line, or seek advice via their website. The Agency will advise them of what precautions they should take and what action is needed in the event of a flood emergency.

Skills Builder 3

(a) Use an atlas to draw a sketch map of the British Isles.

(b) Add arrows to show the fetch that would directly affect the south-west of Ireland, the south coast of England and the coastline of eastern England.

(c) How would these different fetches affect the coastal processes in these three locations?

Activity 3

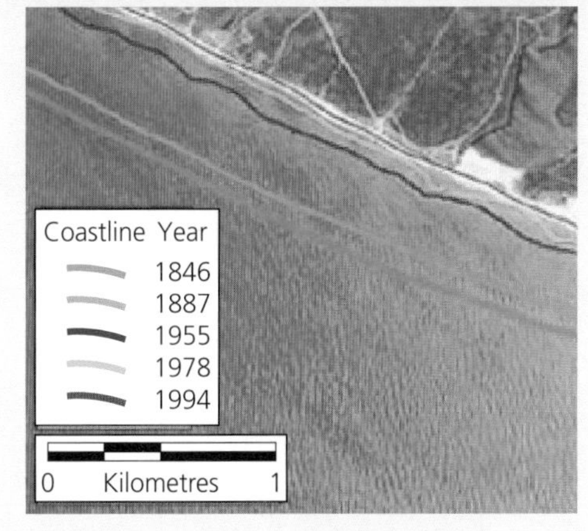

Figure 11: Digital map of Nantucket Island, Massachusetts

Look at Figure 11.

(a) Name and outline two processes by which waves erode.

(b) Explain two possible reasons why cliff recession is happening at a fast rate on this coastline.

ResultsPlus
Watch out!

■ Try not to confuse the words 'prediction' – suggesting what flooding is likely to happen in the future – and 'prevention', which relates to stopping the impacts of any flooding.

Activity 4

(a) Use the internet to find another location that suffers from coastal flooding.

(b) Research how this area tries to predict or prevent coastal flooding.

The Health Protection Agency also provides advice about how flooding can affect people's health and what precautions they should take. The following is an extract from their website:

> *The main threats to health during and immediately after a flood are drowning, and injuries caused by accidents in flowing water.*
>
> *The other main health hazards in floods come from the stress and strain of the event and clean-up.*
>
> *There is also a serious danger posed by carbon monoxide fumes from the indoor use of generators and other fuel-powered equipment, such as driers.*

In the Indonesian Province of Acech, 106 cases of tetanus and 20 deaths were reported after the coastal flooding associated with the tsunami at the end of 2004. This resulted from sea water entering and mixing with sewage in drains while submerging coastal settlements.

In Bangladesh the Coastal Embankment Project has led to the building of many flood walls, and 500 flood shelters have been built – although it is estimated that 10,000 are needed. The local people are now being better educated about what to do in a flood, and flood warning systems are being put in place. In 1997 a flood warning in the Cox's Bazaar area allowed over 300,000 people to be evacuated. As a result, only about a hundred people died in the flood – a much lower figure than in previous, similar floods. There are now laws in place ensuring that the roof of any new one- or two-storey building has to be accessible via an exterior stairway, so that people can escape the rising waters.

The types of hard and soft engineering used on the UK coastline

Coastal areas can be defended against wave erosion in a variety of different ways. The methods used are described as being either **hard engineering** or **soft engineering**. The soft engineering methods are generally considered to be more environmentally friendly. The advantages and disadvantages of all the methods are described in the tables on pages 65 and 66.

Hard engineering methods

Sea wall – a long concrete barrier built at the base of a cliff	
Advantages	**Disadvantages**
It protects the base of the cliffs against erosion because it is made of resistant concrete. Land and buildings behind it are protected. If it is 'recurved', it can reflect wave energy.	It is expensive to build and the cost of maintenance is high. It restricts access to the beach and it may be unsightly.

Groynes – wooden, rock or concrete 'fences' built across the beach, perpendicular to the coastline	
Advantages	**Disadvantages**
These prevent the movement – by longshore drift – of beach material along the coast. The beach can then build up as a natural defence against erosion – and as an attraction for tourists.	They may look ugly and they do not last very long because the wood rots. Sand is prevented from moving along the coast, and places elsewhere may lose their beach and the natural defence it provides.

Rip rap – large boulders of resistant rock	
Advantages	**Disadvantages**
These absorb wave energy and protect weak cliffs behind. They look quite natural.	They can be expensive. They still let some wave energy through. They can restrict access for the very young and the elderly.

Revetments – slatted wooden or concrete structures built at the base of a cliff	
Advantages	**Disadvantages**
These absorb and spread wave energy through slats. They do not interfere with longshore drift.	Regular maintenance is needed. They are quite expensive to install.

Offshore reefs – rock or concrete barriers built on the sea bed a short distance from the coastline	
Advantages	**Disadvantages**
Waves break on the barrier before reaching the coast. This significantly reduces wave energy and allows a wide beach to develop.	They are very expensive to build. They can interfere with the movement of boats.

Skills Builder 4

Look at the table below which shows the costs of constructing and maintaining various methods of **coastal management**.

Method	Construction costs (£ per metre)	Maintenance costs (£ per metre per year)
Rip rap	1,100	5
Revetments	2,000	5
Sea wall	4,000	20
Groynes	2,000	10
Offshore reef	4,500	5
Beach replenishment	2,000	2

(a) Which is the cheapest method to construct, and which is the most expensive to maintain?

(b) Explain how groynes can help reduce rates of cliff recession.

(c) Suggest why the cheapest method would not necessarily be chosen.

Activity 5

With a partner, discuss and decide whether you think hard or soft engineering methods are more suitable to protect busy tourist resorts from coastal erosion.

Soft engineering methods

Beach replenishment – adding sand taken from somewhere else, often offshore

Advantages	Disadvantages
This looks completely natural. It provides a beach for tourists. The beach absorbs wave energy and protects the land or buildings behind. This is quite cheap.	The sea keeps on eroding it away – so it has to be replaced every few years.

Managed retreat – people and activities are gradually moved back from the vulnerable areas of coast

Advantages	Disadvantages
Natural processes are allowed to happen. There is no threat to human safety.	Compensation has to be paid. There is quite a lot of disruption to people's lives and to businesses.

Cliff regrading – making the cliff face longer, so that it is less steep

Advantages	Disadvantages
The angle of the cliff is reduced, making mass movement less likely. This method is relatively cheap.	Other methods need to be used at the base of the cliff to stop it being steepened again by erosion. Properties on the cliff may have to be demolished.

Coastal management – a case study of Swanage

Swanage Bay and Durlston Bay in Dorset have both suffered from some significant coastal erosion during the last century. Rates of erosion in both places are estimated to have been about 40–50 cm per year. You can see the location of these two places in Figure 12.

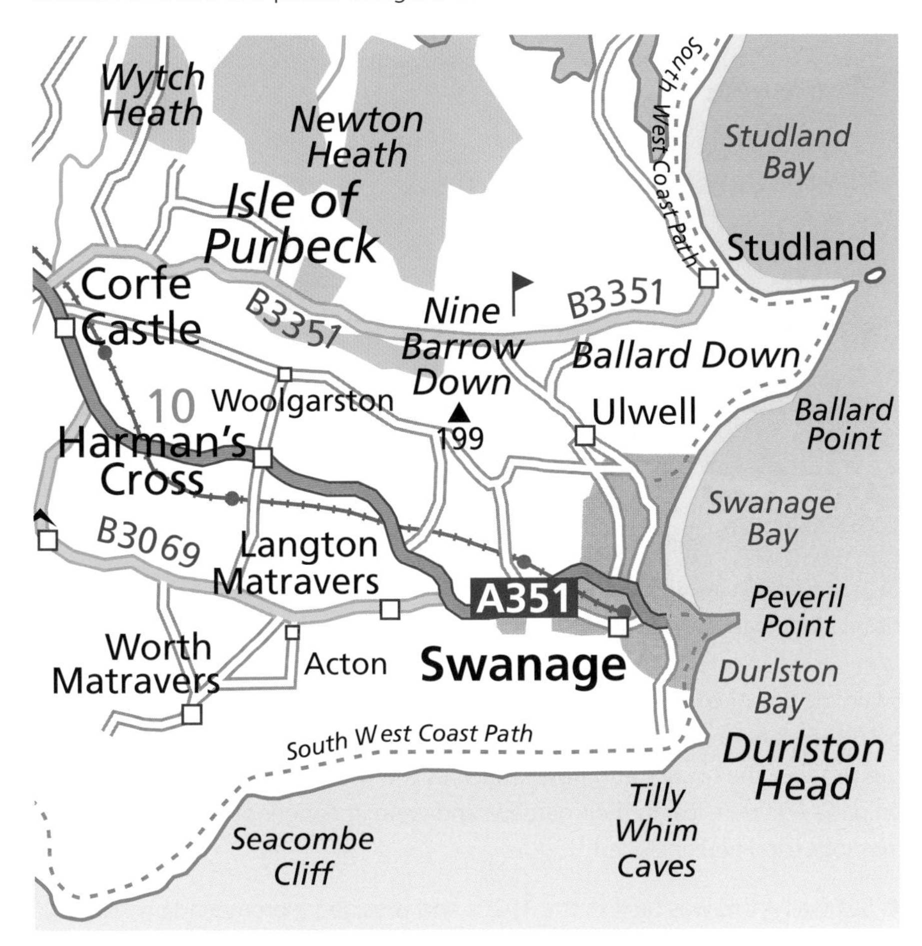

Figure 12: The location of Swanage Bay and Durlston Bay in Dorset

In Durlston Bay, several methods were used to protect the cliffs from erosion and to safeguard the apartments and houses on the cliff top. These cliffs are shown in Figure 13 on page 68. Erosion mainly occurred at one particular point, where there was a major weakness in the rock. The methods included:

- Regrading of the cliff – this extended it forward at the base, making the slope longer and therefore less steep.
- Installing drainage – this removed excess water, so that the slope was not as heavy or as well lubricated after rain.
- Placing rip rap – large granite boulders each weighing about 8 tonnes – at the base of the cliff to resist wave attack.

Objectives

- Know why the coastline at Swanage needs managing.
- Be able to describe the methods of management used.
- Understand why the methods of management have been successful.

Exam Tip

Examination questions sometimes ask about the need for coastal protection. Students who explain why it is necessary, as well as describing the methods used, tend to score high marks.

Activity 6

(a) List the management methods used on the Swanage coastline under the headings 'Hard engineering' and 'Soft engineering'.

(b) Which type of engineering is mostly used here?

(c) Why do you think that is the case?

Answers to examination questions that require a case study should contain some located detail, such as names of places, dates, facts and figures.

Figure 13: Management methods in Durlston Bay

Swanage Bay needed different methods of defence, because the erosion occurred along a considerable length of cliff rather than at one point. On the cliff top, the houses and hotels (such as the Grand Hotel – see Figure 14 on page 69) were losing their gardens and were in danger of collapsing. The methods used here included:

- Sea wall – this was built in the 1920s and provided a promenade (walkway) as well as a barrier to wave attack.
- Cliff regrading – a series of steps were made in the cliff to reduce slope angles.
- Groynes – a series of mainly timber groynes were installed in the 1930s, and eighteen of them have recently been replaced with new ones. These reduced longshore drift and helped make sure that a beach remained in place to absorb the energy of breaking waves.
- Beach replenishment – 90,000 m^3 of sand was dredged from Studland Bay and pumped on to the beach at Swanage. This works with the groynes to ensure a good size of beach.

The cost of the recent new groynes and the beach replenishment was about £2.2m.

You can see some of the methods used at Swanage in Figure 14.

Figure 14: Swanage beach and cliffs

Quick notes (Swanage):

- Different parts of a coastline have different levels of vulnerability to cliff recession.
- Coastlines that have significant human activity need managing.
- Many different methods of management can be used.
- Methods of management can be changed over time.
- Methods of management often work best when used in combination with each other.

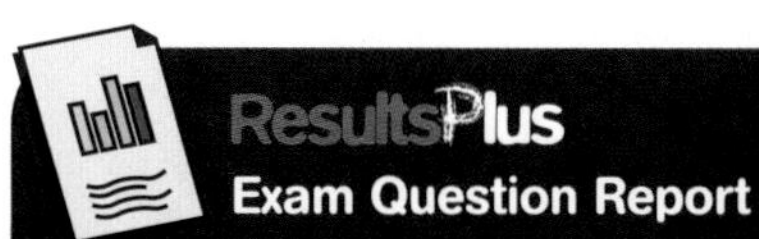

Choose a stretch of coastline or coastal area that you have studied where cliff recession is occurring or has occurred. Explain the management techniques used to control cliff recession in your chosen area. (5 marks, June 2007)

How students answered

Many students answered this question poorly. They often only described the techniques used and did not explain them. The question also required specific techniques actually used in the location studied rather than general comments about any techniques.

37% (0–1 marks)

Most students gave some explanation of how techniques are used to control recession, but their answers usually lacked specific reference to the chosen area. The explanations were often generalised and could relate to many different techniques.

55% (2–3 marks)

Some students answered this question well. They explained at least two techniques and referred to the studied location, giving costs or dates of installation for example. Their explanations were directly related to the named techniques.

8% (4–5 marks)

Activity 7

Read the case study about coastal management in Swanage.

(a) Briefly state why the coastline in Swanage Bay needed managing.

(b) Describe the methods of management that were used.

(c) Explain how these methods were effective.

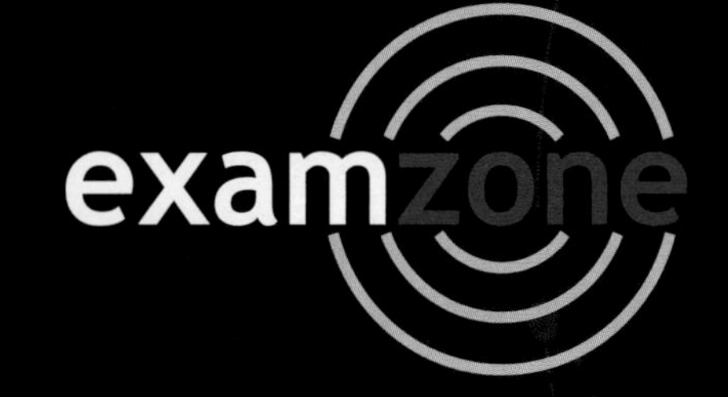

Know Zone Chapter 3 Coastal landscapes

The vast majority of people live within half a day's journey of the coast. The climate is usually less extreme, the farming more productive and the links with the outside world easier. With increasing pressure on them, these environments need more careful management than ever before.

You should know...

- ☐ How to describe cliffs, wave-cut platforms, headlands and bays, caves, arches, stacks and stumps
- ☐ How these landforms are created and the role of different processes in their formation
- ☐ How to recognise these landforms on photographs and maps
- ☐ What longshore drift is and how it affects the coastline both in the features it forms and the related issues, e.g. deposition in estuaries
- ☐ How to describe beaches, spits and bars
- ☐ The processes involved in the formation of beaches, spits and bars
- ☐ How to recognise these landforms on photographs and maps
- ☐ The factors which cause cliff recession, including erosion, weathering processes and mass movement
- ☐ The impact of cliff recession on both the human and natural environments, including insurance claims and loss of land
- ☐ Some examples of these impacts
- ☐ How the effects of coastal flooding are reduced through planning before the event
- ☐ Some examples of these plans
- ☐ How to define hard and soft engineering
- ☐ The main types of coastal defence used on the coastline of the UK, including groynes, sea walls, offshore reefs, rip rap, revetments, beach replenishment, managed retreat and cliff regrading
- ☐ The case study of Swanage to describe and explain the management of the coastal area

Key terms

Backwash
Bar
Coastal flooding
Coastal management
Constructive wave
Deposition
Destructive wave
Erosion
Fetch
Geology
Hard engineering
Longshore drift
Mass movement
Soft engineering
Spit
Stack
Stump
Swash
Vegetation
Weathering

Which key terms match the following definitions?

A The movement of sand along a coast by waves

B The distance a wave has travelled toward the coastline

C The breakdown and decay of rock by natural processes, without the involvement of any moving forces

D The dropping of material that was being carried by a moving force

E An embankment of sand, extending a beach into the open water

F The forward movement of water up a beach after a wave has broken

G A wave that removes more material from a beach than it adds

H An isolated column of rock, standing just off the coast, that was once attached to the land

To check your answers, look at the glossary on page 289.

ResultsPlus
Maximise your marks

Foundation Question: Study Figure 14 on page 69. Sea walls are a method of hard engineering and are constructed to prevent coastal erosion. Describe how sea walls prevent erosion from taking place. (4 marks)

Student answer (awarded 2 marks)	**Examiner comments**	**Build a better answer** (awarded 4 marks)
Coastal erosion happens when waves crash against a cliff.	• ***Coastal erosion happens...*** is correct but the student does not describe the processes.	Coastal erosion takes place when waves hit the cliff, pounding them.
The waves weather away the beach by being big and heavy and pounding the cliff.	• ***The waves weather...*** is partly right but ***weather away*** is the wrong term. The correct points made here and above score a combined 1 mark.	They also use energy to throw rocks at the cliff and this can wear away the cliff.
If you build a sea wall this stops the waves from arriving.	• ***If you build a sea wall...*** is wrong. Building sea walls does not stop the waves from arriving.	Sea walls are built in front of the cliff like a barrier.
If they do they bounce off the wall and go back out to sea.	• The idea of ***bouncing off the wall...*** gets 1 mark.	If they do they bounce off the wall and go back out to sea.

Overall comment: The student would have scored more marks if they had *described* the processes they mentioned and used geographical terms correctly.

Higher Question: Study Figure 14 on page 69, which shows different methods of managing the coastline. For a coastal area that you have studied describe and explain how different types of engineering are used to protect the coast. (6 marks)

Student answer (awarded Level 1)	**Examiner comments**	**Build a better answer** (awarded Level 3)
Hard engineering involves concrete. It is where man tries to stop erosion from happening.	• ***Hard engineering involves...*** This is wrong because it offers a limited view. Concrete is not used in all hard engineering.	The coast either side of Swanage is quite protected by hard engineering which involves building structures to protect the coast.
Common methods are groynes, which trap sand on a beach making the beach wider.	• ***Common methods are...*** is worth 1 mark. This describes groynes, but does not explain why making the beach wider is useful. Also, there is no location provided and the question asked for an area that you have studied.	Sea walls are used in Swanage Bay. Sea walls absorb wave energy but groynes are also used to trap sand, meaning the waves break on the wider beach.
In other places rip rap will be used, which is large rocks wrapped up in wire baskets.	• ***In other places...*** This sentence is partially correct because the student identifies rip rap as a hard engineering method. It receives 1 mark. However, the definition of rip rap is confused with gabions. Again, there is no location provided in the answer.	Further along the coast, rip rap is used in Durlston Bay. These are larger rocks that are put there to break up the waves.
In the picture of Swanage there is a sea wall which protects the cliff by acting like a bumper on a car and absorbing the wave energy.	The last sentence is good. The student refers to the location and explains how the sea wall is used. This receives 1 mark.	In the picture of Swanage there is a sea wall which protects the cliff by acting like a bumper on a car and absorbing the wave energy.

Overall comment: This student could have improved their answer by identifying a location as they were asked.

Chapter 4 River landscapes

Objectives

- Define the key drainage basin terms.
- Be able to describe the change in characteristics of a river and its valley, from source to mouth.
- Understand how processes produce landforms.

ResultsPlus
Exam Tip

Examination questions will often ask you to define one or more of these very important terms – so you must learn them.

River processes produce distinctive landforms

Drainage basin terms

A **drainage basin** is the area of land drained by a river and its tributaries. When it rains, water eventually finds its way into rivers, either by moving across the surface or by going underground and moving through the soil or the rock beneath.

There are a number of important technical terms associated with drainage basins. These are shown in Figure 1, and their meanings are given underneath.

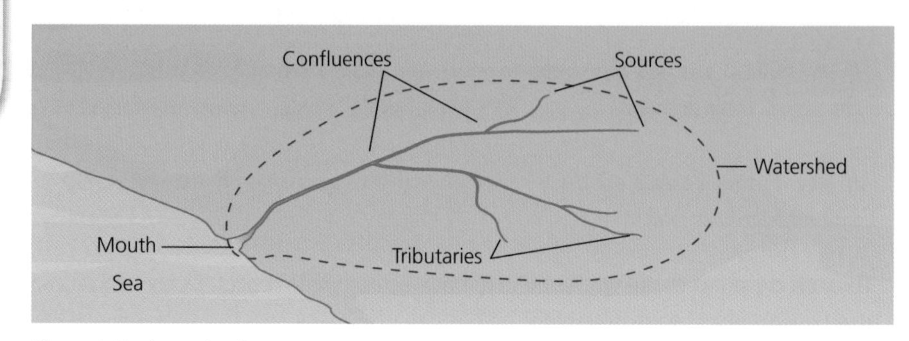

Figure 1: Drainage basin terms

Watershed – the boundary of a drainage basin. It separates one drainage basin from another and is usually high land, such as hills and ridges.
Confluence – a point where two streams or rivers meet.
Tributary – a stream or small river that joins a larger stream or river.
Source – the starting point of a stream or river, often a spring or a lake.
Mouth – the point where a river leaves its drainage basin as it flows into the sea.

The impact of processes on river landscapes

The processes that affect river landscapes are very much like those that affect coastal landscapes (as covered in Chapter 3). The main difference is that rather than waves providing the force for the **erosion** (the wearing away and removal of material by a moving force) processes, it is the flow of water in the river channel.

The **weathering** (the breakdown and decay of rock by natural processes) and **mass movement** (the downslope movement of material due to gravity) processes act on the valley sides, usually making them less steep over time, as material is moved from the top of the slope to the bottom. However, the river itself erodes its own channel, by wearing away the bed and/or banks, as well as possibly eroding the base of the valley side, making it steeper – and perhaps leading to mass movement.

Weathering can happen in many ways. Some common ones affecting river valleys are:

- **Freeze–thaw** – this happens when rainwater enters cracks or gaps in a rock and then freezes if temperatures drop below zero. The water expands as it turns to ice and this exerts pressure on the rock, causing it to break down into smaller pieces.

- Acid rain – all rain is slightly acidic. If the air is polluted by factories and vehicles, it can become very acidic. When rain falls on rocks, the acid in it can react with weak minerals, causing them to dissolve, and the rock to decay.
- Biological weathering – the roots of plants, especially trees, can grow into cracks in a rock and split the rock apart.

River erosion involves several different methods, including:

- **Hydraulic action** – this results from the force of the water hitting the river bed and banks and wearing them away. This is particularly important during high-velocity flow.
- **Abrasion** – this is caused by the river picking up stones and rubbing them against the bed and banks of the channel in the flow. This wears the bed and banks away.
- **Attrition** – any material carried in the river will become rounder and smaller over time as it collides with other particles and the sharp edges get knocked off.
- **Corrosion** – the dissolving of rocks and minerals by river water flowing over them.

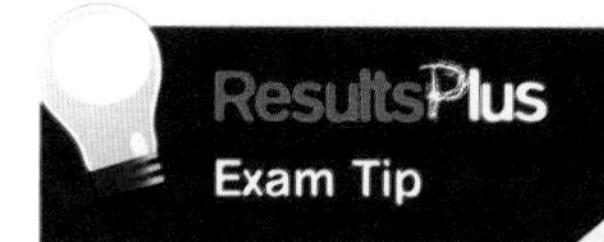

These changes in characteristics from source to mouth will not necessarily be gradual. Sometimes there will be sudden and quite dramatic changes, for example when a tributary joins and the discharge increases significantly. High quality answers to examination questions asking for one or more of these changes to be described might mention this.

There are several different processes of mass movement. In river valleys, two of the main forms of mass movement are:

- Soil creep – individual particles of soil move slowly down slope under gravity and collect at the bottom of the valley sides. The river may then erode this material.
- Slumping – this happens when the bottom of a valley side is eroded by the river. This makes the slope steeper and the valley side material can slide downwards in a rotational manner, often triggered by saturation due to rain, which both 'lubricates' the rock and makes it much heavier.

Change in characteristics from source to mouth

A river changes with increasing distance downstream from its source towards its mouth, as it moves from its **upper course**, to its **middle course** and finally into its **lower course**. The key characteristics of a river are:

- width – the distance from one bank to the other
- depth – the distance from the surface of the water to the bed
- velocity – how fast the water is flowing
- discharge – the rate at which water is moved through the river channel
- gradient – the steepness of the river bed.

Characteristic	Change from source to mouth
Width	increases
Depth	increases
Velocity	increases
Discharge	increases
Gradient	decreases

Formation of landforms

Rivers produce a wide range of landforms, many of which help make river landscapes attractive places for people to visit and enjoy.

Interlocking spurs

Near their source, rivers do not have a lot of power as they are very small. They tend to flow around valley side slopes, called spurs, rather than being able to erode them. The spurs are left interlocking, with those from one side of the valley interlocking with the spurs from the other side. You can see these spurs labelled in Figure 2.

Figure 2: Interlocking spurs

Waterfalls

Where a river flows over bands of rocks of differing resistance, the weaker rock is eroded more quickly and a step may develop in the river's bed. The increased velocity gained by the water as it falls over the step further increases the rate of erosion of the weaker, downstream band of rock. Abrasion and hydraulic action at the base of this step cause undercutting and the formation of a plunge pool. Eventually the overhanging, more resistant rock collapses due to a lack of support, making the **waterfall** steeper. The waterfall in Figure 3 is almost vertical. Over time, repetition of this process means that the position of the waterfall retreats in an upstream direction.

ResultsPlus
Watch out!

- Meanders are not the same as interlocking spurs. Although the river has a winding course in both cases, with interlocking spurs it is flowing around obstructions. This is not the case in meanders.

ResultsPlus
Exam Question Report

The feature in Figure 4 is a meander bend. Explain the changes which might happen to this meander over a period of time. You may use a diagram or diagrams in your answer. (5 marks, June 2007)

How students answered

Most students answered this question poorly. They often described the changes, stating that the bend would get bigger and perhaps become an ox-bow lake. However, they did not explain *why* this happened.

60% (0–1 marks)

Many students gave some explanation of the changes by mentioning erosion and/or deposition and stating where they were happening.

31% (2–3 marks)

Some students answered this question well. They not only mentioned erosion and deposition, but also gave details of specific process mechanisms such as abrasion. They also established a clear sequence in the changes happening.

9% (4–5 marks)

Figure 3: Waterfall on Hills Creek, West Virginia

Meanders

Meanders are large bends in a river's course, found well below the source, often in its lower course as the river approaches its mouth (see Figure 4). They are common landforms found on a river's flood plain. The flow of the river swings from side to side, directing the line of maximum velocity and the force of the water towards one of the banks. This results in erosion by undercutting on that side and an outer, steep bank is formed, called a **river cliff**. **Deposition** takes place in the slower moving water on the inside of the bend, leading to the formation of a gently sloping bank, known as a **slip-off slope**. The cross-section of a meander is, therefore, asymmetrical – steep on the outside, gentle on the inside.

Ox-bow lakes

As meander bends grow and develop, their neck becomes narrower. Eventually the river may erode right through the neck, especially during a flood. Water then flows through the new, straight channel and the old bend is abandoned by the river. The bend gradually dries up and, helped by deposition at the neck by the river, becomes sealed off as a horseshoe-shaped lake, as Figure 5 (on page 76) shows.

Skills Builder 1

Figure 4: A meander bend

Look at Figure 4.

(a) Which of the features labelled A and B is (i) a river cliff and (ii) a slip-off slope?

(b) Describe the appearance and position of feature A.

(c) Explain how river processes have produced these features.

Eventually, many ox-bow lakes dry up completely, and then they may not be so easy to recognise.

Activity 1

Draw up a table to show all of the landforms that are produced by river processes. Use three headings – 'Erosion', 'Deposition' and 'Erosion and Deposition' – putting each landform under the correct heading.

Activity 2

Consider these four river landforms – waterfall, flood plain, meander, interlocking spurs.

(a) Which two are most likely to be found in the upper course of a river near the source?

(b) Describe the appearance and position of a flood plain.

(c) Explain how a waterfall is formed.

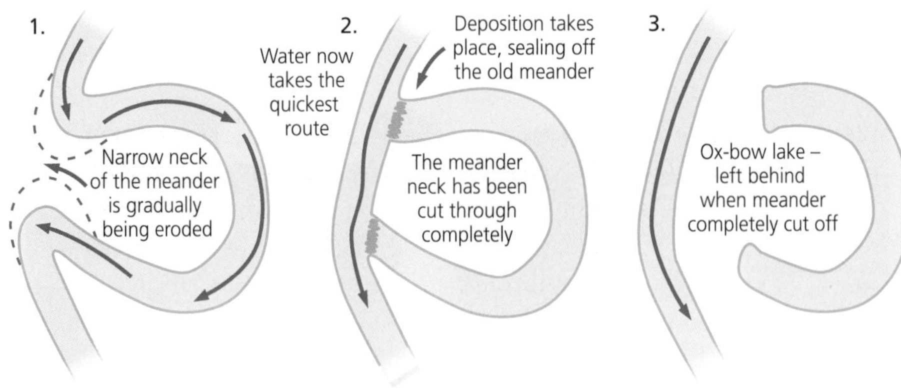

Figure 5: Formation of an ox-bow lake

Flood plains

A **flood plain** is a wide, flat area of land either side of a river in its lower course. The flood plain is formed by both erosion and deposition.

Lateral (sideways) erosion is caused by meanders eroding on the outside of their bends. This makes the valley floor wide and flat. When the river floods, the flood water spreads out on the valley floor, slows down and deposits the sediments it was carrying. This can be seen in Figure 6.

Levees

Levees are natural embankments of sediment along the banks of a river. They are formed along rivers that carry a large load and occasionally flood. In times of flood, water and sediment come out of the channel as the river overflows its banks. The water immediately loses velocity and energy as it leaves the channel and so the largest sediment is deposited first, on the banks. Repeated flooding causes these banks to get higher, forming levees. You can see these alongside the river in Figure 6.

Sometimes natural levees are raised and reinforced by humans to protect settlements, industry and transport links on the flood plain from future floods.

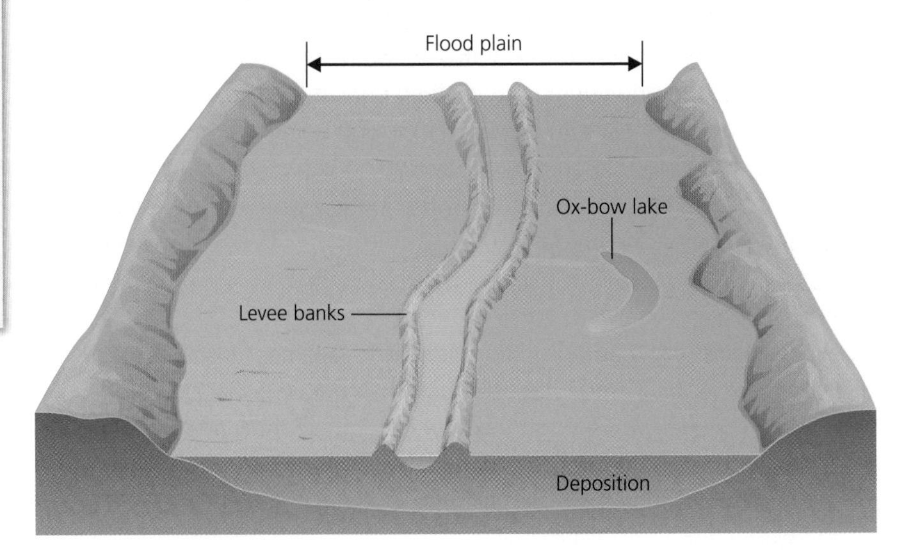

Figure 6: Flood plain and levees

Flooding and flood prevention

Causes of flooding

A river flood is when a river overflows its banks and water spreads out on to the land alongside the river channel. Rivers usually flood as a result of a combination of causes. These are often a mixture of physical and human factors which, together, can result in large amounts of water getting into the river very quickly – so quickly that it fills up and overflows.

Physical factors:

- Intense rainfall – When rain falls too fast to fully allow its **infiltration** into the ground, it flows quickly across the surface and into the channel.
- Snow melt – In some places lots of snow falls during the winter months, but when temperatures rise above zero in the spring all the snow that has built up suddenly melts.
- **Impermeable** rocks – Some rock types, like granite, do not let water enter the ground and so rainwater runs off across the surface into the channel.

Human factors:

- **Deforestation** – Vegetation collects, stores and uses water from a drainage basin. The less vegetation there is, the more water can reach the channel.
- **Urbanisation** – In towns and cities, rainwater will not infiltrate the hard, man-made surfaces like concrete and tarmac and so it runs off into the channel.
- **Global climate change** – Increasing global temperatures, partly due to human activities such as burning fossil fuels, can cause more melting of ice in glaciers and, in places, more rainfall and more frequent storms.

Effects of flooding on people and the environment

There are many different and wide-ranging effects that floods can have on people and the environment (see Figure 7 on page 78). Some of these are very immediate, whilst others can happen much later – and last much longer.

The flooding that affected parts of northern England and the Midlands in the summer of 2007 caused the loss of thirteen lives and over £2 billion worth of damage. In Lincolnshire, 40% of the pea crop was damaged, and in north Wales 30 tonnes of debris and earth blocked the only road out of Barland near Presteigne on 23 July.

Floods in Mozambique in Africa during February 2007 led to 121,000 people being made homeless. In Sudan at least 12,000 livestock and 16,000 chickens were lost, whilst outbreaks of acute diarrhoea killed dozens of people as water supplies became contaminated.

Objectives

- Know the major factors that cause flooding.
- Be able to describe the effects of flooding on people and the environment.
- Understand how the effects of flooding can be reduced.

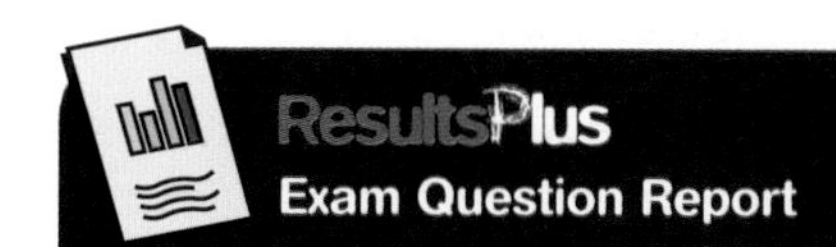

Name and explain two physical factors that cause floods. (4 marks, June 2007)

Most students answered this question poorly. They often only gave answers that related to human factors, such as urbanisation, rather than physical factors, as asked for in the question.

47% (0–1 marks)

Many students gave two causes, such as snow melt or impermeable rock, but did not explain how they caused flooding.

33% (2 marks)

Some students answered this question well. They named two valid physical factors and explained how they caused flooding. This was best done by explaining why water runs quickly across the surface into a river, causing it to rise rapidly and overflow its banks.

20% (3–4 marks)

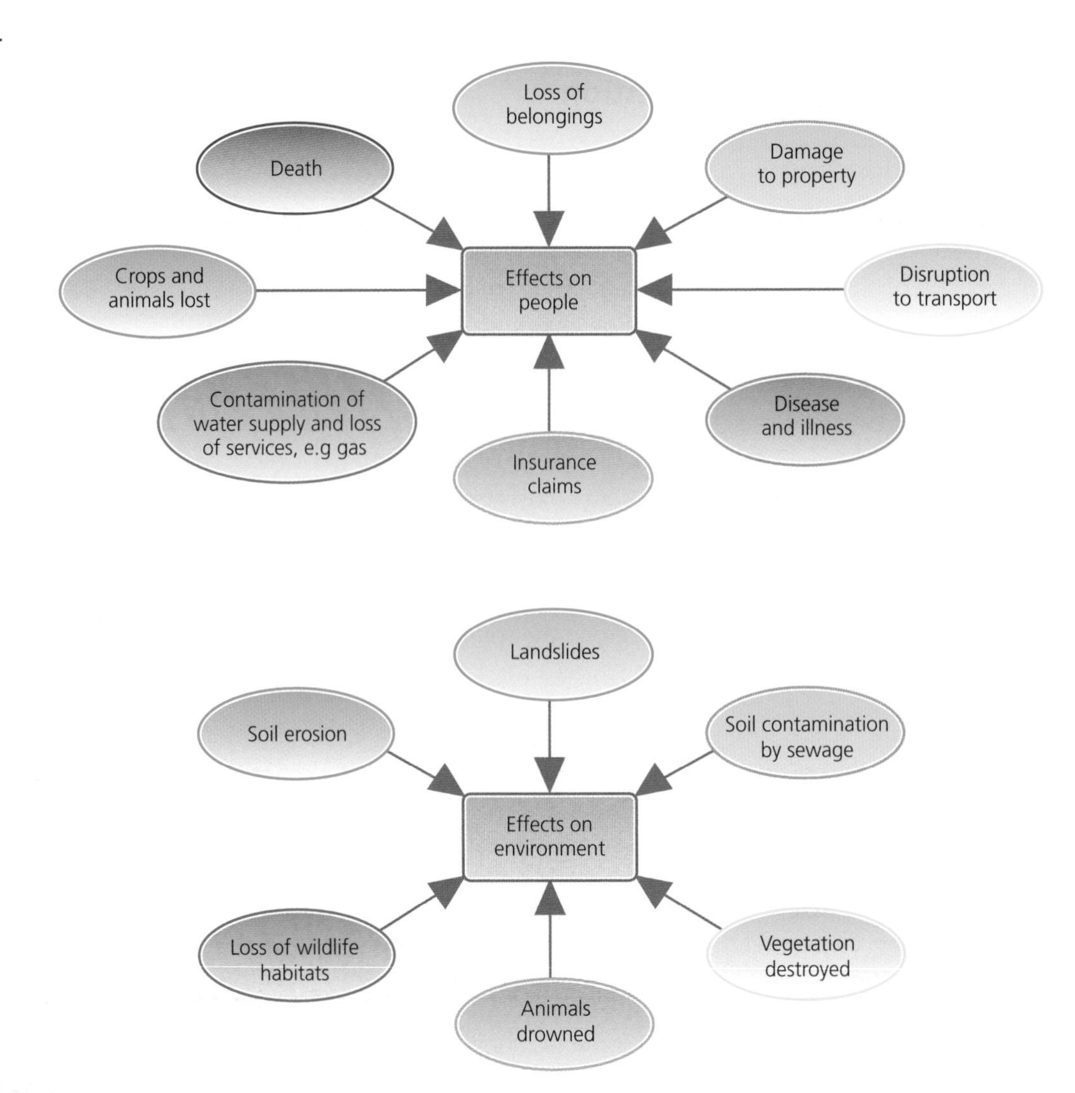

Figure 7: Effects of flooding on people and the environment

The key to a good answer about causes of flooding is to explain why water gets into river channels so quickly. This is usually because it is running across the ground surface, rather than going underground, from where it would take much longer to reach the rivers.

Prediction and prevention of the effects of river flooding

Forecasting

Many rivers, especially those that have a history of flooding, are monitored by the Environment Agency. If river levels rise to potentially dangerous levels, they are able to warn people in areas of risk and, if necessary, evacuate people from their homes to a safer place. They have also produced maps, which are available online, that show areas at different levels of risk.

Computer simulation models are now being used to help predict flooding. In Texas, for example, in response to a prediction from Rice University, the Texas Medical Centre was evacuated before a large flood occurred in 2001.

Building design

There are things that people can do to their property that will make it easier and cheaper to clean up, after a flood.

The Environment Agency advice includes:

- Lay ceramic tiles on your ground floor and use rugs instead of fitted carpets.
- Raise the height of electrical sockets to 1.5 metres above ground floor level.
- Fit stainless steel or plastic kitchens instead of chipboard ones.
- Position any main parts of a heating or ventilation system, like a boiler, upstairs.
- Fit non-return valves to all drains and water inlet pipes.
- Replace wooden window frames and doors with synthetic ones.

In many low-income countries, wooden buildings are built on stilts so that living areas are above the flood risk level (see Figure 8).

Figure 8: House on stilts in Khulna, Bangladesh

Planning

The local government is responsible for giving planning permission for new buildings. In recent years, many new buildings have been built on the flood plains of rivers, putting these properties at significant risk. Now, it is widely accepted that this should only be done if protection measures are put in place. Land use zoning is often used. This means flood risk areas are only used for activities that would not suffer too much from a flood. These uses include parks and playing fields.

Figure 9 (on page 80) shows how the town of Nome in Alaska has planned its land uses so that most of the area at risk of an extreme, '1 in 100 year' flood event is left as open space or is used for leisure and recreation.

Activity 3

Read the news report below on flooding in Wales in October 2008.

FLOODS HIT HOMES, SCHOOL AND LAND

Firefighters have been dealing with calls about flooding in several areas of Wales after heavy rain.

They went to Llanelltyd, in Dollgellau, to help pump out a house with 1ft of water in the ground floor and were in the Lovesgrove area of Aberystwyth.

Firefighters were also called to Ty Newydd Farm at Felin Fach, Lampeter, to help rescue 20 sheep which were stuck in flood water.

There were reports of a flooded cellar at Cyfarthfa High School in Merthyr.

In Aberystwyth, crews spent the night pumping water which severely affected Fron Frith Lodge, and a high-volume pump had to be brought into operation.

http://news.bbc.co.uk/1/hi/wales/7691694.stm

(a) What caused this flooding?

(b) Identify three impacts of the flooding.

(c) How did the firefighters respond to this flood event?

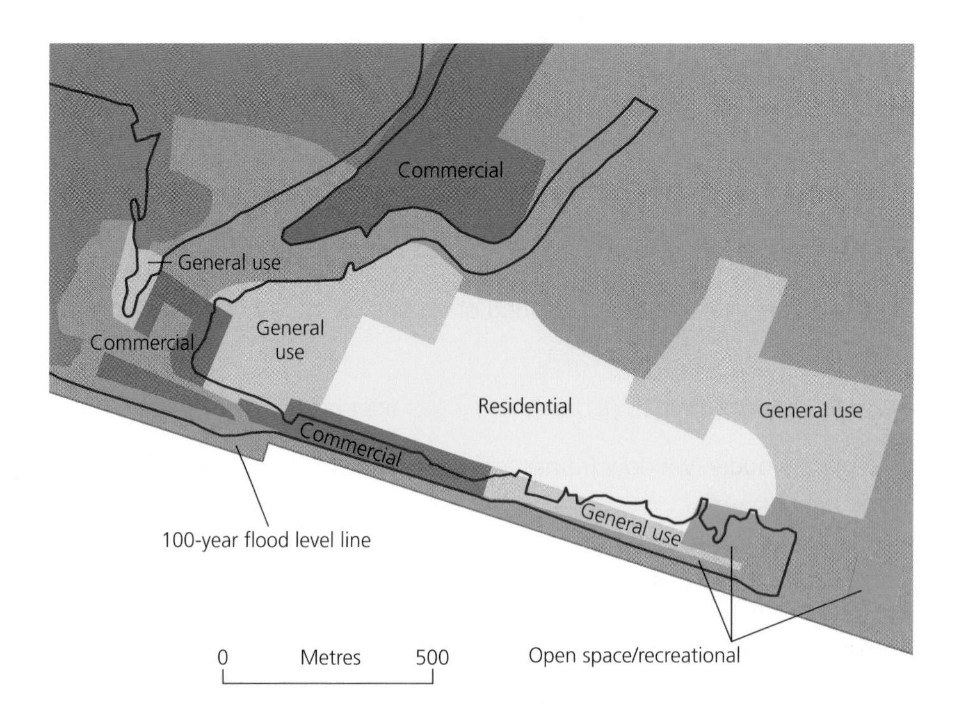

Figure 9: Land use in Nome in Alaska, showing how it has been planned to cope with possible future flooding

ResultsPlus
Watch out!

Despite the efforts made to predict and prevent them, floods can sometimes be unexpected and far more severe than anticipated. You should be aware that high-income countries can generally predict and prevent flooding much better than low-income countries because they have greater levels of technology and they are better able to afford the cost of prediction and prevention schemes.

Activity 4

Produce an A4-sized leaflet that could be distributed to homes in a town in the UK at risk of flooding, explaining what they should do before, during and after a flood.

Education

It is no good local governments or government agencies having emergency plans and strategies if they do not let the local people know about flood risk and what to do in a flood. This is where education is really important. Governments try to educate people by:

- Sending leaflets through the post
- Advertising in newspapers and on television
- Posting information on websites (see below)
- Offering helpline telephone numbers
- Having drills and training exercises.

Here is an extract from the Environment Agency website:

> *The Environment Agency provides flood warnings online 24 hours a day. From this page you can view warnings in force in each of our eight regions covering England and Wales. You can also search for your local area and its current warning status using the panel on the right. The information is updated every 15 minutes.*
>
> *For further information about a Flood Warning in a particular area, you can call Floodline 24 hours a day on 0845 988 1188.*

The National Weather Service in the USA issues advice on what to do before, during and after flood events, such as:

- Store drinking water in clean bathtubs and in various containers – the water service may be interrupted
- Keep a stock of food that requires little cooking and no refrigeration – electric power may be interrupted
- Keep first aid supplies on hand
- Assemble a disaster supplies kit containing: first aid kit, canned food and can opener, bottled water, rubber boots, rubber gloves, Weather Radio, battery-powered radio, flashlight, and extra batteries. (Weather Radios are issued free to all those who live in flood risk areas. These are permanently tuned into the weather service broadcasts.)

Types of engineering used to control rivers

There is a wide range of different engineering methods that can be used to try to control rivers in the UK (Figure 11). Some of these involve **hard engineering**, others involve **soft engineering**. Each method has advantages and disadvantages.

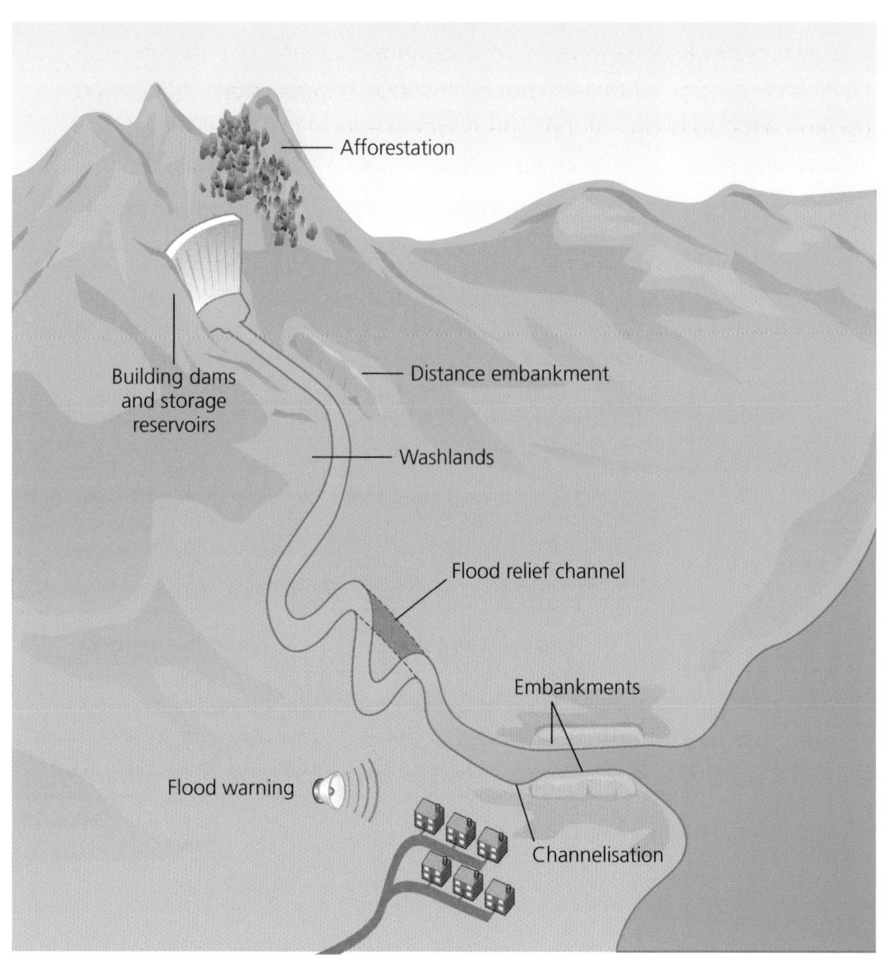

Figure 11: Flood prevention methods (Harcourt and Warren 2001)

Skills Builder 2

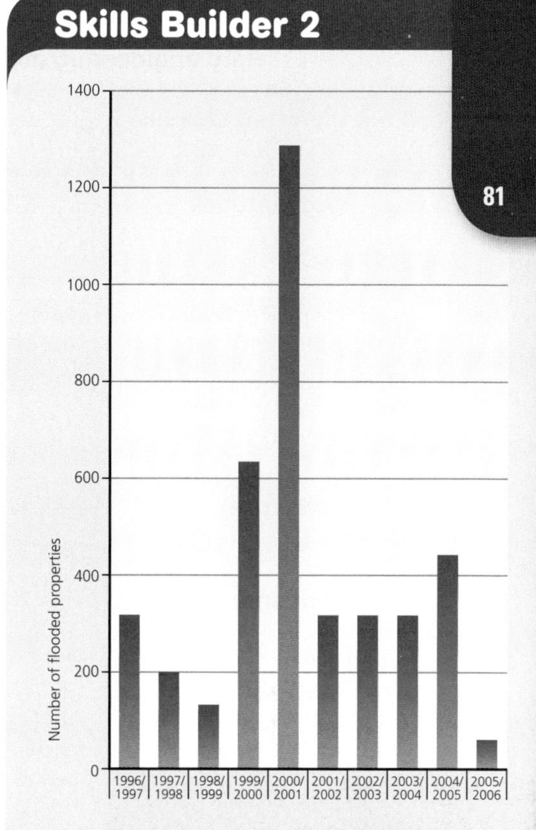

Figure 10: The number of flooded properties in the south-west of England between 1996/1997 and 2005/2006

Study Figure 10.

(a) In which years were the (i) highest and (ii) lowest number of properties flooded?

(b) Suggest two physical reasons why an area may have severe floods.

(c) Explain how any two hard engineering methods can be used to protect properties from being flooded.

Hard engineering methods

Embankments (levees) – these are high banks built on or near riverbanks

Advantages	Disadvantages
They stop water from spreading into areas where it could cause problems, such as housing. They can be earth and grass banks, which blend in with the environment.	Flood water may go over the top. They can burst under pressure, possibly causing even greater damage.

Channelisation – this involves deepening and/or straightening the river

Advantages	Disadvantages
This allows more water to run through the channel more quickly, taking it away from places at risk.	More water is taken further downstream, where another town or place at risk might lie. It does not look natural.

Flood relief channels – extra channels can be built next to rivers or leading from them

Advantages	Disadvantages
The relief channels can accommodate the surplus water from the river so that it won't overflow its banks.	They can be unsightly and may not be needed very often. Costs can be high.

Soft engineering methods

Washlands – these are areas on the flood plain that are allowed to flood

Advantages	Disadvantages
This gives a safe place for flood water to go. Inexpensive and leaves the natural environment unspoilt.	Flood plains cannot be used for other things.

Afforestation – trees are planted in the drainage basin

Advantages	Disadvantages
Trees intercept rainfall and take water out of the soil. This reduces the amount reaching rivers. Wooded areas look attractive and provide wildlife habitats.	The land cannot be used for other activities, such as farming.

Land use zoning – governments allocate areas of land to different uses, according to their level of flood risk

Advantages	Disadvantages
Major building projects are allocated to low-risk areas. Open space for leisure and recreation is placed in high-risk areas because flooding would be less costly for them.	These may not be the best places for the different activities in terms of public accessibility.

Flood warning systems – rivers are carefully watched and if the levels are rising, places downstream can be warned

Advantages	Disadvantages
People living in towns or villages downstream have a chance to prepare or to evacuate.	Sometimes it is not possible to give very much warning and so it is hard to save possessions.

Activity 5

Write a letter to a newspaper arguing either for or against the use of hard engineering methods in flood management.

River management – a case study of Blandford Forum

The River Stour, which runs through part of Dorset, has a long history of flooding – major floods were recorded as far back as 1756. The small town of Blandford Forum – known locally as just 'Blandford' – is on the river and was particularly badly hit by flooding in the latter part of the twentieth century. In 1979 there were two significant floods, the first in May flooded 30 properties, while the second in December flooded 110. The river flowed from the water meadows, over West Street and into Market Place and East Street, flooding the main hotel and several shops. Since then, there have been minor floods in 1990, 1992 and 1994/95, with larger floods occurring in 1999 and 2000. The impact of the 1979 flood can be seen in Figure 12. Under particularly severe conditions some 180 properties in Blandford were at risk.

Objectives

- Know why the River Stour needs managing at Blandford Forum.
- Be able to describe the methods of management used.
- Understand why the methods of management have been successful.

Figure 12: Flooding in Blandford in 1979

Causes of flooding at Blandford

- Upstream of the town the River Stour drains an area of impermeable clay and there is a dense network of streams. When it rains, these streams feed water quickly into the River Stour, causing its level to rise rapidly.
- This area used to be woodland, but is now used for farming – so there are few trees to intercept or take up water when it rains.
- At Blandford there is a very narrow flood plain – so there is little space for flood water to spread on to safely.
- Rainfall has sometimes been very intense. In December 1979, for example, over 5 cm fell in 24 hours.

The flood management scheme

Key

A = Pumping station
B = Washland
C = Flood bank
D = Flood wall
■ = Industry
■ = Blandford

Scale: 1:5,000

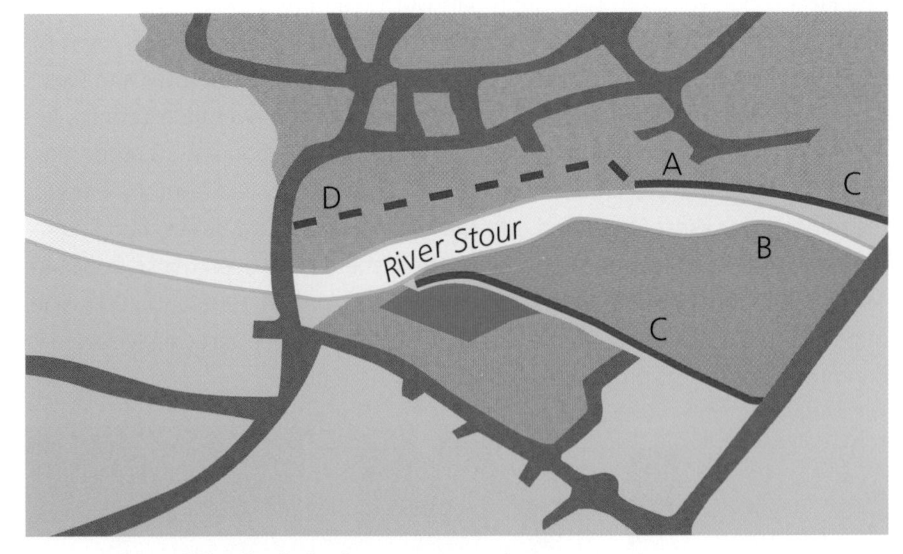

Figure 13: Map of Blandford's flood management scheme

The scheme, shown in Figure 13, which was started in 1986, is designed to protect the town from floods up to '1 in 200 years' events. It involves:

- A flood wall, up to 2.5 m high, on the north bank with a relief channel, 1 m deep, behind it. This protects the houses and shops of the town centre (see Figure 15).
- A flood bank on the south side of the river. This protects a brewery, a new industrial estate and new housing (see Figure 14).
- An area of washland on the south side between the river and the flood bank (see Figure 14). This allows flood water space on which to spread.
- A pumping station at Langton Meadows to pump flood water further downstream.

One important way of considering how successful a flood defence scheme has been is to compare the costs involved with the estimated benefits that have been achieved through having the scheme in place.

Figure 14: Washland and flood bank on the south side of the River Stour at Blandford protect the Hall and Woodhouse brewery

Figure 15: Flood wall on the north side of the River Stour at Blandford, protecting the town centre

This scheme uses a variety of management methods and a combination of hard and soft engineering. It cost £1.45m but is estimated to have already saved the town from over £1.6m worth of flood damage.

Quick notes (Blandford):

- There are a number of reasons why flooding occurs on a river.
- These reasons often act in combination with each other.
- There are a variety of methods that can be used to defend a town against flooding.
- Some methods involve hard engineering whilst others involve soft engineering.
- Methods of management usually work best when used in combination with each other.

Activity 6

(a) List the management methods used in Blandford under the headings 'Hard engineering' and 'Soft engineering'.

(b) Which type of engineering is mostly used here?

(c) Explain how the different methods used work in combination with each other.

Choose a river management scheme that you have studied. Describe the river management techniques that have been used on your chosen river. (5 marks, June 2007)

Most students answered this question poorly. Many did not focus on the question at all and wrote about either the causes or the effects of flooding.

47% (0–1 marks)

Many students answered this question reasonably, describing appropriate management techniques such as afforestation or flood relief channels, but without direct reference to their chosen river.

45% (2–3 marks)

Some students answered this question well. They described at least two techniques and referred to the studied location, quoting details such as the area of land afforested, the cost of construction, the date of installation or the length of the flood relief channel.

8% (4–5 marks)

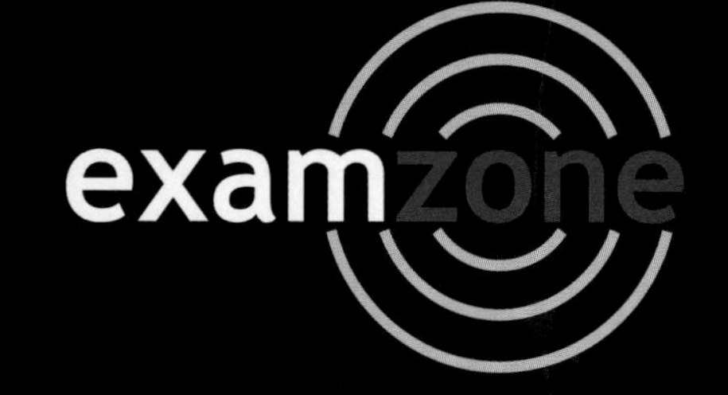

Know Zone
Chapter 4 River landscapes

Rivers shape the world. The majority of people live on or close to major rivers and we all depend on the food grown on their flood plains. We use rivers for transport, power, water, food and recreation. Rivers are also dangerous; more people are killed by river flooding than by any other natural disaster.

You should know...

- ☐ Drainage basin terms such as watershed, confluence, tributary, source and mouth
- ☐ How to recognise these features on diagrams, maps and photographs
- ☐ The processes of erosion and deposition that help form river features, such as interlocking spurs, waterfalls, meanders, river cliffs, ox-bow lakes, flood plains and levees
- ☐ How rivers change in width, depth, velocity, discharge and gradient from source to mouth
- ☐ The changing characteristics of rivers in their upper, middle and lower course
- ☐ How to describe interlocking spurs, waterfalls, meanders, river cliffs, ox-bow lakes, flood plains and levees, and how to explain their formation
- ☐ How to recognise these landforms on diagrams, photographs and maps
- ☐ The general physical and human factors that cause rivers to flood, e.g. intensity of rainfall, urbanisation and deforestation
- ☐ How human and natural environments are affected by flooding, e.g. insurance claims and loss of land
- ☐ How the effects of river flooding are reduced through planning before the event
- ☐ The definitions of hard and soft engineering
- ☐ The main types of defence used on UK rivers, including embankments, channelisation, flood relief channels, land-use zoning, washlands, and flood warning systems
- ☐ The case study of Blandford Forum to describe and explain the management of a river area

Key terms

Abrasion
Attrition
Channelisation
Confluence
Corrosion
Deforestation
Deposition
Drainage basin
Erosion
Flood plain
Freeze–thaw
Hard engineering
Hydraulic action
Impermeable
Infiltration
Interlocking spurs
Levees
Lower course
Mass movement
Meander
Middle course
Mouth
River cliff
Slip-off slope
Soft engineering
Source
Tributary
Upper course
Urbanisation
Washlands
Waterfalls
Watershed
Weathering

Which key terms match the following definitions?

A The point where two streams or rivers meet

B The chopping down and removal of trees to clear an area of forest

C The gradual wearing down of rock materials as they are transported by water, wind or ice

D Areas on the flood plain that are allowed to flood (a soft engineering method)

E The area of land drained by a river and its tributaries

F Using environmentally friendly methods of construction and management to cope with the forces of erosion

G A hard engineering method that involves deepening and/or straightening the river

To check your answers, look at the glossary on page 289.

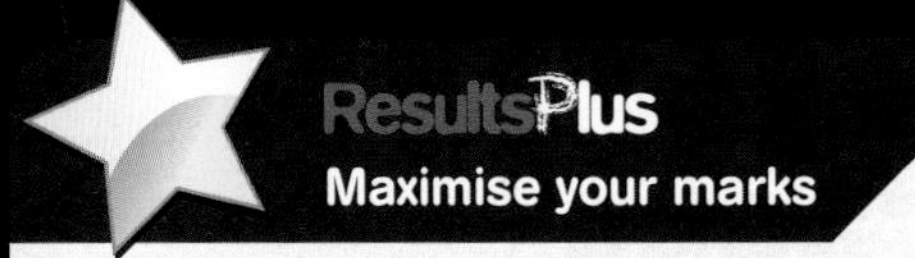

Foundation Question: Explain why some areas are at risk from flooding. You may use evidence from your own case study of river management. (4 marks)

Student answer (awarded 2 marks)	Examiner comments	Build a better answer (awarded 4 marks)
River floods will happen more where there are severe floods because of the really bad weather conditions.	• *River floods will happen...* This scores no marks. The student needs to *explain* why some places are more at risk than others.	In 1979, Blandford, which is sited on or close to a very narrow flood plain, was affected by severe flooding twice. The flooding followed very heavy rain as the water had little space to flow on safely.
Flooding is always a risk for a place, especially if it is not defended properly, which might be the case if it is too hard to do this.	• *Flooding is always...* gains 1 mark. The answer is not precise enough for a second mark. It is not true that flooding is ***always*** a risk.	Lower areas by a river, especially if they have been built on the flood plain, are most at risk from flooding.
When you know a flood is going to happen because it has happened before in that particular area as it is close to the river, you can make the people leave.	• *When you know...* Some areas being more at risk because they are ***close to the river*** scores 1 mark, but more marks would have been gained for using 'lower' rather than ***closer***.	Even after a flood management scheme was developed in 1986, a 1 in 200 year flood would still flood the town. Land-use changes in the future might make the risk of flooding greater.

Overall comment: The student could have improved this answer by being more precise and by using geographical language.

Higher Question: Study Figure 11 on page 81, which shows a number of ways of managing rivers and river flooding. Describe the methods shown and explain how they help protect the population. (6 marks)

Student answer (awarded Level 1)	Examiner comments	Build a better answer (awarded Level 3)
Methods vary to protect the people and they don't always work.	• *Methods vary...* is a general introduction that repeats the question, so scores no marks.	Eight methods shown in the figure, some try to manage the river to reduce risk while others deal with problems after a flood.
Channelisation can make rivers more dangerous. Dams and reservoirs trap water but it is better to plant trees which absorb water.	• *Channelisation can...* is awarded 1 mark. This is an excellent point but channelisation is not fully explained so does not gain further marks.	Some hard engineering methods like flood embankments (or levees) might make flood risk worse. Instead of letting rivers flood, more water goes downstream, making the risk more severe. A compromise is a distance embankment.
Washlands are wetlands that are good for rivers and it is always important to warn people about floods.	• *Washlands are wetlands...* The first part of this sentence means little but there is recognition of the method in the second part which scores 1 mark.	Restoring wetlands absorbs floodwater, although land is lost for farming. Afforestation helps reduce the flow of water to a river and dams allow control of discharge.
This didn't happen at Boscastle, when a sudden storm destroyed the village.	• The final sentence is irrelevant.	Extreme weather events cause floods so flood warnings and sensible land zoning to keep people off the flood plain are needed.

Overall comment: The answer needed more detail and the student should have used the resource more fully.

Chapter 5 Glaciated landscapes

Objectives

- Know the names of the main landforms created by glacial processes.
- Be able to describe the appearance of these landforms.
- Understand how processes produce these landforms.

The impact of glaciation on river valleys

Glaciation is the impact of glaciers on the landscape. A glacier is a large mass of ice that moves under gravity. Glaciers form when snow that falls on the ground during one year does not melt before more snow the next year falls on top of it. If this process is repeated many times over a number of years, then it compresses and compacts the old snow into ice – similar to the way in which you might make a snowball. When enough ice builds up it is able to move slowly downhill, under gravity.

Such build-ups of ice in mountainous areas lead to a general movement of ice from high areas down into the low areas, especially old river valleys. Eventually a large mass of ice develops at the head of such a valley and then moves down the valley under gravity. This is a **valley glacier**, and as it moves it shapes the landscape.

The processes of glacial erosion

One set of processes by which the landscape is shaped is **erosion**. Glacial erosion mainly happens in two, quite different ways:

Plucking	This happens when ice at the base of a glacier melts and becomes water. This water can enter cracks in the rock of the valley floor and sides. If it freezes again, it turns back into ice and the rocks become attached to the glacier. As the glacier moves forward, large pieces of rock are pulled out of the ground and carried along in the glacier.
Abrasion	This mechanism is often described as being like sandpapering. Rocks that are in the base and sides of the glacier rub against the rocks of the valley floor and sides as it moves. This wears away the rocks, often leaving them with scratches.

The effect of freeze–thaw

Freeze–thaw is a **weathering** process that commonly happens during glaciation. This provides much of the rock material found in a glacier. Weathering often happens towards the tops of the valley sides, above the level of the ice. Water enters cracks in the rocks there and freezes. When it freezes it expands and this puts stress and pressure on the rock, gradually widening the crack. After many such freeze–thaw events, the rock may be split apart, as you can see in Figure 1. Fragments may then fall under gravity, by a mass movement process known as rock fall, down on to the glacier below.

(1) Water freezes to ice, expanding the crack

(2) Ice thaws and more water fills the bigger crack

(3) Water freezes to ice, expanding the crack even further

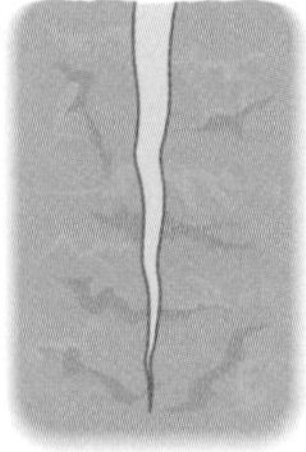

Water fills crack in the rock

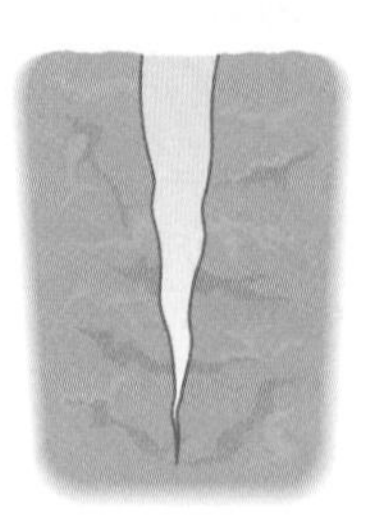

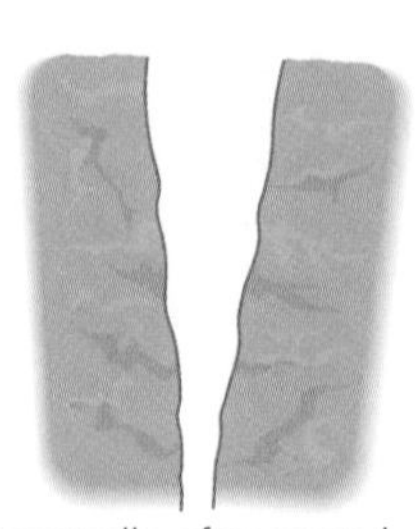

Eventually, after several cycles, the rock breaks apart

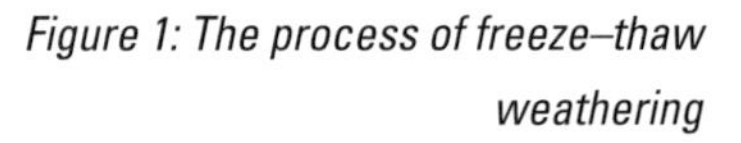

Figure 1: The process of freeze–thaw weathering

This rock material gradually finds its way to the bottom of the glacier, where it plays an important part in the process of **abrasion**, rubbing against the valley floor rocks, scratching them and gradually wearing them away as the glacier moves forward.

This material – a mixture of different-sized rocks and sediments – is eventually left behind in the landscape when the glacier melts. The material is then called **moraine**, and it can be deposited in many different ways to form distinctive types of landforms.

Landforms of glaciated uplands

The landscapes of upland areas that have been glaciated contain many landforms that were produced by erosional processes.

Corries are armchair-shaped hollows. They typically have a steep back wall of bare rock, a deep base and a rock lip at the front. A corrie is formed by ice collecting in a small hollow on a hillside. When the ice is thick enough, it moves out of the hollow, down slope, under gravity. As it moves it causes plucking at the back of the hollow, and the plucked material is then used in the abrasion of the base of the hollow. The ice moves out of the hollow in a rotational manner. The thicker ice in the middle exerts great downward pressure on the base, giving very high rates of erosion. At the front the ice is thinner and so less erosion happens. This means a rock lip is left at the front (see Figure 2).

After glaciation has finished, water can collect in the corrie from rain and from surface flows over the steep slopes around. This can eventually form a small, circular lake in the base of the corrie, because the rock lip traps the water. This is known as a **corrie lake**.

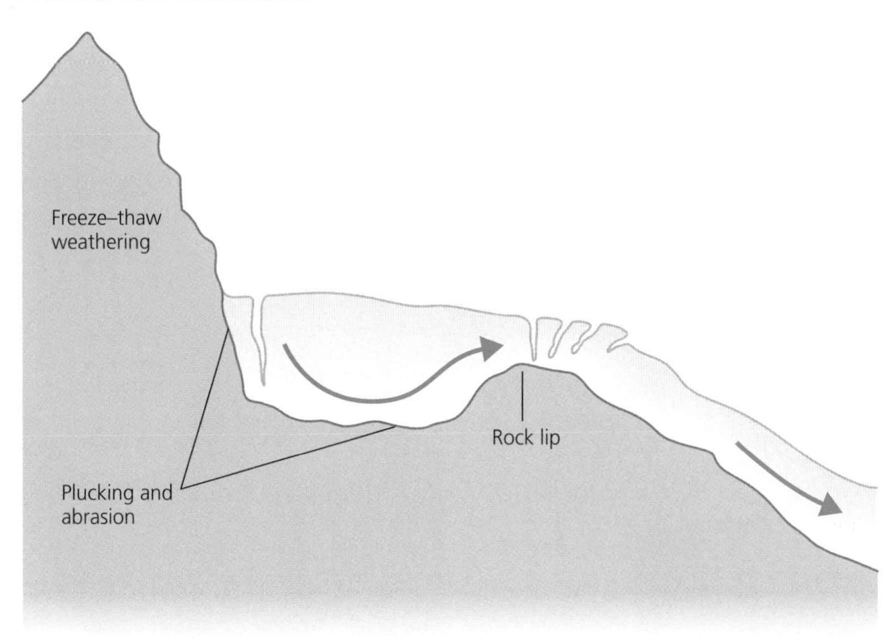

Figure 2: Corrie formation

The water in a corrie lake is not the water left from when the ice melted. This will have long since evaporated or drained away. Recent rain water collects in the hollow to form the lake.

State three features of a corrie. (3 marks)

Basic answers (0–1 marks)
Are not direct or accurate enough, as students misunderstand what 'features' means. Many state processes responsible for the corrie's formation.

Good answers (2 marks)
Give one or two valid features. The steep back wall is the most common answer and some also identify the presence of the corrie lake.

Excellent answers (3 marks)
Appreciate which features can be mentioned and include the rock lip at the front and, occasionally, an overall description of the shape, referring to armchair or amphitheatre shapes.

ResultsPlus
Build Better Answers

Explain the formation of a ribbon lake. (3 marks)

■ **Basic answers** (0–1 marks)
Give a description only with no explanation.

● **Good answers** (2 marks)
Offer some explanatory statements relating to either deposition forming a dam or erosion deepening the valley floor.

▲ **Excellent answers** (3 marks)
Provide a full explanation of processes, such as why erosion deepens a valley at a particular point.

If two corries form 'back to back' on a hillside, the land between them becomes narrowed due to the erosion of their two back walls. Over time, this becomes a very sharp ridge called an **arête**. Weathering happens on this ridge, making it even steeper and sharper. If corries form on three or more sides of a mountain, then the top will become very sharp and jagged due to the erosion of all of their back walls, leaving little to separate them. This jagged top is known as a **pyramidal peak** and weathering also further sharpens its appearance. These landforms are shown in Figure 3.

As a valley glacier moves through a former V-shaped river valley, erosion leads to the valley being widened, deepened and straightened. The result is a **U-shaped valley**. The moving ice has much more erosive power than the relatively small stream or river that previously flowed in the valley. Not only does it make the floor deeper and erode the valley sides to make it wider, but it also erodes through the interlocking spurs that previously existed, turning them into **truncated spurs** as part of the steep valley sides.

The main valley glacier is very powerful and its erosion rates are high because of the large amount of ice that it holds and the pressure that this exerts on the valley floor. Smaller, tributary valleys do not contain as much ice and so do not erode as deeply. At the end of a glaciation, all of the ice melts and the valleys are exposed. The small tributary valleys are left high above the floor of the main valley as **hanging valleys**. Often, they are easily recognised because the streams flowing through them have waterfalls where the water drops down the steep valley sides to reach the floor of the U-shaped valley, as shown in Figure 4.

Ribbon lakes are long, narrow lakes found on the floor of U-shaped valleys. They can be formed in different ways. Some are the result of weak rocks on the valley floor being eroded more rapidly than more resistant rocks around them. The area of weak rock becomes lower and so after glaciation fills with water, forming a long thin lake. They can also form when a ridge of moraine left at the end of the glaciation forms a natural dam and water then collects behind it, forming a lake on the valley floor. A good example is shown in Figure 5 (page 91).

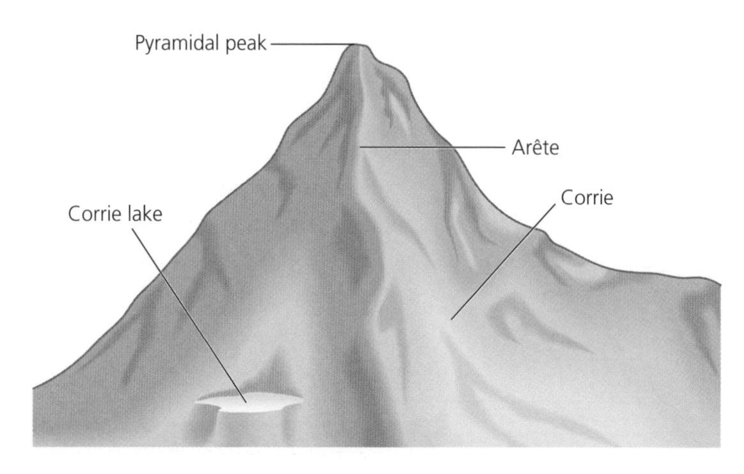

Figure 3: Corrie, arête and pyramidal peak

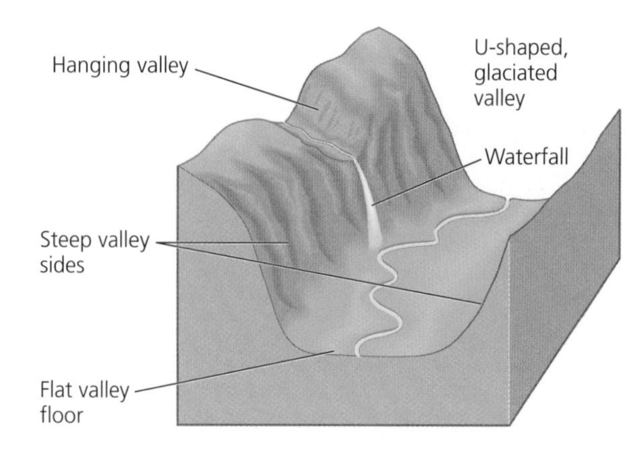

Figure 4: U-shaped, glaciated valley

Figure 5: A ribbon lake in Jasper National Park, Canada

Processes and landforms of deposition

Glacial **deposition** involves the dropping or laying down of rock material that was being carried in the glacier. This mainly happens when the glacier is melting, such as in the summer months or at the end of a period of glaciation when temperatures are rising. This is known as **ablation** deposition. Deposition also happens due to **lodgement**. When glaciers are moving forward, some of the rock material that they are carrying at their base can become pushed and pressed into the existing valley floor by the weight and pressure of the large mass of ice.

The main landforms produced by glacial deposition are the various forms of moraine, and these are generally the result of ablation.

Glaciers contain large amounts of rock material, obtained from freeze–thaw weathering, plucking and abrasion. This material is carried on, in, and at the base of the glacier, as it is moving forward. When the ice melts, this material is gradually lowered on to the valley floor and left behind. Material found along the edges of a glaciated valley after the ice has all melted is called lateral moraine. If two glaciers meet, their lateral moraines join together to become a medial moraine, which is left in a line up the centre of the valley. A terminal moraine is a ridge of material deposited at the front of a glacier, left behind and marking the furthest point the glacier reached down the valley. These different moraines are shown in Figure 7.

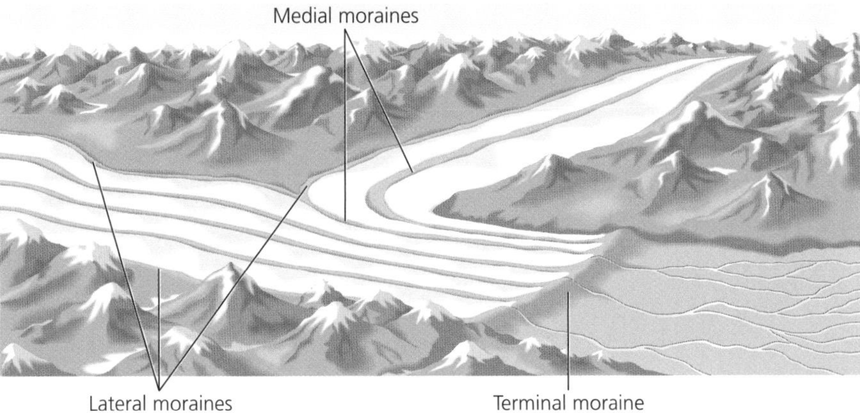

Figure 7: Types of moraine

Skills Builder 1

Figure 6: A glaciated upland

Look at Figure 6.

(a) Which of the features labelled A and B is (i) a pyramidal peak and (ii) an arête?

(b) Describe the appearance of the main valley.

(c) Explain how glacial processes have created this valley.

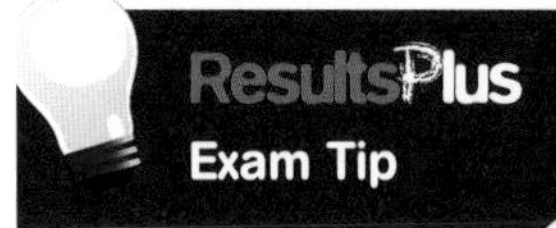

Exam Tip

Lodgement is not an easy process to explain. If something becomes 'lodged' it gets stuck (hence the opposite – 'dislodged'). Lodgement is, therefore, what happens to the material at the bottom of the glacier when it is pressed into the valley floor.

Erratics are large rocks that have been transported by a glacier and deposited in an area of different rock type, so that they now seem out of place. For example, at Norber in the Yorkshire Dales, a number of sandstone boulders were deposited on top of limestone, as you can see in Figure 8.

Figure 8: Sandstone erratics in a limestone area in the Yorkshire Dales

Ground moraine is the name given to mounds of rock material deposited on the valley floor. Sometimes the mounds have distinctive shapes. **Drumlins**, for example, are low, elongated hills formed by lodgement of rock debris into the valley floor. As the glacier passes over the deposited material, it squashes it and streamlines it. Drumlins usually have a steep slope facing up-valley and a less steep slope facing down-valley (Figure 9). Often a group of them are formed together and this is then called a 'swarm'.

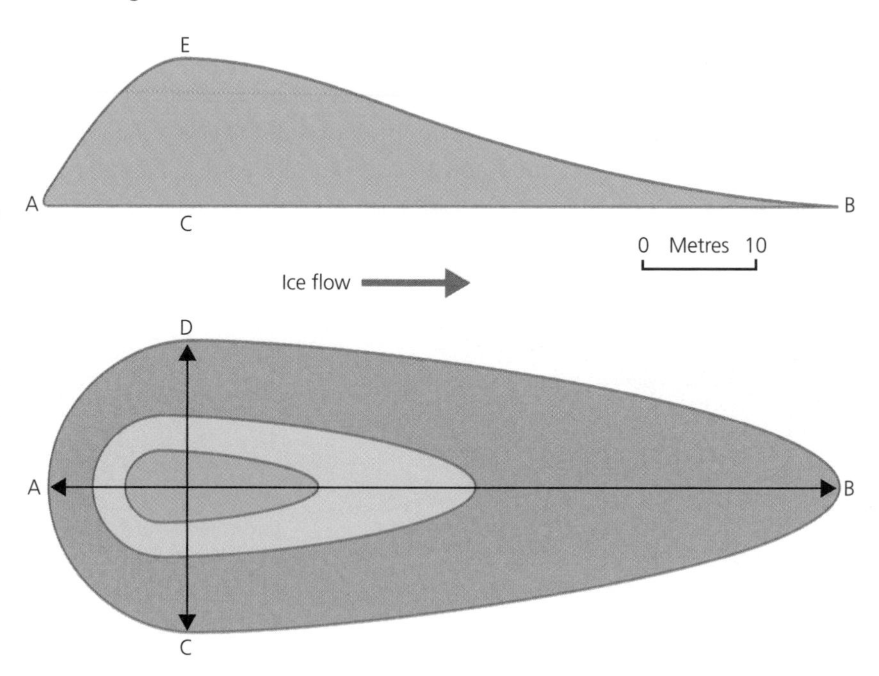

Figure 9: The typical shape of a drumlin. The length of AB is typically a few hundreds of metres. The height of CE is typically tens of metres.

Skills Builder 2

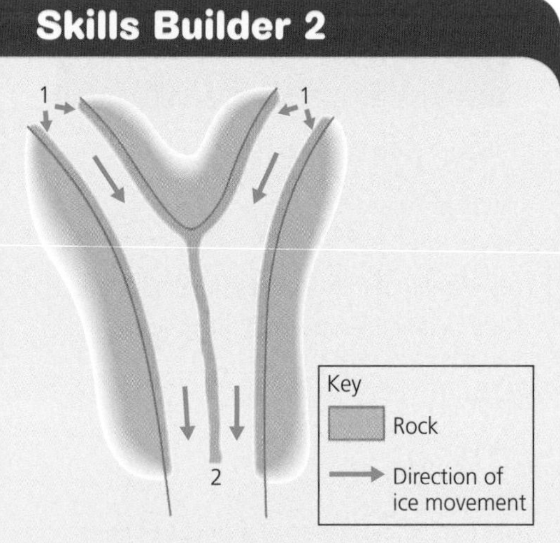

Figure 10: Glacial moraines

Look at Figure 10.

(a) Which of the features labelled 1 and 2 is (i) lateral moraine and (ii) medial moraine?

(b) Describe the typical appearance of a drumlin.

(c) Explain how glacial processes create drumlins.

How people use glaciated landscapes

Objectives

- Know the ways in which people use glacial and glaciated landscapes.
- Be able to describe activities that take place in these landscapes.
- Understand why these uses take place in these landscapes.

Hydro-electric power

There are many human uses of glaciated areas. These include using areas that are currently being glaciated as well as using areas that were glaciated in the past.

Glaciated areas can provide suitable locations for **hydro-electric power** stations. One example of this is the Dinorwig pumped storage hydro power station in Snowdonia, North Wales which opened in 1984. This takes advantage of the close proximity of a corrie lake and a ribbon lake. Water from the 80-metre deep corrie lake, called Marchlyn Mawr, is released through a series of tunnels, falling a height of 750 metres to drive turbines that generate electricity. The water eventually enters the ribbon lake, Llyn Peris, on the valley floor. From here it is pumped back up to Marchlyn Mawr using cheap, off-peak electricity, where it can be held, ready to be released again when there is high demand. This is a very effective way of meeting sudden surges in demand, because electricity is produced within 16 seconds of the water being released from the top lake. The two lakes can be seen in Figure 11.

Figure 11: Dinorwig power station area

Leisure and recreation

Skiing and hiking are popular leisure and recreation activities in glacial areas. Most ski resorts in the Alps are based on skiing on snow that falls in the winter but which melts in the summer. Many resorts still attract visitors in the summer for walking and hiking holidays. However, one area in Austria, around Hintertux, has introduced a ski pass called 'White 5' that gives access to high-altitude glaciers, where the skiing season can stretch from September to June. As a further new attraction, Dachstein near Salzburg has created a 'Sky Walk', a viewing platform at 2,700 m with a partly glass floor that juts out of a vertical rock face (Figure 12). On a clear day, Slovenia and the Czech Republic can both be seen from this 360-degree viewpoint and there are fantastic views of the glaciers and the glacial landscape. The Ice Palace gives visitors a chance to explore the inside of the Dachstein Glacier. In 2006/7 Austria's traditional resorts became more accessible to British skiers thanks to low-cost flying. The launch of an average of seven new daily services from the UK, primarily to Salzburg but also to Klagenfurt and Innsbruck, increased passenger capacity by close to 6,000 a week. In 2007 Austria experienced a 13% increase in British skiers visiting their resorts.

Figure 12: The Dachstein Sky Walk at an altitude of 2,700 m in the Austrian Alps

Activity 1

Complete this paragraph, using some of the words in the box.

Glacial landscapes provide opportunities for human activity. Many visitors go to places like the Alps for skiing and _____ holidays. People can ski on high-altitude _____ even in the summer months. The _____ industry is important to the economy of glacial areas.

Glaciated landscapes provide a different opportunity, using water in glacial lakes for _____ The water is used to turn _____ which generate _____

hydro-electric power		ice skating	
corrie	skiing	electricity	turbines
tourism	glaciers	drumlin	hiking

Skills Builder 3

Read the news report opposite.

(a) How many people visit the Alps each year? What percentage of most resorts' income does not come from skiing?

(b) Describe the likely impact of warming weather on ski resorts.

(c) Suggest how ski resorts might attract more visitors.

Tourism in the Alps is a multi-million pound industry that is severely threatened by climate change. About 80 million people head to the Alps, the majority of them in the winter.

Most resorts rely on skiing for 75 per cent of their income. It is estimated that hotels will lose a quarter of their business if the temperature rises as predicted. Lift companies could lose 35 per cent.

The company Transmontagne, which operates mid-altitude resorts in France, Switzerland, Italy and Slovenia, is currently under a court bankruptcy protection order for six months. Warming weather is seen as a key reason for its financial woes.

http://news.bbc.co.uk/1/hi/world/europe/6932936.stm

Avalanches and their management

The causes and effects of avalanches

Avalanches are significant hazards to people living in or visiting glaciated landscapes. They can occur with very little, if any, warning and they can be devastating in their consequences. Death and injury normally result from suffocation, inhaling snow, frostbite and shock. Damage to transport systems and buildings is caused by the weight of snow landing on them.

The main physical causes of avalanches usually act in combination and include:

- Steep slopes – 30–40 degree slopes have the highest frequency of avalanches.
- Heavy snowfall – which increases the weight of snow on the slope.
- Rapid thaw at the bottom of the snow layer – which can allow the snow to slide.

Human factors that may contribute include:

- Deforestation – because trees help hold snow in place on slopes.
- Skiing on potentially unstable snow – which may disturb the delicate balance that exists on the slope.
- Loud noises – explosions (or possibly even gunfire) can cause vibrations which disturb the delicate balance.

Objectives

- Know the factors that cause avalanches.
- Be able to describe the impact of avalanches.
- Understand how avalanches can be predicted and prevented.

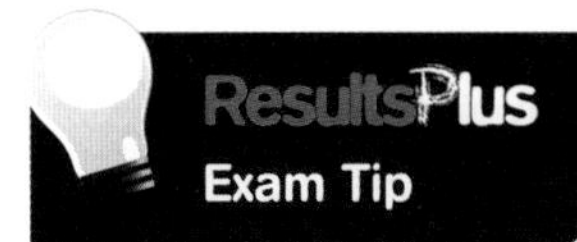

Exam Tip

Good quality examination answers show an awareness that avalanches are usually caused by a combination of different factors acting together.

Case study: The Montroc avalanche, 1999

In February 1999, a major avalanche disaster happened at Montroc in the Chamonix Valley, in the French Alps. This area is a very popular destination for skiers and much work had been done in recent years to increase the number of ski runs available. As a result of the increasing demand, the valley sides had been cleared of many of their trees to allow for the development of more runs. The valley sides were very steep, rising 250 m in a short distance on the eastern side of the valley to the 2,500 m peak of Montagne de Peclerey. There had been very heavy snowfall for the previous four days, and 2 m of snow had fallen in total. At 2.40 p.m. on Tuesday 9 February, a wall of snow 15 m high and 300 m wide broke away from the eastern valley side. The avalanche raced into the valley at Montroc, 250 m below, at speeds of 25 m/sec, demolishing seventeen ski chalets, killing ten people and badly injuring five. You can see this sequence of events in Figure 13. An estimated 90,000 m^3 of snow came down the slope into the village in a few minutes.

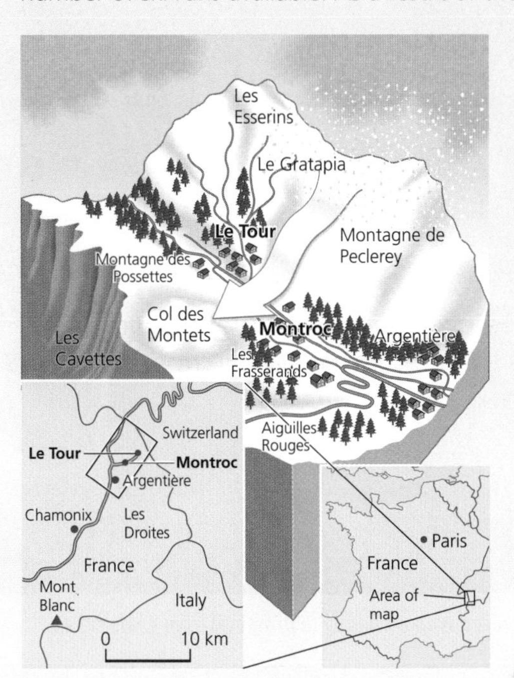

Figure 13: The Montroc avalanche

Case study quick notes:

- There are a number of factors that can combine to cause an avalanche.
- Avalanches can move very quickly.
- Avalanches can cause major loss of life.
- Avalanches can cause significant damage to buildings.

Prediction and prevention of avalanches

The potentially serious impacts of an avalanche can be reduced through efforts to predict their occurrence and by planning before the event. Accurate prediction of avalanches is currently not possible, although the level of the hazard can be forecast. Attempts to predict them use two very different strategies. The first is the use of computer modelling. As the understanding of the causes and the movement of avalanches increases, so more information can be added to sophisticated computer programmes that model how and when an event is likely to occur. The Swiss Federal Institute for Snow and Avalanche Research in Davos has the best system in the world. (It has been estimated that 65% of the population in Switzerland live in areas at risk of avalanches.) The avalanche warning system automatically updates every 15 minutes, using sensors placed on slopes to record the climatic and snow conditions.

The other method is to carefully monitor the condition of the snow as it builds up on a slope. This is done by digging snow pits, several metres across, which reveal a cross-section of the various layers in the snow to try to discover how stable the snow is. This is still the most effective way because the computer models are thought to still be ten years away from full development and use.

Building design is a vital way of planning for avalanches. Roofs are reinforced in order to withstand the weight of snow that could be added during an event. Roads and railways are often covered with concrete canopies (sheds) as they cross slopes at risk of avalanches, as you can see in Figure 14.

Figure 14: A road in the Alps, covered with a canopy to protect it from avalanches

Activity 2

Design an A4-sized poster to put up in an Alpine ski resort, warning people of the dangers of avalanches and how they should try to avoid triggering them.

For an area you have studied, explain the causes and effects of avalanches. (6 marks)

■ **Basic answers** (Level 1)
Give a description only, with no explanation and no reference to the area studied.

● **Good answers** (Level 2)
Offer some explanatory statements about both causes and effects, or a better developed explanation on one. There is some reference to the area studied.

▲ **Excellent answers** (Level 3)
Provide full explanation of both causes and effects with specific detail from the area studied, effectively used as evidence.

Barriers can be constructed on slopes, especially where villages lie further down below. These may be wooden fences, concrete walls or even earth mounds (Figure 15). These can be designed to block, divert or even split the avalanche.

Afforestation can also be used, because the trees help to trap snow on slopes and slow down the movement, should an avalanche happen. But this might be a difficult approach in areas that have many ski runs.

Land use planning is vital. Areas can be classified according to the level of risk, with building being prevented or limited in the areas of greatest risk. Colour-coded zones have been used in Vail, Colorado to successfully plan development in the town.

Controlled explosions can be used to trigger small avalanches that do not cause much impact, rather than allowing the amount of snow to build up to more dangerous levels. This method is often used in locations at risk in the USA, such as the Sylvan Pass in Yellowstone National Park.

Another important aspect of avalanche prevention is education. People who use these areas for skiing need to be educated – not only about the risks they are exposed to, but also about how their actions can trigger avalanches. In particular, those who ski 'off-piste' (away from the marked and compacted runs) have to be taught how to recognise potentially hazardous slopes and how to avoid skiing across slopes whose delicate balance could be disturbed by their actions. Signposts and notices are often displayed to emphasise these messages (Figure 16).

Skills Builder 4

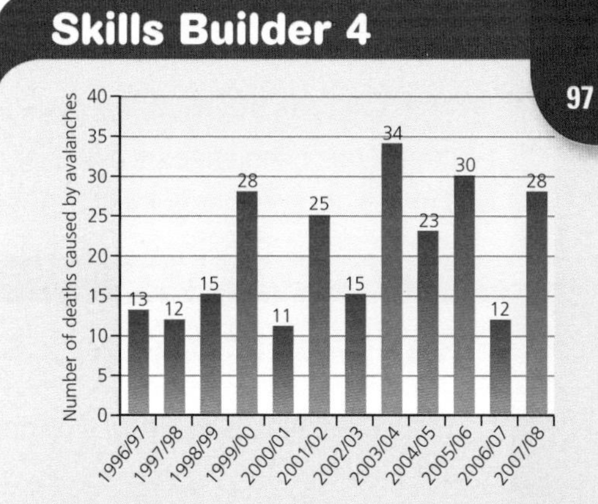

Figure 17: Deaths caused by avalanches in Russia

Study Figure 17.

(a) In which year were there the fewest deaths from avalanches? What was the highest number of deaths in any one year?

(b) State three potential causes of avalanches.

(c) Explain how avalanches can be predicted and prevented.

Figure 15: Barriers constructed on slopes are a method of avalanche protection

Figure 16: A warning sign to keep skiers away from an area where they might cause an avalanche

Know Zone Chapter 5 Glaciated landscapes

10,000 years ago there were glaciers in this country. Ice is a very powerful agent of erosion and much of the landscape of the UK has been affected either by glacial erosion or by deposition. Today, glaciated areas are used to generate energy, and for leisure and recreation.

You should know...

- ☐ The key processes of glacial erosion, including abrasion and plucking
- ☐ The effect of freeze–thaw weathering in providing debris for glacial erosion
- ☐ The key characteristics of landforms of glacial uplands: corries, arêtes, pyramidal peaks, U-shaped valleys, truncated spurs, hanging valleys and ribbon lakes
- ☐ The processes of deposition
- ☐ The landforms created by deposition: moraines, drumlins and erratics
- ☐ How people use glaciated landscapes for energy provision, leisure and recreation
- ☐ The human and physical causes of avalanches
- ☐ How avalanches can be predicted
- ☐ How places can be protected from avalanches

Key terms

Ablation
Abrasion
Arête
Avalanche
Corrie lake
Corries
Deposition
Drumlin
Erosion
Erratics
Glaciation
Freeze–thaw
Hanging valley
Hydro-electric power
Lodgement
Moraines
Pyramidal peak
Ribbon lakes
Truncated spurs
U-shaped valley
Valley glacier
Weathering

Which key terms match the following definitions?

A The use of fast flowing water to turn turbines which produce electricity

B An armchair-shaped depression formed by a glacier in a mountainous area

C A sudden and rapid movement of a mass of snow or ice down a slope

D A tributary valley left high above a main valley that has been deepened by glacial erosion

E A long, narrow and sharp ridge that separates two corries

F Long, narrow lakes found on the floor of U-shaped valleys

G A mixture of different-sized rocks and sediments eventually left behind in the landscape when a glacier melts

H A weathering process which causes rock to break down by the repeated action of water freezing and expanding in cracks

To check your answers, look at the glossary on page 289.

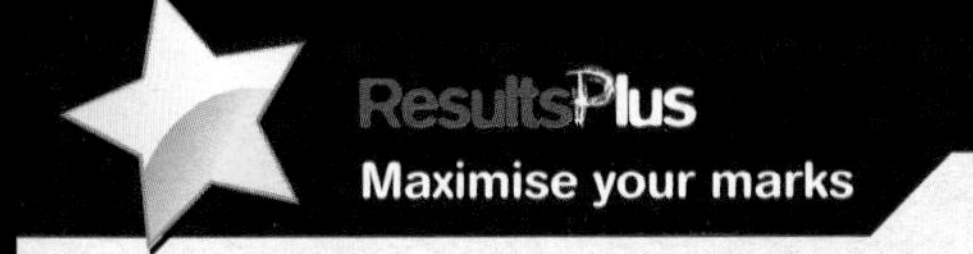

Foundation Question: Name and describe one type of moraine. (3 marks)

Student answer ■ (awarded 1 mark)	Examiner comments	Build a better answer △ (awarded 3 marks)
Chosen type - Terminus Moraine *This is dropped where glaciers stop. This happens when glaciers melt.* *It is made up of lots of freeze-thaw material that comes from higher up on the mountain.*	• ***Terminus Moraine...*** is not quite right, but it is close enough for 1 mark. • ***This is dropped...*** scores no marks because it is about the formation of terminal moraines and the question asks for a *description*. • ***It is made up...*** is not accurate so scores no marks. Although it is fine to write about the material, the detail is not correct.	Chosen type - Terminal moraine. These are lines of material across a valley, almost blocking it. The material is often pushed there or slides off the glacier. It is made up of a mixture of boulders and ground up rock.

Overall comment: Make sure that you read the question and pay particular attention to the command words. In this question they are *name* and *describe*.

Higher Question: Explain how the effects of avalanches can be reduced through planning before the event. (4 marks)

Student answer △ (awarded 3 marks)	Examiner comments	Build a better answer △ (awarded 4 marks)
Prediction of avalanches takes place by using modern techniques like satellites. *Other ways of doing it involve using computers to look at the causes.* *If the snow on a slope is watched very carefully and snow pits are dug then the snow can be seen to be either stable or unstable.*	• ***Prediction of avalanches...*** is not correct and scores no marks. • ***Other ways of...*** scores one mark because it identifies the basic idea of using computers. However, there is no explanation of how computers are used. • ***If the snow on...*** is good and so scores 2 marks. Semi-technical terms such as ***stable*** and ***unstable*** help the answer.	One way of planning is to prepare better for the event by improving building design using canopies and reinforcing roofs. Computers can be used to model the behaviour of snow fields and build up a picture of what makes snow unstable. The Swiss have done much work in this area. If the snow on a slope is watched very carefully and snow pits are dug then the snow can be seen to be either stable or unstable.

Overall comment: It is very important to interpret a question correctly – this question is about *planning*, which obviously involves more than just predicting an event but also preparing for it.

Chapter 6 Tectonic landscapes

Objectives

- Know the names of the major tectonic plates.
- Be able to describe the distribution of earthquakes and volcanoes.
- Understand what causes these tectonic events.

The location and characteristics of tectonic activity

The world distribution of earthquakes and volcanoes

Earthquakes and **volcanoes** occur in many different parts of the world. But their distribution is not random. Figure 1 shows the distribution of earthquakes and Figure 2 shows the distribution of volcanoes. If you study them you will see that there are many similarities. Both tend to occur mainly in narrow bands, although some do fall outside of this general pattern. In many cases the two occur in the same places, although this is not always the case.

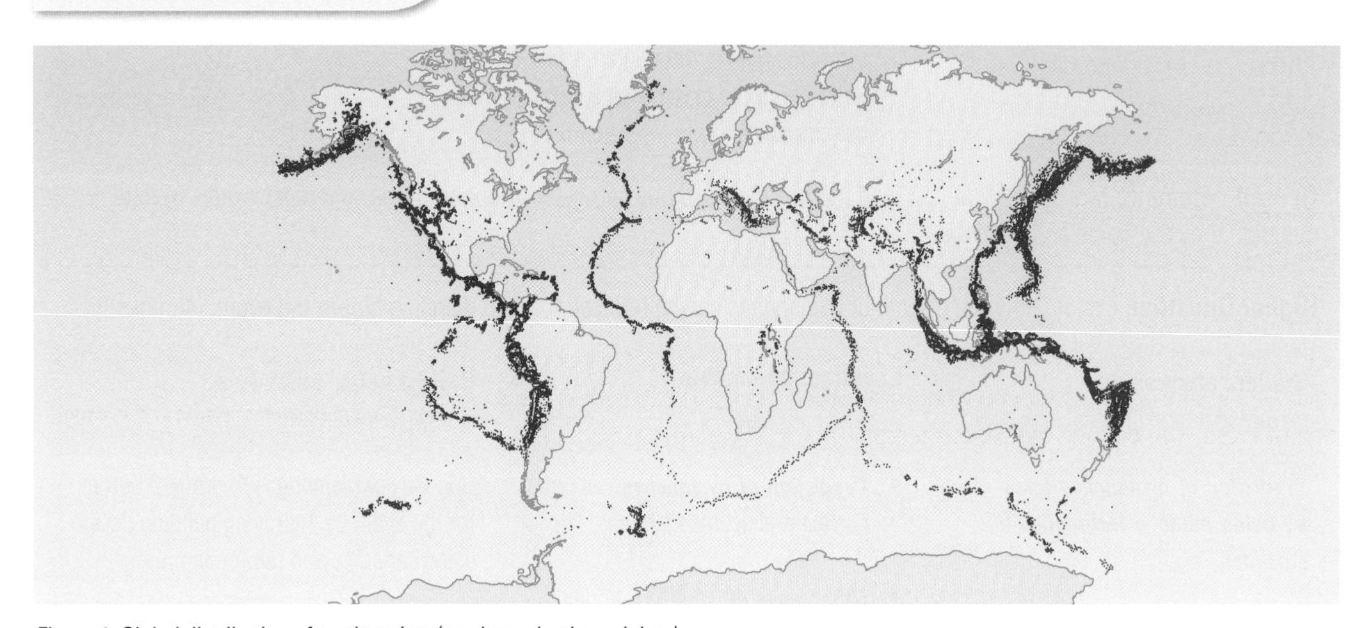

Figure 1: Global distribution of earthquakes (as shown by the red dots)

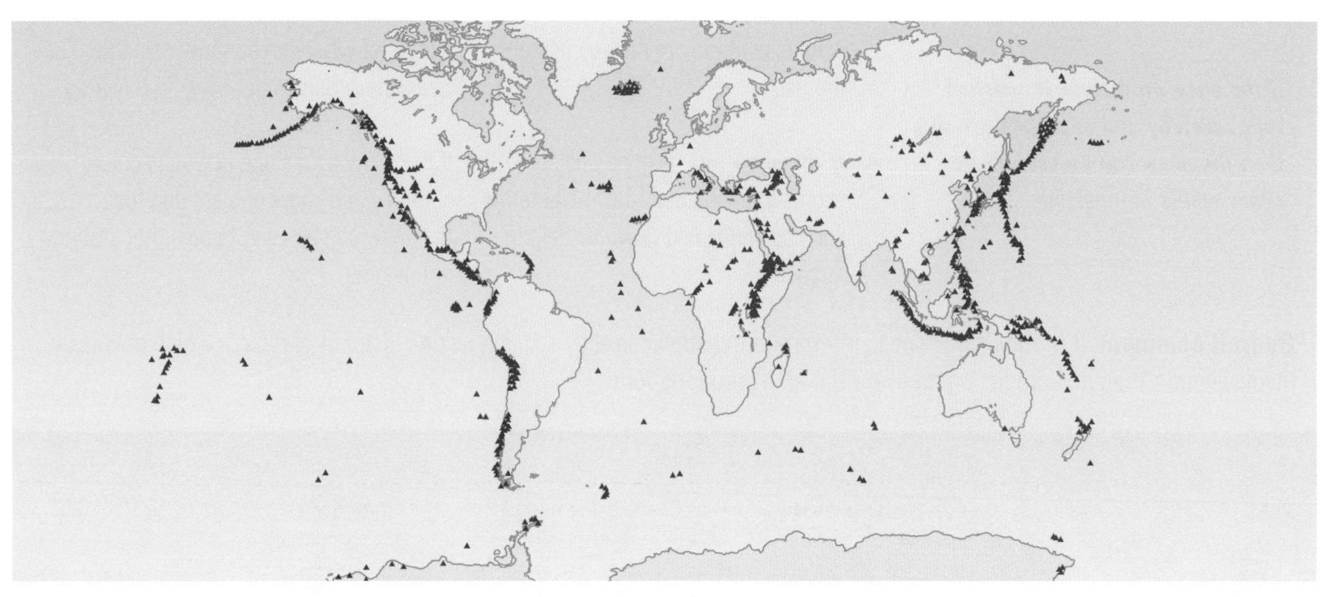

Figure 2: Global distribution of volcanoes (as shown by the red triangles)

Earthquakes and volcanoes tend to occur together in bands that are found:

- In the middle of oceans, such as the Atlantic
- Along the edges of continents, such as the west coast of South America.

There is a particularly strong pattern around the edges of the Pacific Ocean, known as the 'Ring of Fire'.

Earthquakes occur without volcanoes through central Asia. Volcanoes are also found in more isolated clusters, such as the Hawaiian Islands in the Pacific Ocean.

Skills Builder 1

Study Figures 1 and 2 on page 100.

Describe the global distribution of (a) earthquakes and (b) volcanoes.

Why do earthquakes and volcanoes occur where they do?

Earthquakes and volcanoes occur as a result of the movement of tectonic plates. Plate tectonic theory states that the Earth's surface is divided into a number of large and small slabs of rock that together make up the Earth's crust. These plates 'float' on the hot, molten rock – the **magma** – that is in the **mantle** below. These plates move in relation to each other because of **convection currents** in the mantle's magma. These currents are shown in Figure 3. Depending on the direction of the convection currents, the plates can move towards each other, away from each other or slide past each other.

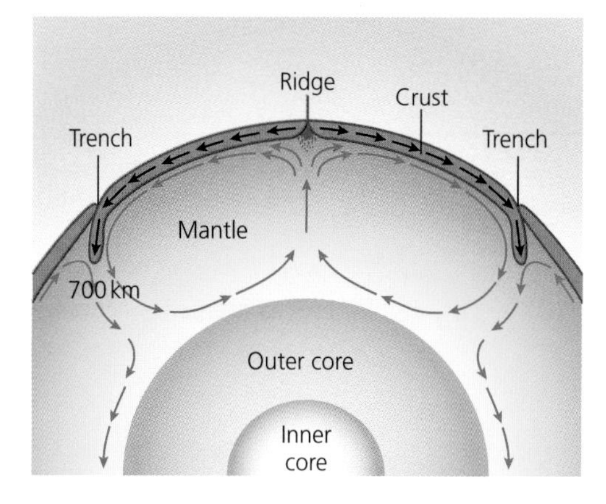

Figure 3: Convection currents in the mantle

There are two types of crust that make up these plates. Continental crust is usually between 25 and 100 km thick and is made of relatively low density material. Oceanic crust, on the other hand, is only 5 to 10 km thick but is made of relatively high density material. Some plates are made only of oceanic crust, such as the Pacific Plate; some are almost exclusively made of continental crust, such as the Eurasian Plate, whilst others are made of both. The African Plate is one of these. You can see the various tectonic plates in Figure 4.

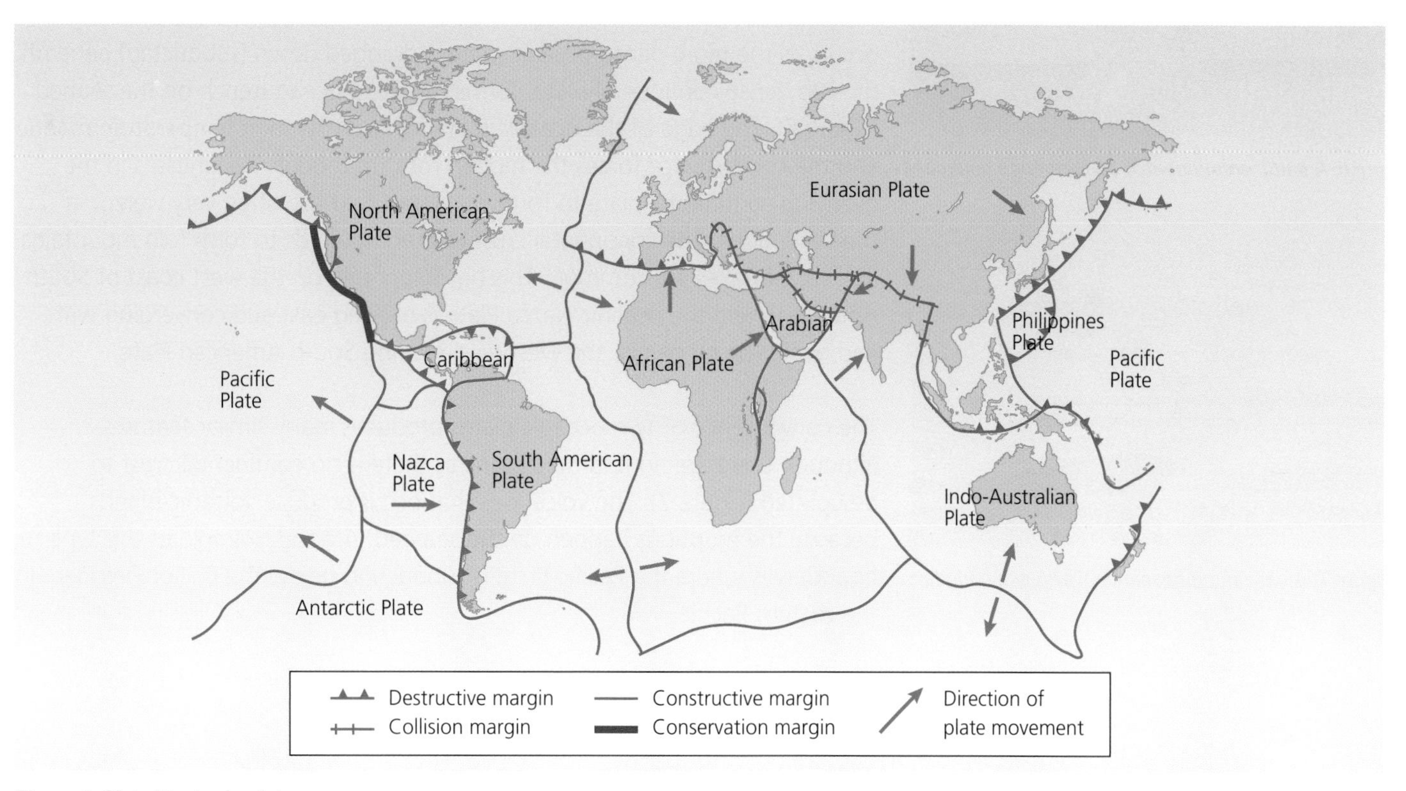

Figure 4: Global tectonic plates

Activity 1

(a) State two differences between oceanic and continental crust.

(b) Explain why tectonic plates move.

The movements of the plates are very slow – typically only a few centimetres a year on average. Sometimes, however, the huge stresses that the plates are put under by the convection currents make the rock suddenly break, causing a brief but violent shaking of the ground above. This is an earthquake.

Volcanoes occur when some of the molten magma from the mantle is able to escape out on to the surface through fractures in the crust. These can occur when plates are pulling apart or pushing together, but not when they are sliding past each other.

Convergent, divergent and conservative plate boundaries

It is because of the movements of the plates that most earthquakes and volcanoes happen where two plates meet, at a plate boundary. There are three types of plate boundary.

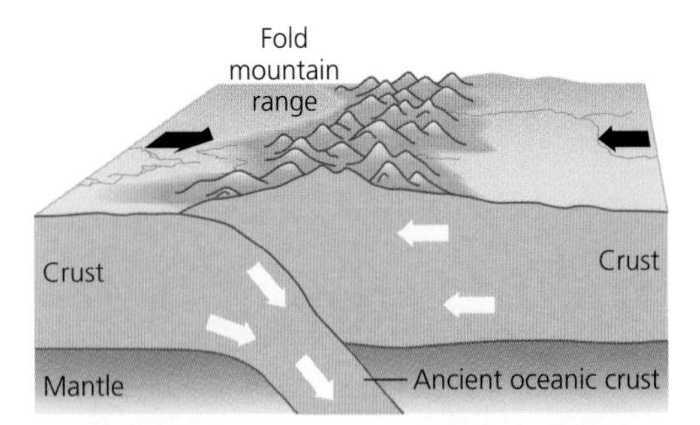

Figure 5: A continental/continental convergence boundary

Convergent plate boundaries are those at which the plates are moving towards each other. This can happen when two continents are converging, when two oceanic plates are converging or when an oceanic plate converges with a continental plate. The features that result from these different types of boundary vary. The convergence of two continental plates is known as a collision boundary (Figure 5). As they are made of relatively low density rock, both plates are pushed upwards by this collision, leading to earthquakes and the gradual formation of a **fold mountain range**. No volcanoes are found here. A good example is the collision between the Eurasian Plate and the Indo-Australian Plate. This has led to the formation of the Himalayas and is now responsible for earthquakes, such as the catastrophic one that struck Pakistan in 2005.

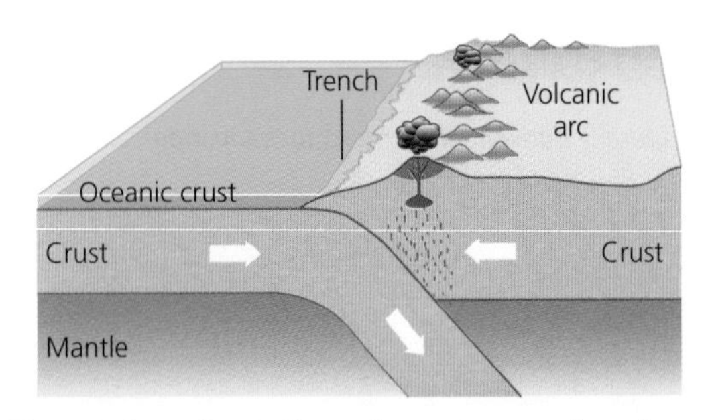

Figure 6: A continental/oceanic convergence boundary

When an oceanic plate converges with a continental plate, **subduction** occurs as the more dense oceanic plate is dragged down (subducted) beneath the less dense continental plate, forming a deep ocean trench on the seabed (Figure 6). The edge of the oceanic plate melts in the high temperature mantle and the convergence forces the molten rock up though weaknesses in the overlying continental plate to form volcanoes that are often very violent in their eruptions. The continental crust is lifted upwards to form fold mountains and earthquakes are common. This has happened on the west coast of South America where the oceanic Nazca Plate is moving east and converging with, and being forced below, the westward-moving South American Plate.

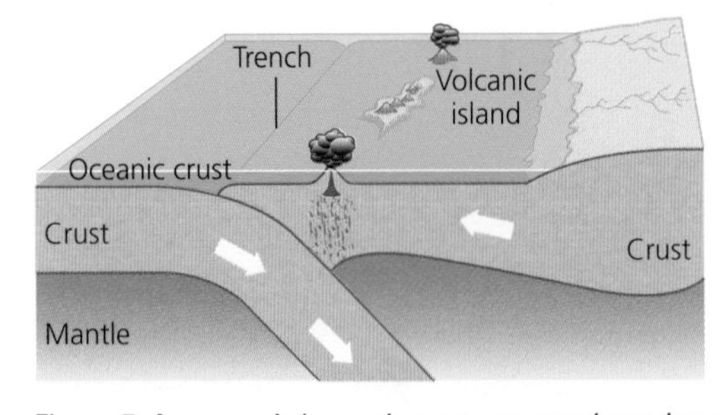

Figure 7: An oceanic/oceanic convergence boundary

The convergence of two oceanic plates produces many similar features, although there are no fold mountains as there is no continental crust to be uplifted (Figure 7). The volcanoes that occur produce volcanic islands because the eruptions happen on the sea bed. A good example of this type of boundary is where the Pacific Plate is subducting below the Philippines Plate in the eastern Pacific.

Divergent plate boundaries are found when two oceanic plates are moving away from each other, such as in the mid-Atlantic (Figure 8). The pressures exerted by the divergence of the plates lead to fractures being formed in the crustal rocks. Rising magma is then able to enter the crust, filling the gaps and creating new sea floor. If a lot of magma escapes on to the surface, submarine volcanoes occur, which may build up over time to form volcanic islands, such as Iceland. The features associated with these boundaries are seen in Figure 8, which shows the mid-Atlantic. Such boundaries can also occur on land when continental plates diverge. This is currently happening in east Africa where a large valley called a 'rift valley' has formed.

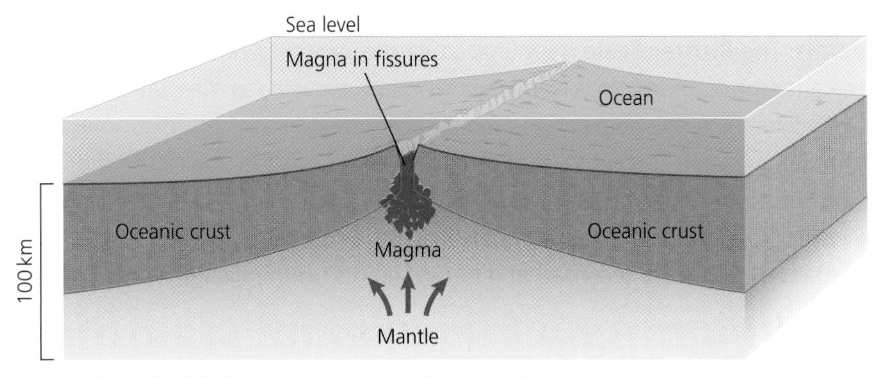

Figure 8: The mid-Atlantic oceanic/oceanic divergence boundary

The final type of plate boundary is one where two plates slide past each other. This is known as a **conservative plate boundary**. No magma is rising in the mantle and no subduction is taking place – so volcanoes cannot occur. The sliding is not smooth, however, and extreme stresses build up in the crustal rocks. When this pressure is eventually released the result is an earthquake. The best example of this situation is along the west coast of North America, where the Pacific Plate is sliding past the North American Plate (Figure 9). Earthquakes occur along fault lines such as the San Andreas fault. Friction between the plates causes them to 'stick' and only when enough pressure has built up will they move again. If this pressure is released frequently and easily then the earthquakes are small. If the pressure builds up over time, the occasional large earthquake will result, such as the one that hit San Francisco in 1989.

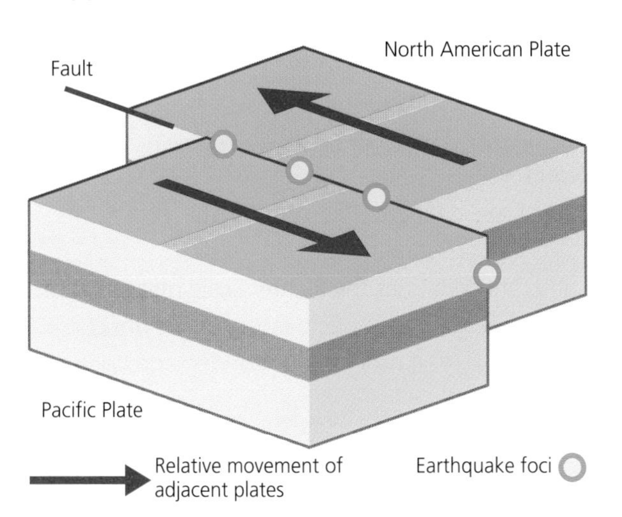

Figure 9: The conservative plate boundary on the west coast of North America

Activity 2

(a) What is subduction?

(a). Why does oceanic crust subduct but continental crust does not?

Activity 3

Choose one example of a divergent plate boundary.

(a) Name the two plates involved and describe the movement of the plates.

(b) Explain why the plates are moving in this way.

(c) Identify the typical landforms produced at this boundary.

(d) Explain how these landforms are produced.

Activity 4

Explain why no volcanoes occur on conservative plate boundaries.

ResultsPlus
Build Better Answers

Explain the processes that take place at divergent plate boundaries. (4 marks).

Basic answers (0–1 marks)
Give description only with no explanation.

Good answers (2 marks)
Offer some explanatory statements relating to either volcanic or earthquake activity.

Excellent answers (3–4 marks)
Provide full explanation of processes, such as how convection currents move plates apart and why magma is able to escape.

Skills Builder 2

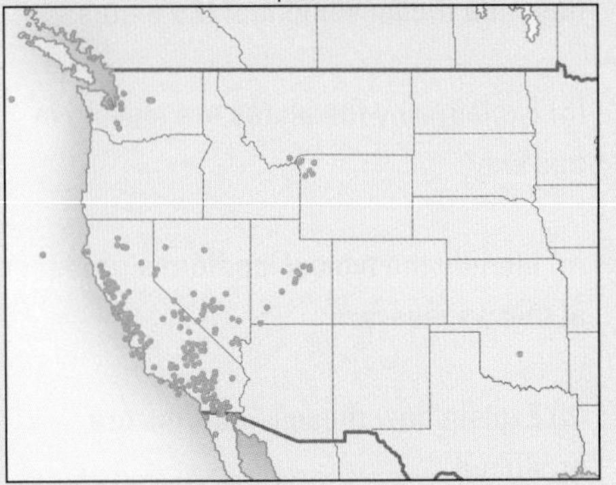

Figure 10: The location of earthquakes in the USA over a 30-day period in 2008 (as shown by the orange circles)

Study Figure 10.

(a) How many earthquakes occurred during this period?

(b) What is the average number of earthquakes per day?

(c) Describe the pattern of earthquakes on the map.

(d) Suggest why so many earthquakes occur in one part of the USA.

Hot spots

In some places, like the Hawaiian Islands in the Pacific Ocean, very high temperature magma rises from the mantle, even though there is no plate boundary there. If the crust is thin or weak, this magma can escape through it to form volcanoes. These locations are called **hot spots**. In Hawaii the volcanoes originally formed on the sea floor, but over time the lava has built up high enough to stick up above sea level to form volcanic islands.

Measuring the magnitude of earthquakes

There are two different scales that are used to measure the magnitude (strength) of earthquakes. Both are useful when studying earthquake events. The **Richter Scale** measures an earthquake's strength according to the amount of energy that is released during the event, as measured by a seismograph. A seismograph is a remarkably sensitive scientific instrument that can measure earthquakes of any strength – from the smallest tremor to those that shatter the surface of the Earth. There is no upper limit to the Richter Scale. The table below shows the sort of damage usually caused by earthquakes that are recorded at different points on the Richter Scale.

Magnitude on Richter Scale	Typical effects
2.4 or less	Usually not felt, but can be recorded by seismograph
2.5–5.4	Often felt, but only causes minor damage
5.5–6.0	Slight damage to buildings and other structures
6.1–6.9	May cause a lot of damage in very populated areas
7.0–7.9	Major earthquake. Serious damage
8.0 or more	Great earthquake. Can totally destroy communities near the **epicentre**

The **Mercalli Scale** is quite different. Whereas the Richter Scale measures the strength of the cause, the Mercalli Scale is a simple measure of the effects themselves – based on what people feel and the amount of damage done (see table on page 105). To avoid confusion, the Mercalli Scale uses Roman numerals, from I to XII.

Mercalli Scale	Defining criteria
I	People do not feel any Earth movement.
II	A few people might notice movement if they are at rest and/or on the upper floors of tall buildings.
III	Many people indoors feel movement. Hanging objects swing back and forth. People outdoors might not realise that an earthquake is occurring.
IV	Most people indoors feel movement. Hanging objects swing. Dishes, windows, and doors rattle. The earthquake feels like a heavy truck hitting the walls. A few people outdoors may feel movement. Parked cars rock.
V	Almost everyone feels movement. Sleeping people are awakened. Doors swing open or close. Dishes are broken. Pictures on the wall move. Small objects move or are turned over. Trees might shake. Liquids might spill out of open containers.
VI	Everyone feels movement. People have trouble walking. Objects fall from shelves. Pictures fall off walls. Furniture moves. Plaster in walls might crack. Trees and bushes shake. Damage is slight in poorly built buildings. No structural damage.
VII	People have difficulty standing. Drivers feel their cars shaking. Some furniture breaks. Loose bricks fall from buildings. Damage is slight to moderate in well-built buildings; considerable in poorly built buildings.
VIII	Drivers have trouble steering. Houses that are not bolted down might shift on their foundations. Tall structures such as towers and chimneys might twist and fall. Well-built buildings suffer slight damage. Poorly built structures suffer severe damage. Tree branches break. Hillsides might crack if the ground is wet. Water levels in wells might change.
IX	Well-built buildings suffer considerable damage. Houses that are not bolted down move off their foundations. Some underground pipes are broken. The ground cracks. Reservoirs suffer serious damage.
X	Most buildings and their foundations are destroyed. Some bridges are destroyed. Dams are seriously damaged. Large landslides occur. Water is thrown on the banks of canals, rivers, lakes. The ground cracks in large areas. Railway tracks are bent slightly.
XI	Most buildings collapse. Some bridges are destroyed. Large cracks appear in the ground. Underground pipelines are destroyed. Railway tracks are badly bent.
XII	Almost everything is destroyed. Objects are thrown into the air. The ground moves in waves or ripples. Large amounts of rock may move.

- Do not confuse the two scales used to measure earthquakes. Remember, the Richter Scale measures the earthquake's strength, whereas the Mercalli Scale measures its impact on people.

Activity 5

Using the internet, compile a list of the ten earthquakes with the highest magnitude that have ever occurred.

The **focus** of an earthquake is its central point – the point deep below the surface (sometimes as much as 600 km down) where the earthquake starts and where the greatest release of energy occurs. Shock waves then spread in all directions. Energy is absorbed by the crustal rocks and that is why the amount of energy and therefore the amount of ground movement decreases with distance away from the focus. The point on the surface directly above the focus is called the **epicentre**. It is the place on the surface that is nearest to the focus and therefore where the greatest effects will potentially be felt. The relationship between the focus and the epicentre is shown in Figure 11.

Skills Builder 3

Study this table of data about earthquakes in 2007.

Location of earthquake	Magnitude on Richter Scale	Number of deaths
Solomon Islands	8.1	52
Coast of central Peru	8.0	514
Southern Sumatra, Indonesia	8.4	25
South of the Fiji Islands	7.8	0

(a) Which earthquakes had (i) the highest magnitude, (ii) the highest death toll?

(b) Suggest four reasons why the earthquake with the highest magnitude did not cause the most deaths.

(C) Explain two differences between the Richter Scale and the Mercalli Scale.

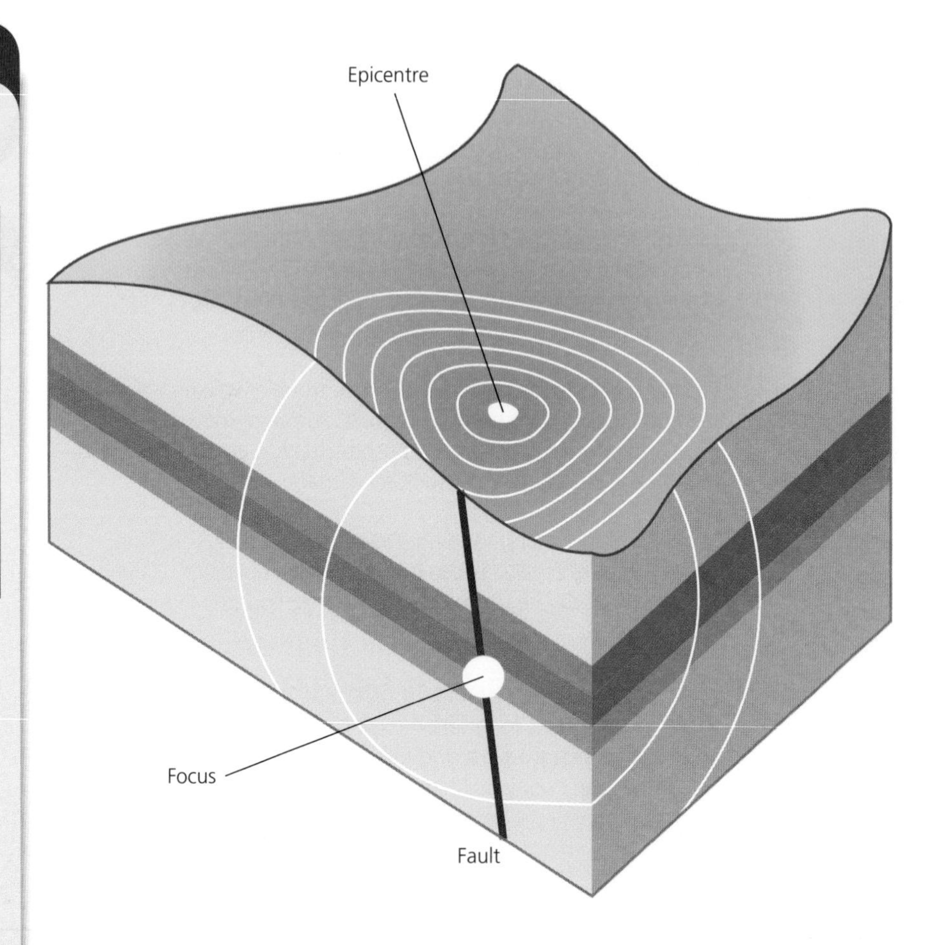

Figure 11: The focus and the epicentre of an earthquake

Managing the effects of tectonic activity

Why do people continue to live in areas of volcanic and earthquake activity?

People continue to live in places that are at risk of earthquakes and volcanoes for a number of reasons – economic, social and environmental. The reasons often differ between places that are at different levels of development.

In high-income countries, people may choose to live in areas that are tectonically active for economic reasons. In California, for instance, there is a significant concentration of high-tech industry, and people are attracted to work there because high-income jobs are available in companies such as IBM and Apple. With 23,000 factories, the industry employs over 700,000 people. All of this is despite the fact that the area is on the conservative plate boundary between the North American and Pacific Plates. Small earthquakes are a daily feature of life, and there were major events in 1989 and 1994.

People are attracted to live and work in this area by the attractive landscape and the pleasant climate. The coast provides beautiful beaches and opportunities for surfing at places such as Half Moon Bay. Summer temperatures can reach 30°C and winters are mild at around 10°C. Rainfall is low, typically 300–400 mm per year, with very little falling in the summer. The standard of living is high, with people enjoying high quality services such as education and medical care.

Volcanoes are attractive to tourists, with many visiting active areas such as Iceland, creating a demand for jobs in hotels and as tour guides. It is estimated that tourism provides over 5,000 jobs and that tourism accounts for about 5% of the country's earnings. People may make conscious decisions to live and work in these places despite the risks.

Volcanoes can provide good environments for farming because some volcanic lava weathers to produce very fertile soils. In Sicily, for example, many farmers grow olives, grapes and citrus fruits on the slopes of Mount Etna. In the Philippines, many people lived on the slopes of Mount Pinatubo before the 1991 eruption because they could grow good crops of rice on its fertile soils (and in this low-income country there were few other opportunities for them to produce enough to survive). In many cases, it is also possible that people may be ignorant of the risks.

Objectives

- Know the reasons why people live in areas of volcanic and earthquake activity.
- Be able to describe the effects of a named volcanic eruption or earthquake.
- Understand how people try to predict and prevent the effects of volcanic eruptions and earthquakes.

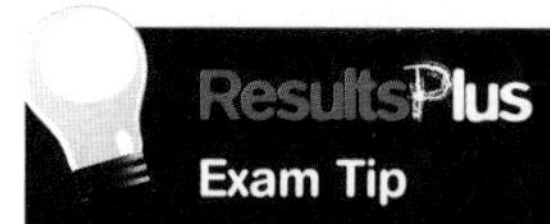

Good quality examination answers usually show an awareness that the reasons why people live in tectonically active areas may be very different in HICs, compared to LICs, due to greater education, mobility and choice.

Skills Builder 4

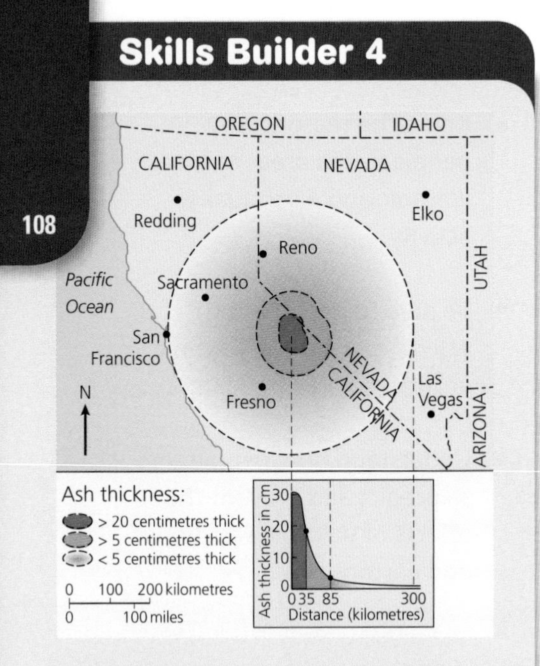

Figure 13: The potential hazard zones for an eruption of the volcano in Long Valley, USA. (The last major eruption was 760,000 years ago.)

Study Figure 13, which shows the likely thickness of volcanic ash to cover the surrounding area after an eruption.

(a) How thick is the ash expected to be at its maximum?

(b) What is the maximum distance from the eruption that is predicted to be covered by ash?

(c) Describe the predicted pattern of ash thickness with increasing distance from the eruption.

Figure 14: Ash-covered ground in Plymouth, the capital of Montserrat

Case study – The causes and effects of the Montserrat volcanic eruption, 1995

Montserrat is one of the Leeward Islands in the Caribbean. It is a British overseas territory – and it lies on a convergent plate boundary. As the two plates merge, the oceanic plate is forced down (subducted) under the continental plate. This increases pressure, which triggers earthquakes. At the same time, heat produced by friction melts the descending crust to form molten magma (as was shown in Figure 6 on page 102). The hot magma tries to rise to the surface, and when it succeeds it will form a volcano such as the one at Soufriere Hills. Between 1995 and 1997 this volcano erupted huge quantities of lava, ash and **pyroclastic flows** – high speed clouds of hot gases, ash and rock fragments. The area affected is shown in Figure 12.

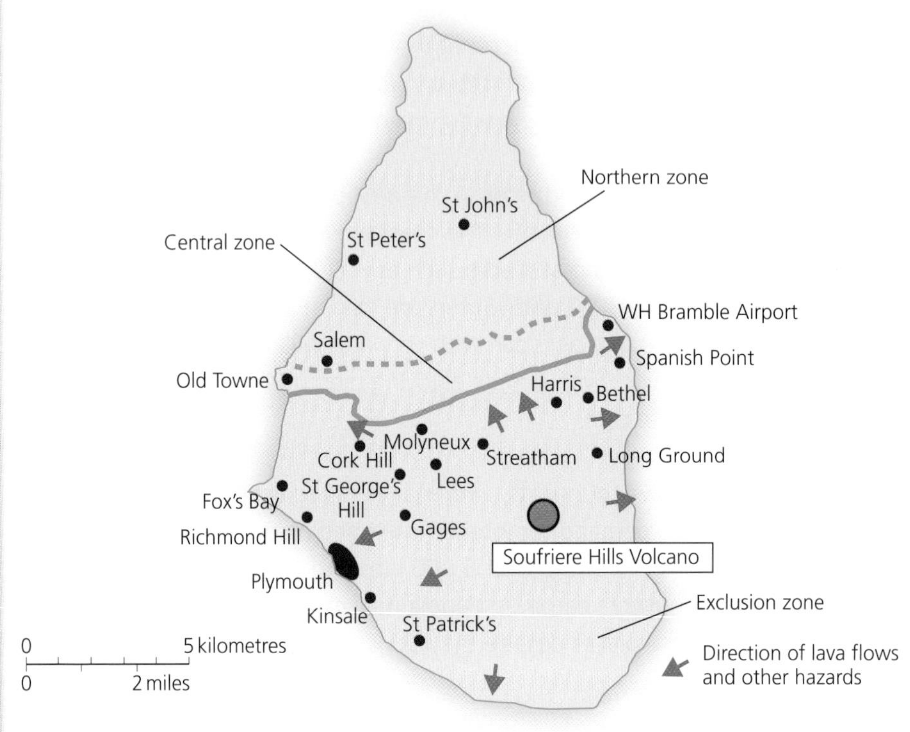

Figure 12: Map of Montserrat

Effects of the eruption

- Twenty-three people died.
- Two-thirds of the island was covered in ash.
- Half of the population were evacuated to the north of the island.
- Half of the population eventually left the island.
- Plymouth, the capital, became an abandoned 'ghost town' (see Figure 14).
- Crops and animals were destroyed by layers of volcanic ash.
- Fires were caused in forests by volcanic gases igniting.
- Rivers flooded because they were blocked by ash.
- The port and the airport were closed, badly affecting the tourist industry.

The human response

- £41 million in aid was donated by the British government.
- Money was given to individuals to help them move away.
- There were riots in the streets because the locals felt that the British government were not doing enough to help.
- The Montserrat Volcano Observatory was set up to monitor the volcano.
- A risk assessment was undertaken to help the locals prepare for future eruptions.

Case study – The causes and effects of the Bam earthquake, 2003

At 5.26 a.m. on 26 December 2003, an earthquake shook a large area of the Kerman province in Iran (Figure 15). The epicentre of the devastating earthquake was 10 km south-west of Bam City. The magnitude was estimated as 6.5 on the Richter Scale with a depth of focus of 8 km. A large fracture in the crust, called the Bam Fault, runs from north to south through this area and earthquakes are quite common. In the northern part of the area, there were four major earthquakes between 1980 and 1998.

The earthquakes occur as the result of stresses generated by the movement of the Arabian Plate northwards against the Eurasian Plate. The plates are converging at a rate of approximately 3 cm per year. Before the 2003 earthquake, such stress built up that eventually the crust moved both horizontally and vertically along the Bam faultline, with catastrophic effects.

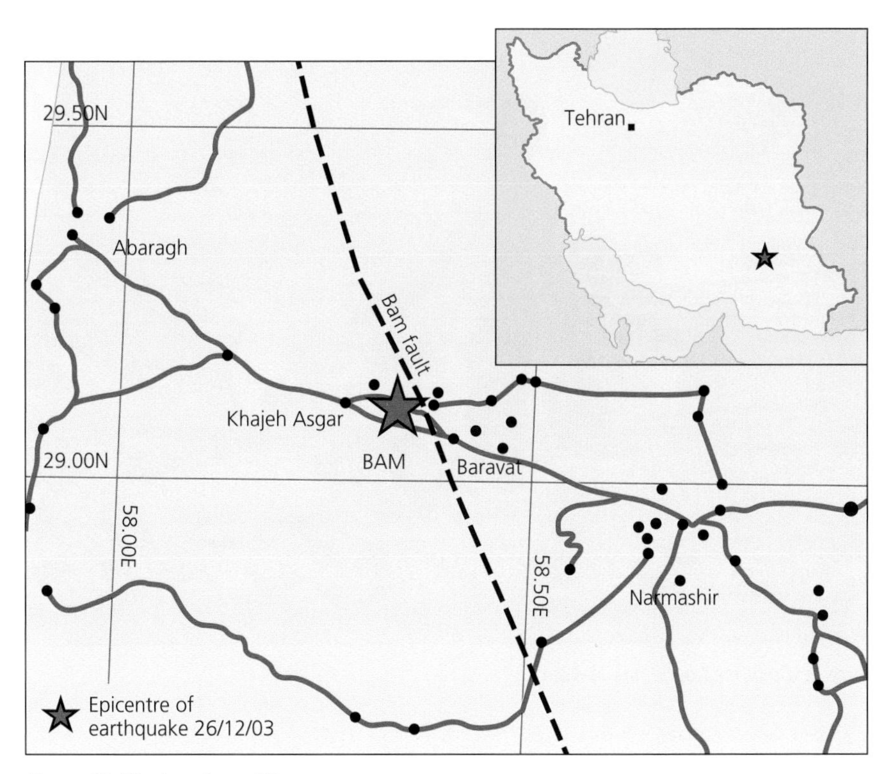

Figure 15: The location of Bam

Quick notes
(Montserrat volcanic eruption):
- Volcanic eruptions can be caused by the activity at convergent plate boundaries.
- Volcanic eruptions have significant effects on the local people, both in the short term and the long term.
- Volcanic eruptions have significant effects on the environment.

Activity 6

Write an eye-witness report for a newspaper on the Montserrat eruption and its effects.

Activity 7

Look at the Mercalli Scale on page 105, and then read the details of the impact of the Bam earthquake. Discuss with a partner what point on the scale you would place this earthquake.

Effects on the people

- 25,000 people were killed.
- 50,000 people were injured.
- 100,000 people were made homeless.
- 25,000 people were treated for trauma.
- Two hospitals collapsed.
- Roads and electricity and telephone lines were damaged.

These effects were so great because people were in bed in their homes at that time in the morning. Most of the deaths and injuries were caused by homes collapsing (see Figure 16). However, it could have been worse – minor tremors the night before had encouraged many people to sleep outside.

Effects on the environment

- Fissures (cracks) formed in the ground.
- There were landslides and rock falls.
- The ground collapsed above underground irrigation channels.
- The lack of irrigation led to the death of many date and palm trees.

Figure 16: Collapsed buildings in Bam

The human response

The short-term emergency response saw the streets of Bam lined with 92,000 tents and thousands of homeless keeping warm with the help of 200,000 blankets, 56,000 items of clothing and 51,000 oil heaters. In addition, over 400,000 ready-to-eat meals were provided, along with bread, rice, sugar and other food items. Later, Finnish and Norwegian Red Cross workers set up a 250-bed emergency hospital.

In the short term there was much anger amongst the local people. Although the Iranian Red Crescent arranged more than 8,500 relief workers, they felt that aid was not arriving fast enough and they suspected that there was much corruption amongst local officials over its distribution. Fresh drinking water was not available until a month after the earthquake.

There were also long-term responses to the earthquake. The government decided almost immediately that the city should be rebuilt. The plan to rebuild took six months to draw up, and again the people were angry at the long delay. Rebuilding is now well underway and modern technology is being used to make houses more earthquake-resistant than the old mud-brick style.

Prediction and prevention of the effects of volcanic eruptions and earthquakes

Volcanic eruptions

Volcanoes can be monitored by a variety of methods to try to predict possible eruptions. The methods include

- Using GPS technology to check the bulging of the volcano as magma approaches the surface
- Gas sampling, because changes in gas composition indicate the activity levels in the magma underground
- Geothermal monitoring from space, to record changes in heat as magma approaches the surface
- Seismic monitoring, to 'listen' to the rising blobs of magma as they force their way upwards and cause small earthquakes
- Looking at the historical records of past eruptions, to see if there is a regular pattern that might be repeated.

Activity 8

Imagine you were a local resident living in Bam at the time of the earthquake. Describe how you would have felt afterwards and explain why you would have had these feelings.

Quick notes (Bam earthquake):

- Earthquakes can be caused by the activity at convergent plate boundaries.
- Earthquakes have significant effects on the local people, both in the short term and the long term.
- Earthquakes have significant effects on the environment.

ResultsPlus
Exam Question Report

Choose an earthquake you have studied. Explain how people responded in the short term and in the long term. (5 marks, June 2006)

How students answered

Many students answered this question poorly and did not score many marks. They tended to write about the impacts of the earthquake rather than the response of the people.

30% (0–1 marks)

Most students made some reference to the response, but it was often only related to the short term, not the long term as well. Little detail was given about the chosen example.

59% (2–3 marks)

Some students answered this question well. They explained both short-term and long-term responses and illustrated their answer well by using detail from their chosen example.

11% (4–5 marks)

Figure 17: Managing a lava flow on Mount Etna

ResultsPlus
Watch out!

Be careful not to over-generalise about volcano monitoring. The ability of a country to predict volcanic eruptions will depend greatly on the level of technology available for monitoring. HICs like the USA have much higher levels of technology available to them than LICs like Mexico, although they may lend them their support.

Good quality examination answers usually show an awareness that at present it is not possible to predict earthquakes. The Tokyo Earthquake Prediction Centre is the best in the world, but it can only give residents a maximum warning of 9 seconds once an earthquake occurs offshore on the nearby plate boundary.

In Mexico, the eruption of Popocatépetl in 2000 was successfully predicted by monitoring of the small earthquakes that often precede eruptions. The nearby town was evacuated and no one was killed.

Once an eruption has happened, it may be possible to prevent it from having major effects on the local people. Lava flows, in particular, can be managed and controlled in various ways:

- Spraying the lava with water to cool it down, making it solidify and stop flowing
- Putting concrete or rock barriers in the path of the lava, diverting the flow away from villages
- Setting off explosives, to divert the lava flow
- Digging ditches, to divert the flow away from areas at risk.

These strategies are often used on the slopes of Mount Etna, Sicily, to protect villages such as Zafferana. They were widely used in the 1991–93 eruptions and were quite successful (Figure 17).

Earthquakes

It is very much harder to predict earthquakes, partly because they don't happen at a visible single point, like volcanic eruptions. They can occur anywhere along the length of the plate boundary. Methods used include:

- Noting strange animal behaviour, because this is often reported before earthquakes
- Monitoring electrical discharges, because there is some evidence that these rise before an earthquake
- Recording minor tremors, because some major earthquakes are preceded by small foreshocks.

Because so little warning can currently be given, a lot of planning needs to be undertaken in order to limit the effects of earthquakes:

- Providing education for local residents, who need to be taught how to prepare and react. They need to have an emergency kit ready, for example, including a torch, bottled water, first aid kit and tinned food.
- Designing buildings that can withstand minor earthquakes – by having flexible steel frames, for instance, which sway as the ground moves. Shutters can be used to cover windows that might shatter, and counter-balance weights can be used in the roof to return buildings to an upright position after they have swayed. Rubberised foundations can be used to absorb the energy of an earthquake.

- Planning regulations are also important, because they determine how high buildings may be and what land uses can take place in areas at risk. Sufficient medical and fire services should be provided, and staff should be fully trained in how to deal with the hazards.

In November 2008, Los Angeles had the largest earthquake drill in history. It cost $2 million and began with radio and television warnings, which instructed more than 5 million people to get under sturdy furniture and to stay indoors with their arms over their heads. This was designed to prepare the people for the 7.5 or greater magnitude earthquake that is predicted to occur in the next thirty years.

The Transamerica Pyramid in San Francisco (Figure 18), which had been built to withstand earthquakes, swayed more than 30 cm but was not damaged in the 1989 California earthquake. Its triangular design gave it a wide, stable base and its steel frame allowed it to sway.

Figure 18: The Transamerica Pyramid in San Francisco

Activity 9

Research other design features of earthquake-resistant buildings:
(a) in HICs and (b) in LICs.

Activity 10

Produce a small leaflet to give to local residents of Los Angeles advising them how to prepare for an earthquake.

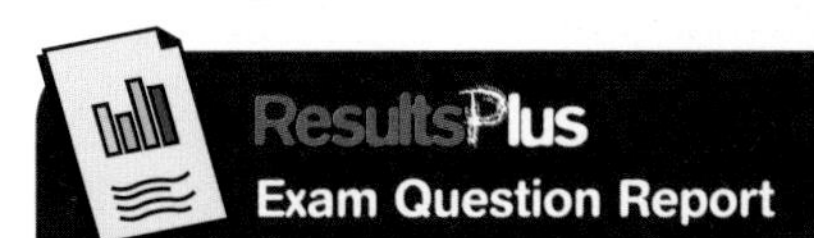

Explain two precautionary methods that have been used on buildings to limit the damage caused by earthquakes. (4 marks, June 2007)

How students answered

Most students answered this question poorly and did not score many marks. In some cases they did not refer to methods used on buildings, but wrote about trying to predict earthquakes.

42% (0–1 marks)

Some students gave one or two valid methods. However, they did not always explain how this limited the damage. Flexible steel frames and rubberised foundations were common answers.

22% (2 marks)

Many students answered this question well. They correctly identified two methods and explained how they limited damage, e.g. the flexible steel frame allowing the building to sway rather than crack.

36% (3–4 marks)

Know Zone
Chapter 6 Tectonic landscapes

The Earth is not 'dead'. Plate movements cause earthquakes and volcanic activity is found in many areas. We need to measure and understand these very powerful forces both for our own personal protection and so that we can prevent widespread damage to property.

You should know...

- ☐ The global distribution of earthquakes and volcanoes
- ☐ The reasons why earthquakes and volcanoes are only found in certain areas, and the role of plate boundaries in explaining this
- ☐ How to draw a cross-section of the main plate boundaries – convergent, divergent and conservative
- ☐ How we measure earthquake activity
- ☐ How we can locate the epicentre of earthquakes
- ☐ The economic, social and environmental reasons why people live in areas of tectonic activity
- ☐ Examples that help us understand these reasons
- ☐ One case study of either an earthquake or a volcanic eruption showing the causes of that event and the consequences for the people and the environment
- ☐ Some examples of how planning helps reduce the impact of both earthquakes and volcanic activity

Key terms

Conservative plate boundary
Convection currents
Convergent plate boundary
Divergent plate boundary
Earthquake
Epicentre
Focus
Fold mountain range
Hot spot
Magma
Mantle
Mercalli Scale
Pyroclastic flow
Richter Scale
Subduction
Volcano

Which key terms match the following definitions?

A Where two tectonic plates slide past each other

B The point on the surface directly above the focus of an earthquake

C Where two tectonic plates move away from each other

D A scale from I to XII, used to indicate the impact and effects of an earthquake

E A place where very hot magma rises from within the mantle to the crust, but not necessarily at a plate boundary

F Circulating movements of magma in the mantle caused by heat from the core

G An open-ended scale indicating the strength of an earthquake, as measured by a seismograph

H The process by which one tectonic plate is dragged down beneath another by convection currents

To check your answers, look at the glossary on page 289.

Foundation Question: Describe two characteristic features that you may find at convergent boundaries. (4 marks)

Student answer ■ (awarded 1 mark)	**Examiner comments**	**Build a better answer** △ (awarded 4 marks)
The margin is very hot where new rock comes to the surface. *Rock goes down into the mantle and forms a very big depression.*	• ***The margin is...*** is not a feature and is not specific to convergent boundaries. This part of the answer does not score any marks. • ***Rock goes down...*** scores 1 mark because it is partially correct. However, it could be improved by using geographical language. A ***very big depression*** is not quite the same as an ocean trench.	Volcanoes form where magma comes to the surface. The heavier crust descends under the lighter plate. This forms a long and deep ocean trench.

Overall comment: The student could have improved their answer by correctly using geographical language.

Higher Question: Describe the effects of a volcanic eruption. Use a named case study. (4 marks)

Student answer △ (awarded 3 marks)	**Examiner comments**	**Build a better answer** △ (awarded 4 marks)
The eruption took place in 1995 when the volcano in the Soufriere hills erupted. *Lots of people had to leave because of the eruption. Some people died as well and everything like the airport closed.* *The ash destroyed crops and fires were caused by volcanic gases igniting, especially in the south of the island. Much of the ash is poisonous and the forests will take years to recover.*	• ***The eruption took...*** This introduction is not relevant to the question. The location is not actually identified as Montserrat. This part of the answer does not score any marks. • ***Lots of people...*** scores 1 mark because it is a valid point. However, it lacks the detail needed to further illustrate the point. • ***The ash destroyed...*** scores 2 marks because it includes some detail of environmental effects that are also located.	In Montserrat the eruptions of 1995 were very violent, with huge quantities of pyroclastics ejected. Twenty three people died and 50% of the population left the island. Ports were closed. The ash destroyed crops and fires were caused by volcanic gases igniting, especially in the south of the island. Much of the ash is poisonous and the forests will take years to recover.

Overall comment: Located detail is important in questions that ask for specific case-study information.

Chapter 7 A wasteful world

Objectives

- Know the different types of waste.
- Be able to describe the differences between HIC and LIC waste production.
- Understand what causes these differences.

Activity 1

Describe the different sources of waste shown in Figure 1.

ResultsPlus
Exam Tip

Good answers to questions about patterns should recognise that there are often exceptions or anomalies that do not conform to the general pattern. This is true in waste production. Japan, for example, only produces 400 kg per person per year, even though it is one of the richest HICs.

Types of waste and its production

Waste is defined by the European Union as 'any substance or object that the holder discards, intends to discard or is required to discard'. It is also referred to in different places as rubbish, trash, garbage or junk, depending on the type of material. In the UK, the government classifies waste as either 'hazardous' (waste that poses a risk to human health) or 'non-hazardous'. Waste comes from a variety of different sources. As you can see from Figure 1, the main source is from the construction and demolition of buildings. But in this chapter we will be focusing on domestic (household) waste.

The differences between LICs' and HICs' waste production

We all produce waste of one sort or another and it has become an increasingly important issue in many countries as they decide how it should be disposed of.

There are many different types of domestic waste. Figure 2 shows the main types of waste produced by households in the USA. This pattern is fairly typical for HICs, with paper and cardboard – much of it newspapers and packaging – being the main component. Food waste is also significant, much of it being leftovers from meals.

HICs produce more waste than LICs – typically about five times as much. It is estimated that in LICs the average weight of waste produced per person is 100–220 kg per year, whereas in HICs it is 400–800 kg per year. The top waste-producing countries are Ireland (800 kg/person/year), Norway (780), USA (760), and Denmark (725). The UK (600 kg/person/year) comes tenth.

It is much harder to establish accurate waste-production figures for LICs, but there are estimated figures for Laos (237 kg/person/year), Vietnam (182), Philippines (146) and Thailand (73).

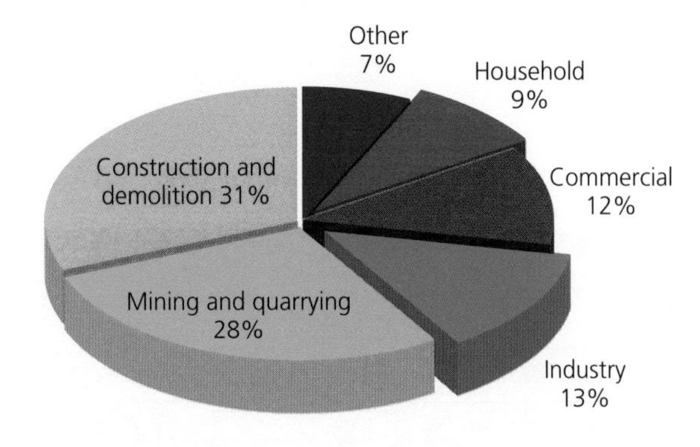

Figure 1: Sources of waste in the UK

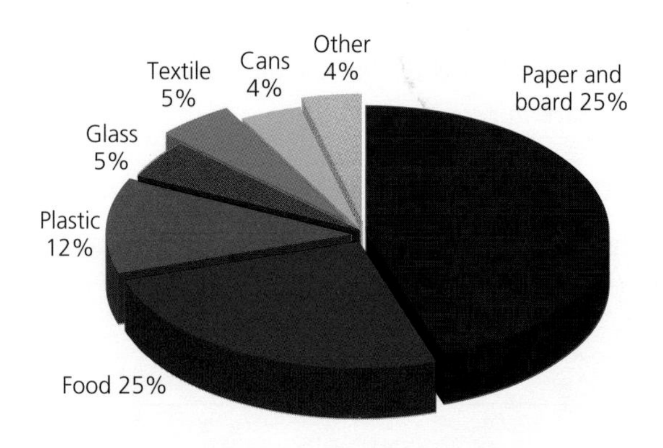

Figure 2: Types of domestic waste in the USA

Wealth and increasing waste

As countries become more wealthy, the people have a greater demand for consumer products, as they become part of what is known as the **consumer society**. They buy more items and they replace them more frequently. This not only leads to products becoming waste, perhaps before they have completely lost their usefulness, but it also means more packaging that then needs to be disposed of. The tendency for people to buy things and then throw them away has led to the creation of what we can call a **throw-away society**.

In LICs, not only is much less waste produced, but the content is also rather different. By far the biggest component is food waste, as can be seen from the data for Dhaka in Figure 3. (Although the actual amount of food wasted is small, the percentage is high as they produce so little waste in total.) Much less waste is produced in LICs because:

- Consumer purchases are very limited because of low incomes.
- Little packaging is used on products, so there is limited plastic waste.
- Many people cannot read, so fewer newspapers are sold.
- Disposable nappies and single-use drinks containers are rarely used.
- Practical recycling for personal use is widespread because people cannot afford to buy new products.

Activity 2

Describe the differences in waste production between the USA and Bangladesh, as shown in Figures 2 and 3.

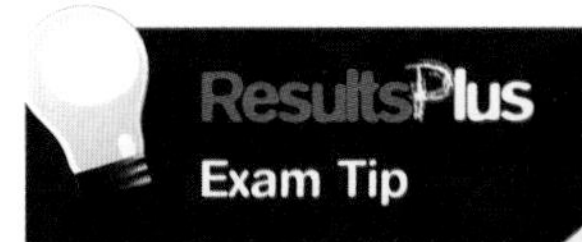

Good answers to questions about differences between countries point out that there are often differences within countries too. There are major differences in Bangladesh between the waste produced by the urban population and that produced by the rural population.

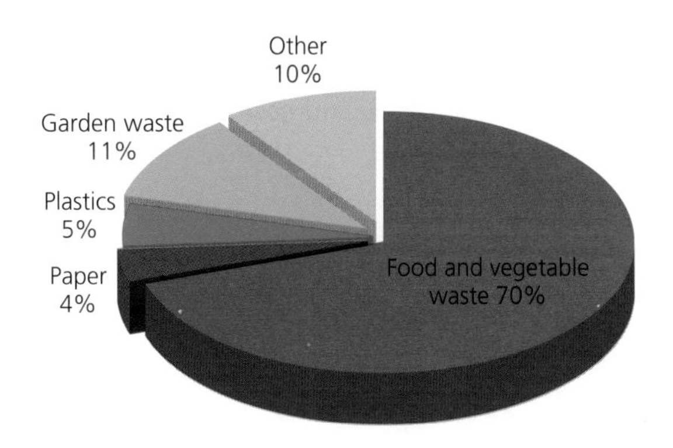

Figure 3: Types of domestic waste in Dhaka, Bangladesh

The difference between HICs and LICs is well illustrated by the amount of waste paper they produce, as shown in the table on the right.

Country	Waste paper (kg/person/year)
USA	293.0
Japan	239.0
Germany	205.0
Poland	54.0
Indonesia	17.0
Bangladesh	1.3

In the UK, the biggest component of waste at the beginning of the twentieth century was ashes from coal fires. The move away from coal fires in the 1960s saw the disappearance of coal waste. Today the main bulk of domestic waste is paper, cardboard, metals, glass and plastics.

ResultsPlus
Exam Tip

Good answers to questions about harmful waste should provide a specific example of something that is harmful. For instance, an example of a harmful metal from a computer.

ResultsPlus
Build Better Answers

Describe and explain differences in waste production between HICs and LICs. (4 marks)

Basic answers (0–1 marks)
Give descriptions of the differences only with no explanation.

Good answers (2 marks)
Offer clear descriptions with some explanatory statements relating to the level of development, perhaps focusing on the amounts of waste produced.

Excellent answers (3–4 marks)
Provide clear descriptions and full explanations of the differences in the amounts and types of waste produced.

In LICs, the rural areas produce less waste than the urban areas because they have lower per capita incomes. Urbanisation and rising incomes are the two most important factors that lead to waste generation because they cause increased demand for resources.

For example, in Bangladesh, the rural population generates only 55 kg of waste per person per year, while the urban population generates 150–180 kg.

Different types of domestic waste in HICs

HICs not only produce much more domestic waste than LICs but they also produce different types of waste. Items that typically find their way into HIC waste include:

- Electronic goods – including mobile phones which are often discarded not long into their potential life.
- White goods – domestic appliances such as washing machines and fridges.
- Packaging – including plastic bags.

In the UK we throw away 15 million mobile phones per year – that is 1,700 per hour. Mobiles contain harmful metals such as lead, mercury and cadmium – and there are thought to be over 5 billion of them in the world today.

The UN estimates that a global total of 50 million tonnes of electronic waste is being produced each year. (A sign of our throwaway society is that twelve years ago the average lifespan of a PC was ten years, but now it is just three years.) Electronic waste contains many harmful metals and gases, as you can see in Figure 4.

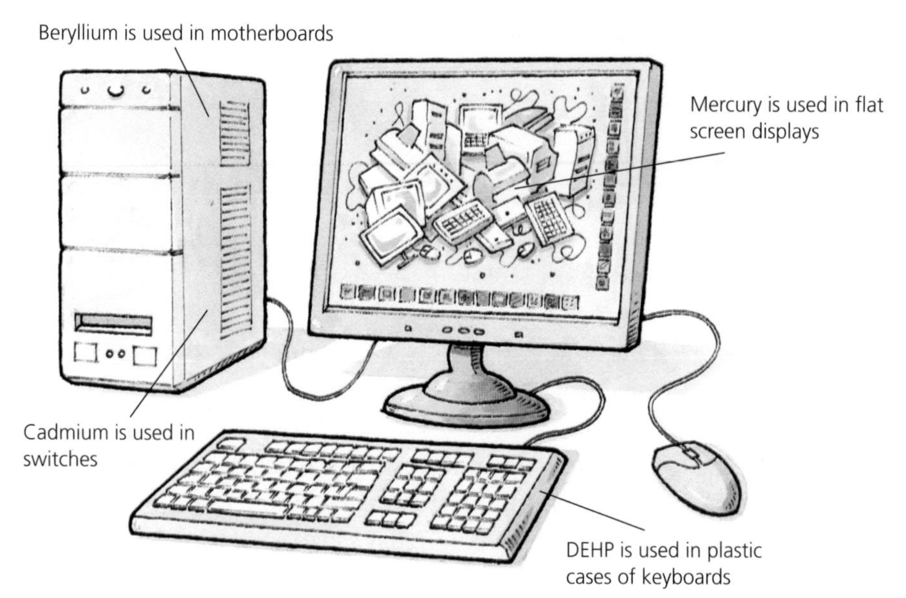

Figure 4: Harmful waste from a computer

Packaging

Packaging can be defined as materials used for the containment, protection, handling, delivery, and presentation of goods. Packaging can be divided into three broad categories:

- Primary packaging: the wrapping or the containers that are handled by the consumer.
- Secondary packaging: the larger cases or boxes that are used to group quantities of primary-packaged goods for distribution and for display in shops.
- Transit packaging: the wooden pallets, the cardboard and plastic wrapping and the containers that are used to enable the loading, transport and unloading of goods.

The UK produced 10.5 million tonnes of packaging waste in 2007, of which 70% was for food and drink. Much of the primary food and drinks packaging is dirty and contains residues from its contents.

Paper and cardboard are the most widely used packaging materials in terms of weight, as can be seen in Figure 5. They account for 43% by weight of all packaging and are used to pack 25% of goods. Plastic packaging accounts for just 20% of the weight of all packaging – but 53% of goods are packaged in plastics. Because of its low weight and relative strength, plastic is one of the most energy-efficient, robust and economic materials available. Unfortunately it is very hard to dispose of.

Mixed-material packaging can in some cases have the benefits of being more resource- and energy-efficient than single-material packaging, but combining materials makes subsequent recycling difficult. An example of this type of packaging is the 'Tetra Pak' which typically consists of 75% paper, 20% polyethylene and 5% aluminium foil.

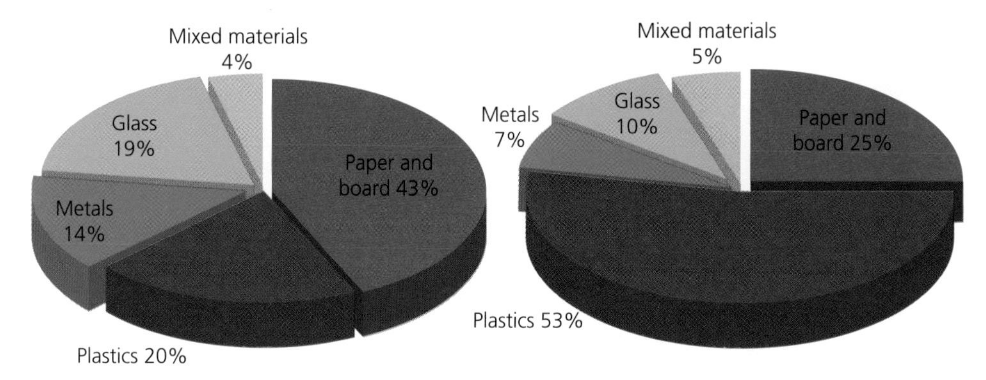

Figure 5: The materials in packaging waste

Activity 3

Study Figure 5.

(a) Describe the composition of packaging waste (i) by weight (ii) by percentage of packaged goods.

(b) State the main differences between the two.

(c) Suggest two reasons for these differences.

Although plastic packaging forms the highest percentage of packaging used, it is often very light and so it is not the highest in terms of weight.

Skills Builder 1

Study this table of data, showing the different types of domestic waste produced in 2006 in one country.

Type	%
Garden	24%
Kitchen	18%
Paper & board	16%
Glass	8%
Plastic	8%
White goods	6%
Other	20%

(a) Draw a divided bar chart to show the data.

(b) Which is the type of waste with the highest percentage?

(c) Which type of waste is three times as high as white goods?

(d) Does this country seem to be an HIC or an LIC?

(e) Suggest two reasons for your answer.

Objectives

- Know how waste is disposed of and recycled.
- Be able to describe how recycled waste is used.
- Understand waste and recycling policies.

Recycling and disposal of waste

Recycling on a local scale

Recycling is good. It stops our rubbish going to landfill. It saves energy – thereby reducing greenhouse gases. It saves resources, and it doesn't cost us anything. It means that as members of the public we can help make our country and our world more sustainable. Recycling is an important part of the 'waste hierarchy' (Figure 6) which the UK government uses to show the order of priority that should exist when dealing with waste.

Recycling reduces the demand for raw materials, lessening the impact of their extraction and their transportation. Activities such as mining, quarrying and logging can destroy the natural environment and precious local wildlife habitats. Although some materials for recycling still need to be transported around the UK, their movement will have less impact than transporting raw materials from often remote locations in other parts of the world.

Recycling uses less energy than producing goods from virgin material and also results in fewer emissions. Using less energy is vital because burning **fossil fuels** for energy produces carbon dioxide, a greenhouse gas that contributes to global warming. In addition, recycling reduces the need for waste to go to **landfill** or **incineration**. Figure 7 shows the symbol used on lots of packaging to show that it can be recycled.

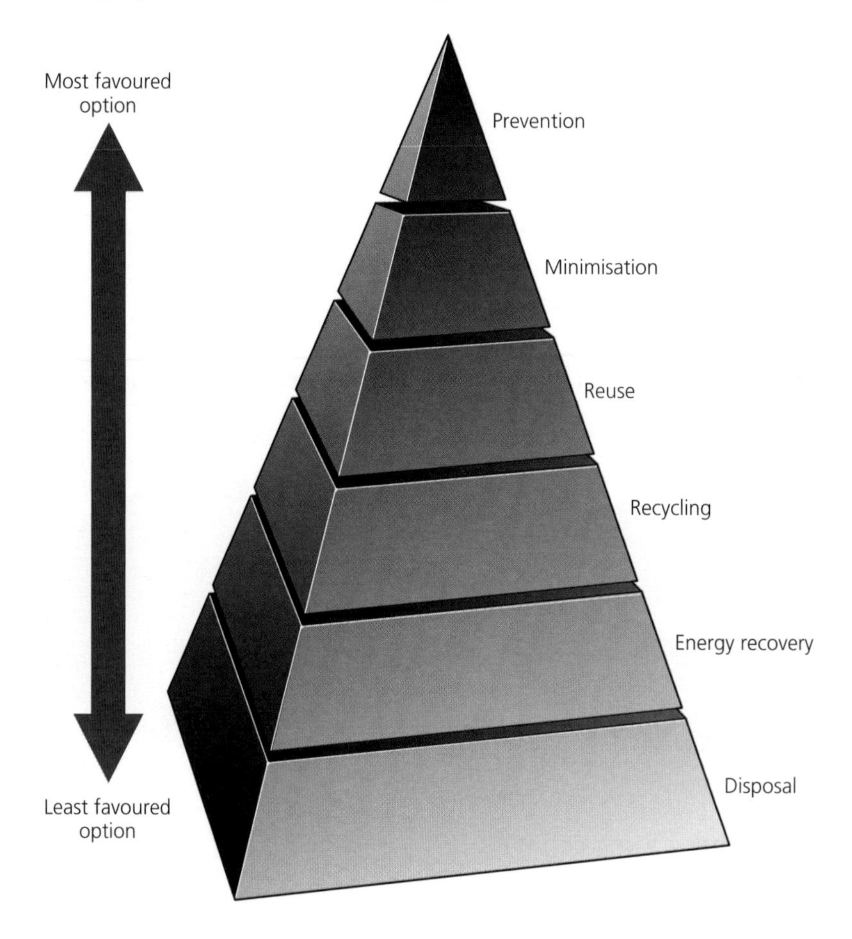

Figure 6: The waste hierarchy

Case study – the London Borough of Camden

Camden is a borough in north London with a population of 228,000 living in 92,000 households. In 2004, the council was collecting 85,000 tonnes of household waste each year. The proportions of the different types of waste are shown in Figure 8.

The council developed a waste strategy in 2006 with some very specific targets to be achieved by 2010, including 35% of household waste to be recycled. This was to be achieved by:

- Increasing participation in household recycling to 60%
- Increasing the amount of recycling collected from housing estates by 10%
- Increasing the amount of recycling collected from schools by 10%
- Recycling 70% of rubbish brought to the Regis Road Recycling Centre
- Providing all parks and open spaces (where practical) with recycling banks
- Putting single recycling banks, which allow three materials to be recycled together, at all tube stations
- Providing a fortnightly garden waste collection for the whole borough
- Trialling a kitchen waste collection scheme
- Introducing a scheme, either fines or incentives, to encourage more people to recycle.

So far the council have improved their recycling significantly. Currently, they provide three refuse collections a week, two of which are for non-recyclable waste. The third collection is a doorstep recycling collection or estate recycling collection. This allows residents to recycle paper, cardboard, glass, plastic bottles and aluminium cans. This service is contracted to a company called Veolia.

The Regis Road Centre allows residents to take for recycling all of the above items as well as: hard/rigid plastics (garden furniture and toys), hardcore and rubble (six sacks maximum per customer), paper-based food and liquid cartons (Tetra Paks), oil, batteries, light bulbs, metals and all electrical goods. If items are too bulky to be taken to the centre, such as white goods, they can be collected by the council, free of charge. The borough also has smaller recycling centres at Summers Lane and Hornsey Street. Recycling rates at Regis Road were 58% in 2004, but reached 70% in 2008.

Figure 7: The recycling symbol

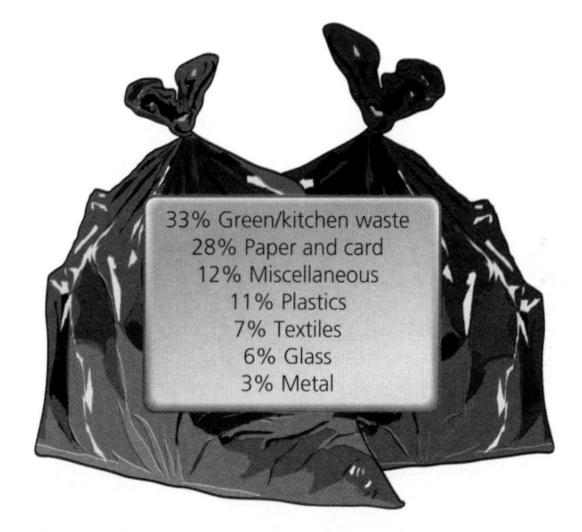

Figure 8: The different types of waste in Camden, 2004

Activity 4

Look at some of the packaging used on products bought by you and your family. List the different materials used and identify which can be recycled.

Activity 5

Look at each of Camden's recycling targets (page 121). Put each under one of three headings: 'Target achieved', 'Target not achieved' and 'Not known'.

How successful has Camden been so far in trying to meet its targets?

The trial kitchen waste collection scheme is already operating, alongside a separate garden waste collection. The scheme started in August 2008 and only operates in a limited area in the north of the borough. The table below shows what the scheme covers.

Kitchen waste	Garden waste
All cooked and raw foods including meat, fish, fishbones, bread, pasta, rice, vegetable and fruit waste, eggs, cheese and teabags.	Grass cuttings, leaves, bark, prunings, dead flowers and twigs.

Residents are provided with a small container for indoor use, called a kitchen caddy, and then a choice of a 240 litre, 120 litre or 23 litre bin which is kept outside the house and is emptied weekly. Residents are also provided with compostable bags to line the containers. Residents do not have to take part in the trial. In the garden waste scheme some households do not have the space to store a container, and so the council is looking into supplying reusable garden waste sacks. The use of either sacks or bins for garden waste will make the service easier to use and there will be no need to book garden waste collections in the future.

Large recycling bins (Figure 9) have been provided in the three parks in Camden where it was thought to be appropriate. These are operated together with an organisation called Rewind Recycling.

Figure 9: Camden's recycling bins for public spaces

Camden has used a team of recycling advisers to encourage increased recycling in schools and housing estates. The work in schools has been successful and recycling rates have been increasing by 10% per year. The incentives were also tried on the estates, but these did not really work and rates have not yet increased very much.

By 2007 the rate of recycling in Camden was 27%, a significant increase from the 17% rate in 2003/04, and moving towards the 2010 target of 35%.

After residents put their waste materials in their outside container or in a recycling bin the materials are taken to a central depot where they are sorted, bulked up and baled for onward transportation. Usually, even if materials are separated fully by the householder, there is still some further sorting to be undertaken because there is likely to be a small amount of contamination by other materials. The sorting is done at a depot called a 'Materials Reclamation Facility' (MRF). There are two types of MRF – 'clean' and 'dirty'.

Clean MRFs only accept recyclables that have already been separated from normal refuse – though they may arrive as a mixture (glass and cans together, for example). Dirty MRFs accept mixed rubbish from households or businesses. The simplest sorting techniques at MRFs are manual, employing people to pick out materials from a raised conveyor belt. Mechanical sorting systems, however, have improved considerably over recent years, and continue to develop.

ResultsPlus
Exam Tip

A good way of looking at how successful recycling schemes are is to see whether they have achieved their targets or not. Camden has already met some of its targets ahead of the planned timescale.

The bales are sent to 'reprocessors' such as paper mills, glassmakers or plastic reprocessing plants where the material is processed for use in other applications or processed directly into a new product. Some materials, such as aluminium and glass, can be recycled indefinitely, as the process does not affect their structure. Other materials, such as paper, require a mixture of reprocessed waste and new raw material to manufacture a new product. With materials such as plastic, the waste is converted into a pellet which is then used in the manufacture of a recycled or part-recycled plastic product. Over 3,000 different products made from recycled materials are available in the UK, ranging from wine glasses and bags to fleeces and greetings cards. In the recession of 2008/09, the market for waste materials fell significantly so the value of waste fell too. This meant that local authorities such as Camden were earning less money for their waste. For example, the value of waste paper fell from £70 per tonne to £10 per tonne in December 2008.

Quick notes (the London Borough of Camden):

- Waste disposal is the responsibility of local councils.
- Local councils are increasingly addressing recycling issues.
- Recycling can take place in a range of ways.
- Recycled material has many potential uses.
- Local councils set targets for recycling.

Case study – how one HIC – Germany – disposes of its waste

Germany produces about 60 million tonnes of domestic waste each year and, like all countries, its three main disposal options are:

1. Landfill

2. Incineration

3. Recycling.

Landfill

At the most basic level, land filling involves putting waste in a hole in the ground and covering it with soil or rock. Today, the engineering of a modern landfill is a complex process, typically involving lining and capping individual cells or compartments into which waste is compacted and then covered to prevent the escape of polluting liquid or gases. In newer landfill sites, systems are installed to capture and remove the gases and liquids produced by the rotting rubbish.

Over recent years Germany has used landfill for getting rid of much of its waste and this has been possible because of the country's geology. Mineral extraction and quarrying left large holes in the ground which were 'restored' by filling with waste. In addition, the underlying geology often provided naturally impermeable ground conditions, allowing the waste to be buried with less risk of liquids seeping out and polluting groundwater. Because of the geological conditions, therefore, landfill was relatively cheap. In particular, radioactive waste from Germany's nuclear power plants is increasingly disposed of in this way, with sealed containers placed in holes several hundred metres underground. The repository at Konrad, for example, is located in a pre-existing iron mine at the exceptional depth of 1,000 metres. The use of these old mines has reduced the need for the waste to be shipped to specialist nuclear waste recycling plants, such as Sellafield in the UK.

But suitable sites are running out, and Germany now exports 1.8 million tonnes of non-hazardous waste per year to countries such as Spain and China, who have spare capacity for landfill and see it as a way of earning money for their economies.

Skills Builder 2

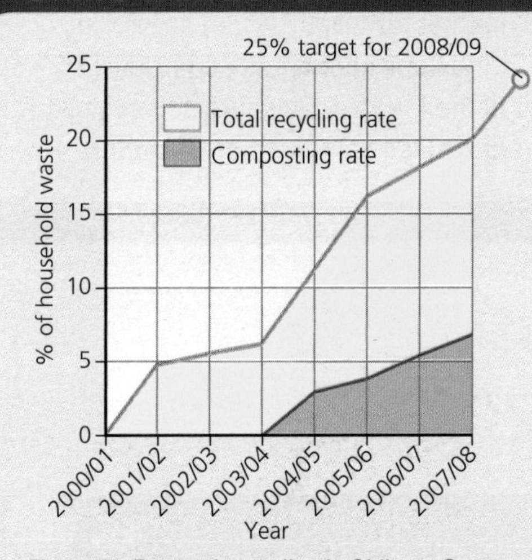

Figure 10: Rates of recycling in Oldham, Greater Manchester

Study Figure 10.

(a) What percentage of household waste was recycled in 2001/02?

(b) Describe the changes in recycling rates over the whole time period shown.

(c) Is Oldham on track to meet its 2008/09 target?

(d) Explain your answer (c).

(e) Explain two ways in which local authorities such as Oldham can try to increase their rates of recycling.

Incineration

Incineration is the burning of waste. Incineration may be carried out with or without 'energy recovery', which means that energy is produced from the burning process. Without energy recovery, it is a form of disposal, like landfill – although it uses less space.

Energy can be recovered from waste either by 'direct waste incineration' (burning the unsorted waste in 'mass burn incineration plants') or by burning bricks or logs of refuse-derived fuel (RDF) which have previously been formed by drying, sorting and shredding general waste. In Germany, there are 68 incinerators, such as the one at Herten in Schleswig-Holstein. They have a total capacity of 18 million tonnes per year and most of them burn RDF.

The technology to burn waste has developed significantly over the past fifty years and incinerators are now much cleaner than they used to be. The energy released from burning the rubbish is often used to generate electricity, but even greater benefits can be gained by using the heat directly – to heat nearby housing or offices, for example.

However, despite improvements in the operation of incinerators, there is strong public concern about their health effects. And from a resource point of view, incineration may not be the best way to deal with waste. Even if energy is obtained through the process, incineration may be a waste of valuable resources. There is opposition to incineration in Germany because the RDF incinerators are exempt from the carbon emissions laws, which were designed to reduce the scale of climate change.

Recycling

Germany has very strict recycling laws, which means that people are very careful to recycle and to do it properly. Unfortunately, the country's own recycling centres can only deal with less than a third of the material collected. This means that Germany has to ship the majority to countries like Denmark and pay for it to be recycled there.

ResultsPlus
Exam Tip

Waste disposal is big business and is increasingly carried out by private companies who are trying to make a profit. This has led to an international trade in waste which is highly competitive, although the recession of 2008/2009 has led to a decline in the market for waste. When answering questions about how countries dispose of their waste, try to provide evidence of it being sent to other countries.

When answering questions about the benefits of recycling, you should point out that it would be better not to produce as much waste in the first place.

Figure 11: Recycling in Germany

One of the most successful recycling initiatives has been the Green Dot scheme, which originated in Germany and now operates in more than twenty European countries. Manufacturers pay for a licence and put a Green Dot logo on their products, to show that they are contributing to the cost of recovery and recycling. (Manufacturers who don't join the scheme must recover all their recyclable packaging themselves – which is just not possible in most cases.) The more packaging there is, the dearer the licence – so the system encourages manufacturers to cut down on their packaging. This has led to less paper, thinner glass and less metal being used, thus creating less waste to be recycled. The net result in Germany has been a significant decline in waste of about one million tonnes per year. However, the scheme is expensive. It costs the German taxpayers $2.5 billion a year – more than $30 per person, an amount that approaches the taxes or fees already paid for regular waste-collection services.

In Bavaria, in southern Germany, residents are enthusiastic recyclers, each collecting an average of 322 kg of recyclable waste per year. Only 1% of this ends up as landfill with 67% being recycled and 32% incinerated.

Quick notes (Germany's waste disposal):

- Germany disposes of waste by landfill and incineration.
- Recycling rates are much higher in Germany than in the UK.
- Germany sends waste for disposal and recycling to other countries.
- The costs of the Green Dot scheme are very high.

Skills Builder 3

Draw a pie-chart to show how Bavaria deals with its recyclable waste.

Suggest reasons why some countries are more successful than others in recycling waste. (4 marks)

■ **Basic answers** (0–1 marks)
Only give a description of differences in recycling, with no explanation.

● **Good answers** (2 marks)
Offer some explanatory statements relating to the role of government and/or level of economic development.

▲ **Excellent answers** (3–4 marks)
Provide full explanations of how some governments, such as Germany, encourage recycling by education, legislation and financial incentive.

Objectives

- Know the difference between renewable and non-renewable fuels.
- Be able to describe the advantages and disadvantages of different types of renewable and non-renewable fuels.
- Understand the global-scale distribution of energy surplus and energy deficit.

ResultsPlus
Watch out!

Renewable energy is not the same as renewable fuel. 'Fuel' refers to something that is burnt to produce energy – and does not, therefore, include renewable energy sources such as tidal, solar and wind power.

ResultsPlus
Exam Tip

Wood and biomass fuels are only renewable if equal or greater amounts are replanted to replace that being used.

Sources and uses of energy

Renewable fuels

Renewable fuels are the fuels produced from renewable resources, as substitutes for fossil fuels. In practice, these are what we call 'biofuels' (although the production of hydrogen from renewable resources may be a future development). Biofuels are any kind of fuel made from living things, or from the waste they produce, including:

- Bioethanol, biodiesel or other liquid fuels made from processing plant material or waste vegetable oil
- Wood, wood chippings and straw
- Pellets or liquids made from wood
- Biogas (methane) from animals' excrement
- Syngas, a mixture of carbon dioxide, carbon monoxide and hydrogen.

The main biofuels being used at present are bioethanol and biodiesel. These are blended to produce fuel for engines and motors. A variety of crops can be used in their production, including rapeseed, palms, sugar cane, soy and maize.

The European Union issued a directive in 2003 calling for biofuels to meet 5.75% of transportation fuel needs by the end of 2010. Because of their advantages, future targets are likely to be higher, but there are some concerns about their use too.

Advantages

- Using biofuels reduces the emission of greenhouse gases in comparison to burning fossil (non-renewable) fuels. Although burning biofuels releases carbon dioxide, growing the plants absorbs a comparable amount of the gas from the atmosphere. A recent UK government report declared that biofuels reduced emissions by 50–60%, compared to fossil fuels.
- The sources of biofuels will never run out.

Disadvantages

- Farming and processing the crops uses energy and, depending on what is grown and how it is treated, this can make the total production and use of biofuels just as polluting as non-renewable fuels.
- According to the World Conservation Union, using ethanol rather than petrol reduces total emissions of carbon dioxide by only about 13% because of the pollution caused by the production process, and because ethanol gets only about 70% of the mileage of petrol.
- Asian countries may be tempted to replace rainforest with more palm oil plantations, some critics say.

- If increased proportions of food crops such as corn or soy are used for fuel, that may push prices up, affecting food supplies for less prosperous people. Food prices are already increasing. With just 10% of the world's sugar harvest being converted to ethanol, the price of sugar has doubled. The price of palm oil has increased 15% over the past year, with a further 25% gain expected in 2010.
- With much of the western world's farmland already consisting of identical fields of single crops, the fear is that a major adoption of biofuels will reduce habitats for animals and wild plants still further, leading to reduced biodiversity.

Non-renewable fuels

Non-renewable fuels are the fuels – like coal, oil and natural gas – that cannot be remade or 'regrown', because it would take millions of years for them to form again. They exist in a fixed amount that is gradually being used up – much of it to generate electricity. In practice, the non-renewable fuels are what we call the 'fossil fuels' (although uranium, used in nuclear power, is also non-renewable). Coal, oil and natural gas are called fossil fuels because they were formed long ago from the remains (fossils) of living organisms. The Earth's population cannot rely on using them in the long term because the reserves are becoming depleted.

Advantages

- Fossil fuels are a relatively cheap way of generating electricity.
- Coal is relatively easy to transport because it is a solid.
- There are large deposits of coal left that are reasonably easy to obtain, some of it being close to the surface.

Disadvantages

- When fossil fuels are burned they produce carbon dioxide, which is a greenhouse gas and a major contributor to global warming.
- Burning coal without first purifying it also contributes to the production of smog (smoke and fog), which is harmful to health.
- Transporting oil around the world can produce oil slicks, pollute beaches and harm wildlife.
- Some sources of coal are deep below the ground, as in the UK. They can be difficult, costly and dangerous to mine.
- In the end, all the non-renewable fuels will run out.

Skills Builder 4

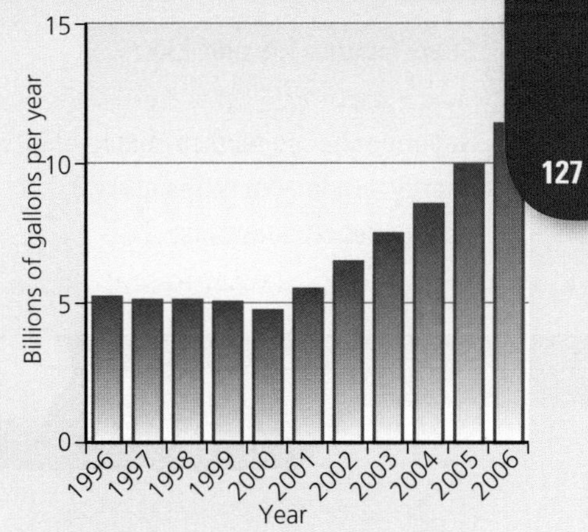

Figure 12: World bioethanol production, 1997–2006

Study Figure 12.

(a) How much bioethanol was produced in 1996?

(b) By how much did bioethanol production increase between 2005 and 2006?

(c) Describe the changes in bioethanol production over the whole time period shown.

(d) Describe how ethanol is used as a 'renewable fuel'?

(e) What are the disadvantages of using renewable fuels such as ethanol?

Activity 6

For one renewable and one non-renewable fuel, list their advantages and disadvantages. Which one seems to have the greater advantages?

Renewable fuels	Non-renewable fuels
Bioethanol	Coal
Biodiesel	Oil
Wood products	Natural gas
Biogas	Uranium (nuclear fuel)
Syngas	

Skills Builder 5

Study Figures 13a and 13b.

Describe the global distribution of (a) energy surplus countries and (b) energy deficit countries.

ResultsPlus
Exam Tip

Good answers to questions about patterns might note that the pattern is not very strong or clear. Patterns can be hard to spot sometimes.

The distributions of energy surplus and energy deficit

Countries that produce more energy than they use are said to have an '**energy surplus**'. Those that use more than they produce have an '**energy deficit**'. Figure 13 shows the world distribution of energy surplus countries and energy deficit countries. (The rest of the world's countries are approximately in balance.) You will see that there is something of a pattern, with HICs generally having a deficit – most of Europe and the USA, for example – but with some exceptions like Canada. Many LICs have a surplus, including those in south-east Asia and north Africa, but others have a deficit, including India.

In trying to explain this distribution, a number of possible reasons need to be considered:

- What energy resources does a country have?
- How much energy does a country use?
- How economically developed is the country?
- How hot is the climate of the country?

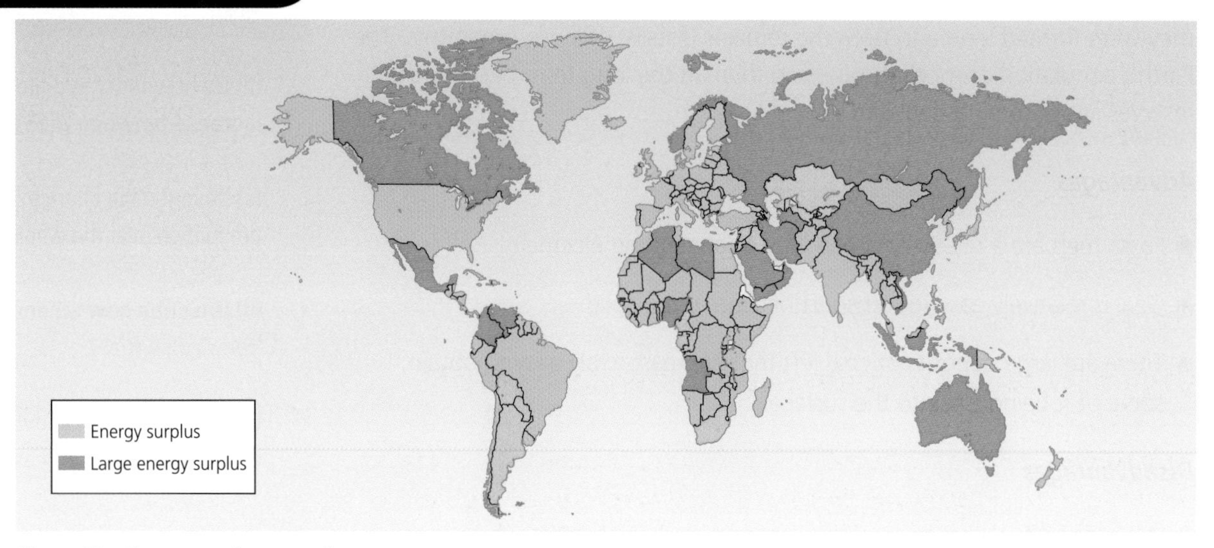

Figure 13a: Energy surplus countries

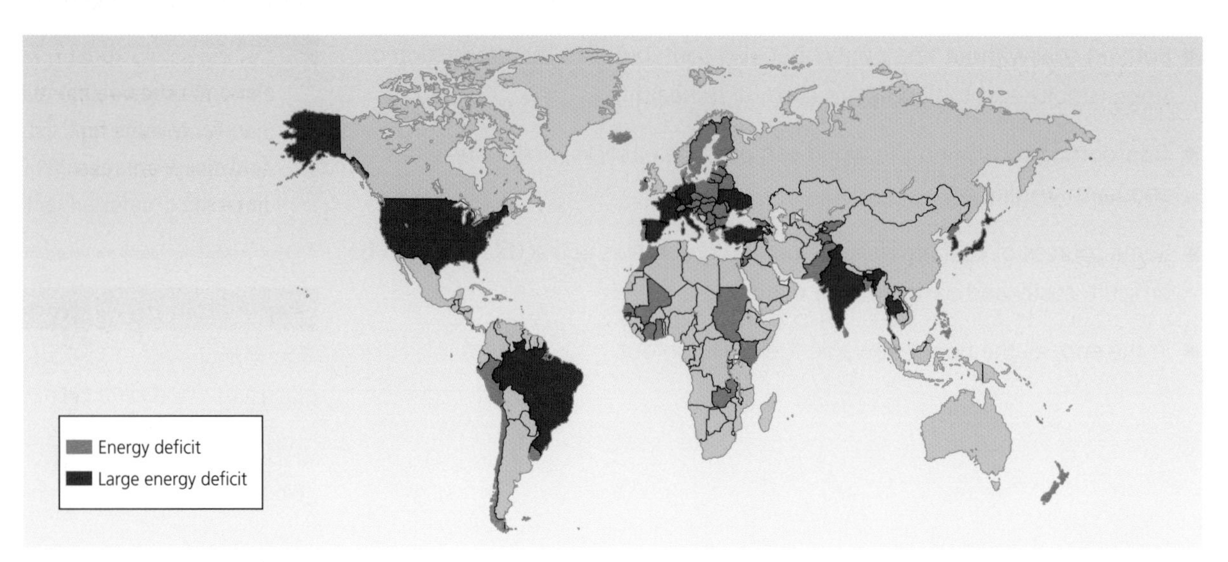

Figure 13b: Energy deficit countries

Management of energy usage and waste

Objectives

- Know the ways in which energy is being wasted.
- Be able to describe the carbon footprints of countries at different levels of development.
- Understand the possible solutions to energy wastage.

How energy is being wasted

Energy is wasted in homes through a combination of carelessness and inadequacies in building design. The percentage of energy lost from homes in different ways is shown in Figure 14. This might include:

- Leaving lights on when they are not needed
- Leaving phone chargers plugged in after the phone has been charged up
- Leaving electronic equipment on 'standby'
- Not having double glazing
- Not having loft insulation
- Not having hot water tank insulation
- Not having cavity wall insulation
- Having thermostats set too high
- Leaving doors and windows open.

Many of these points also apply in factories. In addition, energy may be wasted by:

- Poorly serviced or poorly maintained machinery
- Vibrations in machinery
- Leaks from compressed air valves.

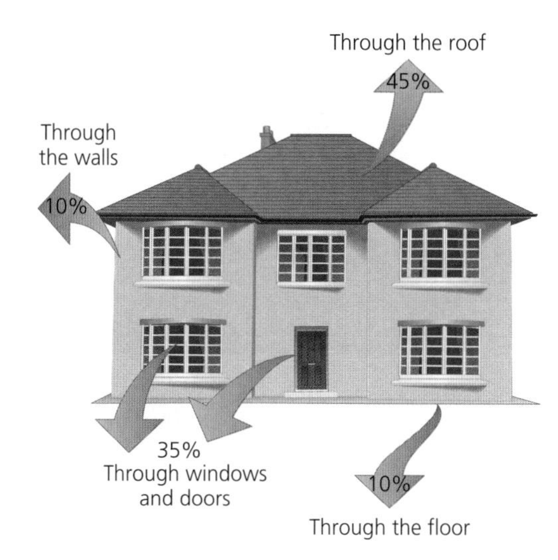

Figure 14: Domestic energy waste

Carbon footprints for countries at different levels of development

A **carbon footprint** is a measure of the impact our activities have on the environment and, in particular, their impact on climate change. It relates to the amount of greenhouse gases produced in our day-to-day lives through burning fossil fuels for electricity, heating and transport, etc.

The carbon footprint is a measurement of all the greenhouse gases we are individually responsible for producing. It is expressed as kg of the equivalent carbon dioxide:

- The average worldwide carbon footprint is about 4,000 kg
- The average for the industrial nations is about 11,000 kg
- The average footprint for people in the UK is 9,700 kg
- The worldwide target to combat climate change is 2,000 kg.

Activity 7

List the points about how energy is lost from homes under two headings: 'Carelessness' and 'Building design'. Which would seem to be the main reason for energy loss in homes?

Activity 8

Draw a simplified version of the house in Figure 14. Add labels to show how heat loss by each 'escape route' could be reduced.

The relative importance of the contributions to a person's individual footprint is shown in Figure 15. Your individual footprint consists of primary and secondary footprints. The primary footprint is calculated by dividing the amount of energy used in your house by the number of people in your house, added to the energy used in your journeys. The secondary footprint involves recreational activities and the energy used to supply you with your goods and services. In practice, this is very difficult to work out accurately. However, good estimates can be made, and you can try it for yourself on websites such as www.carbonfootprint.com.

Figure 16 shows the carbon footprints of a range of countries. The graph clearly suggests that HICs, such as the USA and UK, have much higher footprints than LICs like Bangladesh and Tanzania. Although this graph does not show many countries, it is a good representation of the overall pattern.

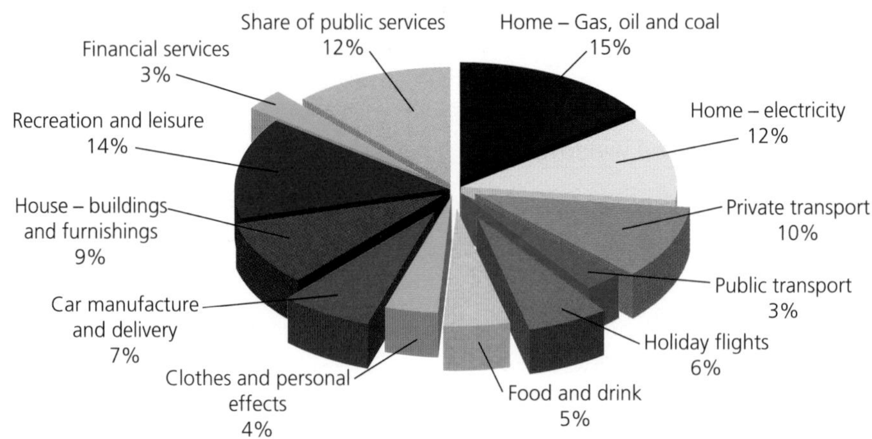

Figure 15: Contributions to individual carbon footprints

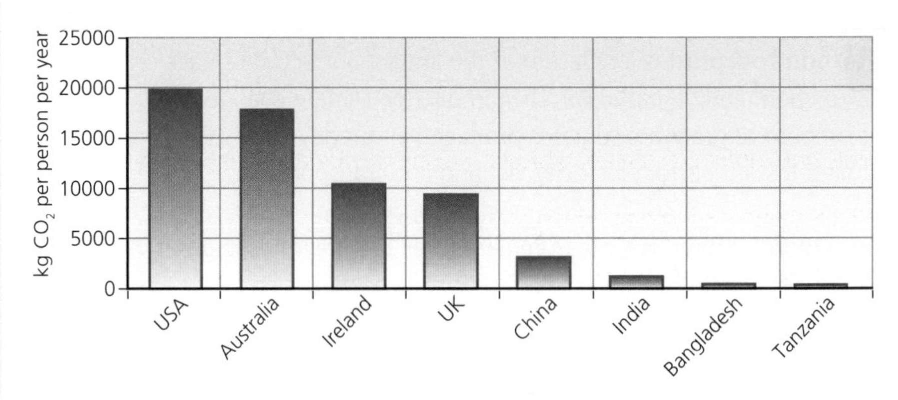

Figure 16: The carbon footprints (per person per year) of selected countries

Possible solutions to energy wastage in the UK

Domestic scale solutions

The main solution to energy wastage in homes is for greater **energy efficiency** to be included in building design. In the UK, all new homes have to be designed and constructed to meet high energy efficiency standards. And owners of older homes can take fairly simple (and relatively cheap) steps to make their homes more energy efficient.

ResultsPlus
Watch out!

■ Do not confuse carbon footprint with ecological footprint. This is a different measure based on the much wider impact that people have on their environment, not just their carbon production.

ResultsPlus
Build Better Answers

With the use of examples, explain how the level of development of a country affects its carbon footprint. (6 marks)

■ **Basic answers** (Level 1)
Give a description of carbon footprint only, with no explanation.

● **Good answers** (Level 2)
Offer some explanatory statements relating to level of development, perhaps relating to standards of living and the use of energy in homes.

▲ **Excellent answers** (Level 3)
Provide full explanation of different reasons, perhaps also recognising that production and transportation of consumer goods is a major contributor.

The table below shows the steps recommended by organisations such as the Energy Saving Trust. Figures are based on average homes.

Method	Effect	Cost (approx.)	Benefits (per year)
Hot water tank jacket	Reduces heat loss and energy use	£12	Saves £20 on bills
Reduce central heating temperature by 1°C	Less heat used	Nil	Saves up to £70 on bills
Cavity wall insulation	Reduces heat loss through walls	£500	Saves about £90 on bills
Floor insulation	Reduces heat loss through floor	£90	Saves £45 on fuel bills
Loft insulation	Reduces heat loss through roof	£750	Saves £110 on fuel bills
Condensing boiler	More efficient way of supplying hot water	About £1,000	Saves £200 on fuel bills
Double glazing	Reduces heat loss through windows	£5,000	Saves £90 on fuel bills
Switch off appliances, rather than leaving them on standby	Reduces electricity use	Nil	Saves £37 on electricity bill
Fit energy-saving light bulbs	Use less electricity than normal bulbs	About £3 each	Saves £50 on electricity bill

The UK government is encouraging energy saving by offering grants as well as advice and guidance. Under the Warm Front Scheme, grants of up to £2,700 for heating and insulation may be available. Some homes needing oil central heating may receive a grant of up to £4,000. Applicants must be in receipt of one of a range of qualifying benefits and be owner-occupiers or private renters. The government is also encouraging energy companies to promote energy efficiency. British Gas, for example, currently offers free cavity wall insulation to any of its customers who are on certain government benefits.

Newly built homes in the UK typically include energy-efficient features such as:

- Polystyrene insulation under a concrete floor slab.
- Thick carpets and underlay.
- Engineering quality hard bricks.
- Thermalite high-performance blocks for the foundations.
- 130 mm insulating lightweight concrete blocks for the internal walls.
- 75 mm wall cavity, filled with mineral wool insulation.
- Double glazed windows filled with argon gas.
- Full draught proofing.
- Foil-backed plasterboard on the ceiling.
- 200 mm of fibreglass loft insulation.

Two examples of this type of design policy can be seen in Milton Keynes, Buckinghamshire. In Giffard Park, thirty-six flats and 'starter' houses were built by Giffard Park Housing Cooperative in 1984. The challenge was to achieve a 60% reduction in space-heating fuel requirements for no more than an additional £500 in construction costs. This was achieved partly through the use of solar heating. In Two Mile Ash, four timber-framed 'superinsulated' houses were built in 1985. These had extremely low space-heating fuel requirements, achieved not by solar methods or special heating appliances but by heavily insulating the fabric, paying great attention to air tightness and introducing controlled ventilation, based on Finnish designs and construction methods.

Local scale solutions

A number of local projects are taking place in the UK to try to provide greater energy efficiency for whole communities, as well as for individual households.

One such project is the Eastcroft District Heating Scheme in Nottingham, shown in Figure 17. Operated by Waste Recycling Group (WRG), the scheme involves burning 150,000 tonnes of waste material each year in an incinerator. This produces steam which is used to supply heat to around 1,000 homes in the St Ann's area, as well as the Victoria shopping centre, an ice rink, Nottingham Trent University and government buildings and offices. It also generates electricity for about 5,000 homes. WRG had submitted plans to increase the amount of waste burnt to 250,000 tonnes per year, but following opposition it is now drawing up a completely new plan to expand the whole facility with a new, more modern design. Nottingham currently sends over 100,000 tonnes of waste to landfill sites outside of the city, so much of this could be burnt at the new facility.

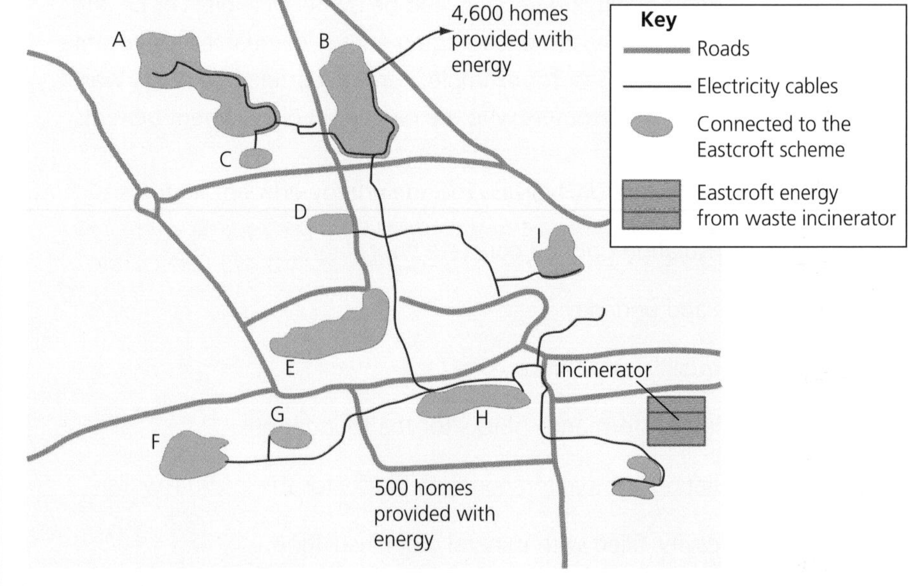

A – Nottingham Trent University
B – Victoria shopping centre
C – Theatre Royal
D – Old Market Square shop
E – Broadmarsh shopping centre
F – Inland Revenue
G – Magistrates' court
H – Capital One offices
I – Ice stadium

Figure 17: The Eastcroft District Heating Scheme

Nottingham is a low-performing council in terms of recycling, but this scheme has reduced the amount of waste going into landfill by about 150,000 tonnes per year. The plant also recycles metal from the incinerator which is left behind after burning. About 3,000 tonnes of iron and steel is reclaimed per year and used in the construction of girders. Even the ash produced is recycled, being used in road construction at other sites. The incinerator operates at very high temperatures – between 850°C and 1,100°C – to ensure complete combustion, and the emissions are then filtered before being released into the atmosphere.

There is opposition to the scheme because gases are released into the atmosphere during the process and because the ash produced contains poisonous metals, such as mercury and lead. Some critics also believe that the scheme encourages the production rather than the reduction of waste. Much of the waste burnt is industrial rather than domestic and some is brought in from surrounding cities, such as Leicester. Nottingham Friends of the Earth estimate that the plant costs the local taxpayers £1m per year. They also believe the emissions are not monitored or regulated properly.

National scale solutions

Current government policy on energy efficiency is related to its plans for reducing greenhouse gas production. As far as business and industry is concerned, the government offers 80% reduction on the Climate Change Levy to companies that meet their energy efficiency targets. They also fund the Carbon Trust which offers advice and training to businesses on issues such as energy efficiency. A video has been produced as part of this process which companies can use to educate their workforce. Also 'smart' energy meters are being trialled to allow businesses to monitor their energy use more effectively and easily.

In terms of building design and energy efficiency, the government have incorporated high standards of energy efficiency in the new regulations that cover any building plans. Energy Performance Certificates have been introduced to show how energy efficiency can be improved. For existing homes, £1.8m of funding has been provided to support grants under the Warm Front scheme which enables homeowners to improve the energy efficiency of their property.

The government itself operates many buildings and offices, and in 2008 it made a commitment to reduce its own energy use by 15% by 2010 and 30% by 2020.

Another major strategy is that of education. This includes educating homeowners and designers about energy efficiency as well as providing educational materials and support for schools through the funding of organisations such as the Energy Saving Trust.

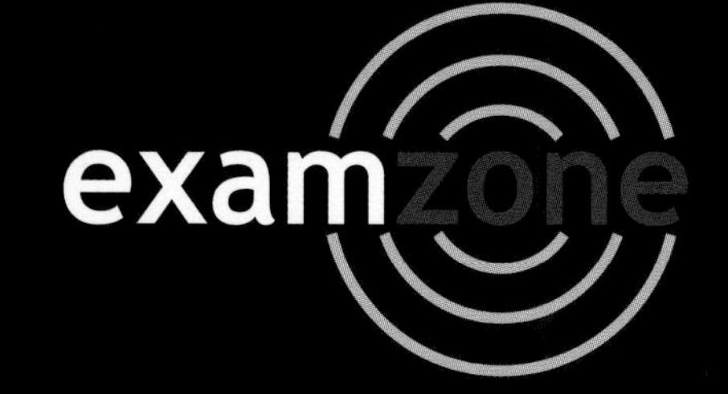

Know Zone
Chapter 7 A wasteful world

As countries grow and develop they create more and more waste. Managing that waste is a problem because although some can be recycled successfully, other methods are less environmentally friendly.

You should know...

- ☐ The variations in waste production from country to country
- ☐ How the level of development of countries helps explain these variations
- ☐ How consumer societies have led to greater waste creation
- ☐ The differences between types of consumer waste, including packaging and electrical goods
- ☐ How waste is recycled
- ☐ A case study of the different ways in which recycling is carried out
- ☐ A case study of waste disposal by an HIC
- ☐ The advantages and disadvantages of renewable fuels
- ☐ The advantages and disadvantages of non-renewable fuels
- ☐ How energy is distributed globally, with some nations having a surplus while others have a deficit of energy
- ☐ How energy is being wasted both in homes and in workplaces
- ☐ How to calculate carbon footprints
- ☐ Possible solutions to energy wastage
- ☐ Some of the policies being used to try to save energy in the UK on both a local and a national scale

Key terms

Carbon footprint
Consumer society
Energy deficit
Energy efficiency
Energy surplus
Fossil fuels
Incineration
Landfill
Non-renewable fuels
Recycling
Renewable fuels
'Throw-away society'

Which key terms match the following definitions?

A Disposal of rubbish by burying it and covering it over with soil

B A situation when the production of energy exceeds the use of energy

C Processing waste materials so that they can be used again

D A wealthy society in which people tend to dispose of goods once they are finished with, rather than reusing or repairing them

E Non-renewable resources that can be burned – such as coal, oil or natural gas – that have been formed in the Earth's crust

F Destruction by burning, e.g. of waste materials

G Combustible sources of energy – like biofuels – that can be regrown or regenerated

H A measurement of all the greenhouse gases we individually produce, expressed as tonnes (or kg) of carbon dioxide equivalent

To check your answers, look at the glossary on page 289

ResultsPlus
Maximise your marks

Foundation Question: Describe advantages and disadvantages of renewable fuels. (4 marks)

Student answer ■ (awarded 1 mark)	**Examiner comments**	**Build a better answer** △ (awarded 4 marks)
Renewable fuels are not likely to run out very quickly, which is the big advantage. *They don't really produce all that much energy because it isn't cheap to make energy in this way.*	• *Renewable fuels are...* is awarded no marks. *Very quickly* makes this part of the answer incorrect as renewable fuels are never likely to run out. • *They don't really...* receives 1 mark. A second mark would be given if it was supported by an example.	Renewable fuels do not run out which is an advantage. An example is wood that re-grows as long as it is replanted. Wind power and solar power are both renewable, but neither are reliable because it is not always windy or sunny. This is a disadvantage.

Overall comment: Always re-read your answers to make sure they are precise. Examiners want to reward good answers so make it easy for them!

Higher Question: Explain why the amount of waste produced by countries tends to increase as they become wealthier. (6 marks)

Student answer ● (awarded Level 2)	**Examiner comments**	**Build a better answer** △ (awarded Level 3)
Because people are wealthy they don't care about the environment so they treat it badly. *Wealthy people often have cars which pollute the air and create other waste like oil.* *Many goods today have lots of packaging. Wealthy countries buy lots of goods and all this packaging, such as cardboard boxes and tins, just gets thrown away, creating waste.*	• *Because people are...* is neither true nor really relevant because waste is not mentioned. No marks are awarded here. • *Wealthy people often...* This is awarded 1 mark. It needs some detail about car ownership or further explanation about the pollution. • *Lots of goods today...* is awarded 2 marks because an idea is presented along with an explanation of it.	HICs are consumer societies and a lot of goods are purchased. Making these goods obviously creates waste at each stage; for example, cars that are made of steel, plastics and lots of other materials. Many goods are transported long distances, creating waste products. Making electricity creates waste, fuel oil needs to be disposed of and, in some countries, nuclear waste has to be made safe. Many goods today have lots of packaging. Wealthy countries buy lots of goods and all this packaging, such as cardboard boxes and tins, just gets thrown away, creating waste.

Overall comment: Be careful not to confuse waste with pollution and environmental damage in general.

Chapter 8 A watery world

Objectives

- Know the different sources of water.
- Be able to describe the differences between HIC and LIC water consumption.
- Understand how level of development affects water consumption.

Skills Builder 1

Study Figure 1.

Describe the differences in water consumption between HICs and LICs.

ResultsPlus
Exam Tip

Answers to questions about differences should provide evidence in the form of data. In Activity 1, for example, it would be good to say '14% of water usage in HICs is domestic, whereas it is only 4% in LICs'.

Water consumption and sources

The differences between LIC water consumption and HIC water consumption

The three main ways that we use water are:

1. **Agriculture** – especially in **irrigation**
2. **Industry** – for cooling machinery, in food and drink manufacture, for energy production, etc.
3. **Domestic** – in our homes, for drinking, washing, etc.

Average **water consumption** per person is found by dividing the total water used in a country by the number of people living there. This does not show how much water each person directly uses themselves, because it includes water used in industry and agriculture as well as domestic use. There are significant differences between water consumption in HICs and in LICs. HIC water consumption is very high. On average people use about 1,200 m^3 per year each. This is about three times as much as in LICs, where it is typically about 400 m^3 per year. (A cubic metre is 1,000 litres.)

There are also big differences in what this water is used for. Figure 1 shows the percentage of water used in each of the three main categories of use (agriculture, industry and domestic) in HICs and LICs.

There are some very clear differences that can be seen in these two pie-charts. LICs use most of their water in agriculture and relatively little in industry and domestic use. HICs use most water in industry and agriculture, but with domestic use still significant.

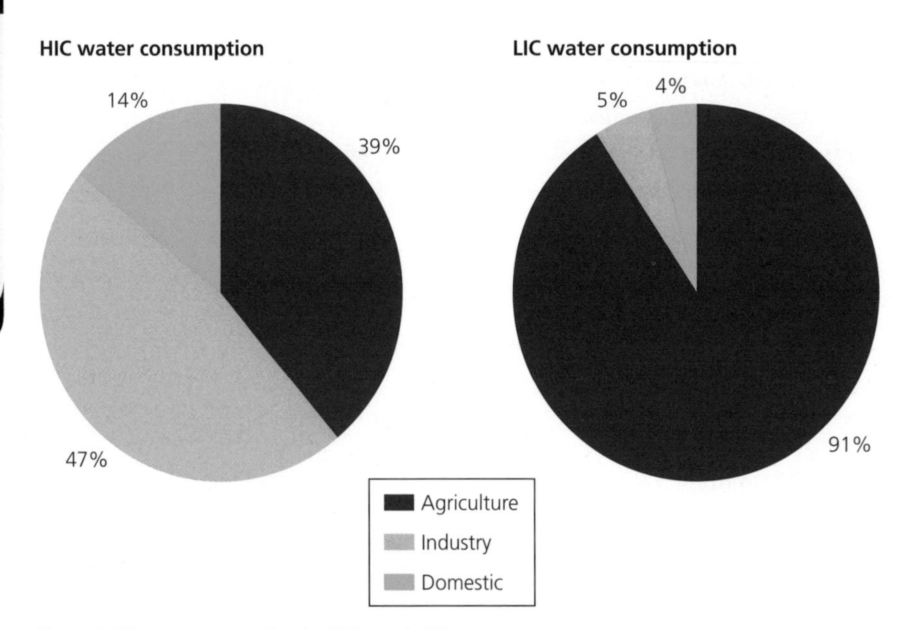

Figure 1: Water consumption in HICs and LICs

The main reasons for the differences are related to the income level of the countries. Agricultural use in LICs is often very inefficient, with a lot of water being added to fields by irrigation channels and by the flooding of fields with water, as in Figure 2. A lot of this water is wasted as it runs off the surface, drains away or evaporates.

In HICs, irrigation tends to be more efficient and more targeted with less being wasted. Sprinklers and drip feeds are used to supply just enough water to exactly the right places at the right time, as shown in Figure 3.

Industrial uses are also very different. There is much less industry in LICs and many industrial units are small-scale cottage industries that use relatively little water in their production methods. Large-scale factories in HICs have a high use of water, especially in the cooling of machinery and equipment.

Domestic use is also different. HIC homes have a piped water supply and have baths, showers and flush toilets. Many have washing machines and dishwashers. Some even have swimming pools. In LIC homes, the water is often brought in manually from wells or communal taps. As it is in short supply it is used carefully to minimise waste. Washing of clothes and dishes is done by hand and bathing often takes place in rivers.

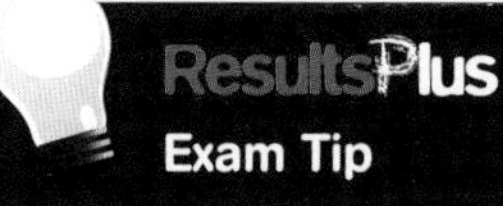

Exam Tip

In examinations you may be asked to describe what is shown in photographs. Use descriptive words to describe shape and appearance. Words such as straight, winding, wide, narrow, high and low may be useful.

Activity 1

Explain why water consumption differs between HICs and LICs in terms of the percentage used for: (a) industry, (b) agriculture and (c) domestic categories.

Figure 2: Irrigation channel

Figure 3: Irrigation by sprinkler

Wealth, development and water consumption

As countries become wealthier and more developed, so their demand for water increases. Figure 4 shows the rapid increase in water consumption in Asia, a continent where many countries are developing very quickly.

As countries develop, growing levels of industrialisation and greater use of machinery increase the demand for water for industrial uses, which is boosted further as the higher standards of living increase the consumption of food and drink. The demand for domestic water grows too, with more and more labour-saving devices such as dishwashers and washing machines being used. There are more cars to be washed, and gardens to be watered. Higher standards of living also lead to greater numbers of swimming pools and hot tubs being installed in homes. Higher standards of personal hygiene lead to the creation of a 'showering society' in which people shower in the morning and again later in the day, perhaps before going out for the evening. These activities can use large volumes of water (see table below left).

The growth of the leisure and tourism industry results in more water being used in water parks, spas and swimming pools, as well as in watering the growing number of golf courses that are being opened. Figure 5 shows a highly watered golf course in a very dry environment near Las Vegas, USA.

Figure 4: Increasing water consumption, by continent

The typical volume of water used for household activities

Activity	Litres used
Washing machine (per wash)	80
Dishwasher (per wash)	35
Drinking (per day)	1.5
Cooking (per day)	8
Washing clothes by hand (per wash)	25
Washing up (per wash)	6
Taking a bath (per bath)	80
Flushing the toilet (per flush)	13
Personal washing (per day)	8
Cleaning teeth (per brush, tap off)	1
Cleaning teeth (per brush, tap left on)	5

Figure 5: Watered golf course near Las Vegas

Activity 2

1. Keep a record of water usage in your home on a typical Sunday. Calculate the total amount of water used and the average usage per person.

2. Compare your results with those of others in your class.

Reservoirs, aquifers and rivers

On a local scale, the three main sources for our water supply system are **reservoirs**, **aquifers** and rivers.

Reservoirs

Reservoirs are artificial lakes created as a source of water supply, by building a dam across a valley and allowing it to flood. The water stored behind the dam can then be used to supply local towns and cities. Kielder Water (Figure 7) is a large man-made reservoir in Northumberland. It is the UK's largest artificial lake by capacity (200 billion litres) and it is surrounded by Kielder Forest, the largest man-made woodland in Europe. It was planned in the late 1960s to satisfy an expected rise in demand for water to support a booming UK industrial economy. A large dam was built between 1975 and 1981 which is over 50 m high and over 1 km long. It was opened by the Queen in 1982.

It took two years for the valley to fill with water completely once construction was completed. Springs ensure that the water level always remains high, regardless of the prevailing weather. This means that while the south of England is often forced to implement drought strategies and hosepipe bans, north-east England always enjoys plentiful water supplies.

The main towns and cities served by the reservoir, which is owned by Northumbrian Water, include Newcastle, Sunderland and Middlesbrough. Kielder Water is also one of the region's major tourist venues, attracting over a quarter of a million visitors a year, who come to enjoy the wide range of leisure opportunities on offer such as Leaplish waterside park.

Figure 7: Kielder Water

Skills Builder 2

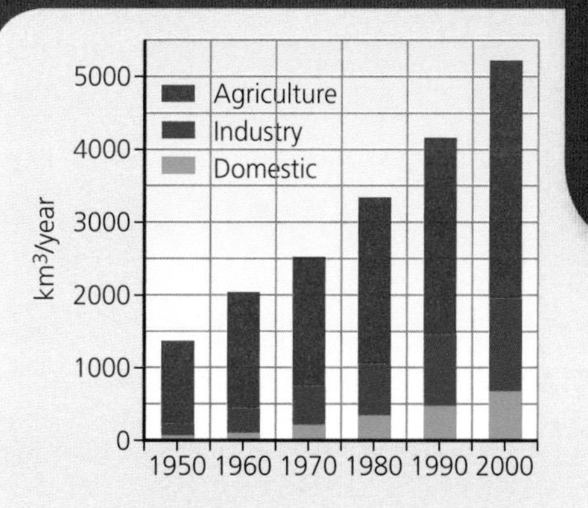

Figure 6: World water consumption 1950–2000

Study Figure 6.

(a) Describe the change in total consumption between 1950 and 2000.

(b) Suggest the main reason for this change.

(c) State the amount of water consumed by agriculture, industry and domestic uses in 2000.

(d) Estimate the percentages of water consumed by agriculture, industry and domestic uses in 2000.

ResultsPlus
Build Better Answers

Describe and explain the global pattern of water consumption. (4 marks)

■ **Basic answers** (0–1 marks)
List places with high/low consumption.

● **Good answers** (2 marks)
Offer a clear description with some explanatory statements relating to level of development, perhaps focusing on total water consumption.

▲ **Excellent answers** (3–4 marks)
Provide a clear description and full explanation of the pattern, referring to differences in level of development and types of usage.

Aquifers

Deep layers or large bodies of rock that can hold substantial quantities of water are known as aquifers. These are often porous rocks – such as chalk – that let water in and then hold the water in tiny holes. The water can be extracted by drilling wells or boreholes down to the aquifer and using a pump or, in some cases, by allowing the water to rise to the surface under its own pressure. Traditionally, chalk aquifers around the River Thames basin were used to supply water to London. The chalk hills of the Chilterns, the Berkshire Downs and the North Downs were all important sources of water. The average rainfall in these places is about 750 mm per year and about half of this enters the chalk. The rest is lost to evaporation or runs off across the surface. The water that enters the chalk moves underground, eventually making its way into nearby rivers or emerging as springs on the ground surface. The amount of water stored in the chalk under London is more than twice that held in Kielder Water.

Up to the 1960s, water extracted from the chalk for industry caused water levels in the centre of London to drop by 65 metres. In the 1960s, about 480 million litres per day were being extracted. As industrial activity declined, this reduced to about 380 million litres per day and remained fairly stable until the 1990s. As a result of the reduced extraction, the water levels in the chalk started to rise again, causing concern about the flooding of deep basements and tunnels. Another concern associated with rising groundwater levels beneath London is the risk of the ground around deep foundations for tall buildings becoming softer.

Extraction has since been increased to stabilise the water levels. The groundwater level rise began to slow down naturally towards the end of the twentieth century, as the water level depression began to fill up. With increased extraction, water levels have stabilised.

ResultsPlus
Exam Tip

It is good to have up-to-date facts and figures when writing about places. Books go out of date quickly – even good books like this one! Try looking up water usage in London on the internet and see if things have changed since this book was written.

Rivers

Water supply can also be obtained directly from rivers. This is the case in Florida where water has been taken from the Peace River (Figure 8) since 1980. This is done to ensure adequate water supplies for an ever-growing population of more than 750,000 people in the local region.

Every day, an average of 70 million litres of water is supplied to local customers. This water is treated at a treatment plant near Fort Ogden which has the capacity to treat up to 90 million litres per day. Treated water is injected into an aquifer, which acts as a natural underground storage tank, and is then recovered as needed. This process, known as aquifer storage and recovery, is an ideal method for meeting seasonal water demands. It allows water to be withdrawn from the river during 'wet' months, and then stored for use during 'dry' periods when river levels are low.

Work has recently started on a new 24 billion litre storage reservoir and the capacity of the treatment plant is being doubled. This work is designed to meet the increasing demand for water in the area over the coming years.

Figure 8: Enjoying the Peace River in Florida

Water surplus and water deficit on a world scale

Whether a place has a **water surplus** or a **water deficit** depends on the relationship between the water it receives from rainfall and the water it loses through evaporation and transpiration. Many places have a rough balance between the two. But some parts of the world receive more than they lose – meaning that they have a surplus of water. And other places have such high rates of evaporation and transpiration (caused by high temperatures) that they can actually lose more water than they receive during a given time period – so they have a water deficit. Figure 9 shows the distribution of places with a water surplus or a water deficit.

If you compare Figure 9 with a global rainfall map in an atlas you will see that not all the places with high rainfall have a surplus. Some places in central Africa, for example, near the Equator, have high rainfall totals but their high temperatures mean that they do not have a water surplus.

Activity 3

Using examples, describe the three main sources of water supply.

Skills Builder 3

Study Figure 9.

Describe the global pattern of water surplus and deficit.

The relationship between rainfall and water deficit or surplus is not a strong one (or an easy one to see). It may be helpful to point this out in an examination answer.

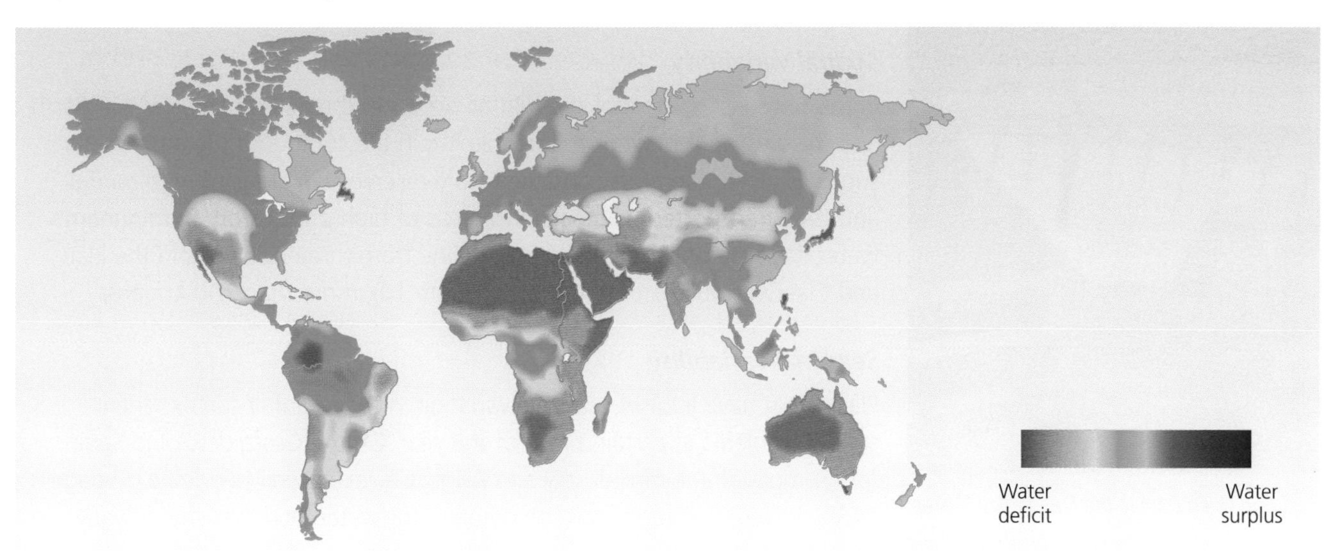

Figure 9: Global water surplus and deficit

Objectives

- Know the different water supply problems in HICs and LICs.
- Be able to describe HIC and LIC water supply problems.
- Understand what causes these problems.

Do not confuse spatial variation (differences from place to place) with seasonal variation (differences from season to season).

Water supply problems

Water supply problems in HICs

There are a number of problems that HICs face in supplying water to meet the needs of their populations:

- Quality
- Spatial variability
- **Seasonal variability**
- Loss through broken pipes.

Quality

The quality of drinking water in the UK is monitored by the Drinking Water Inspectorate, which follows the EU's 'Water Framework Directives'. The Inspectorate regularly and widely tests the quality of water reaching British homes, and in general the quality is high. In 1992, 1.7% of their tests in England found water to be below the quality expected, but by 2007 this had fallen to less than 0.1%.

One of the reasons for the improved quality of drinking water is the work done by the Environment Agency in trying to reduce the use of certain fertilisers. When farms use nitrogen-based fertilisers it can lead to high levels of nitrates in local river water and groundwater. The Agency's work has been particularly successful in the Lake District and in the Norfolk Broads (which has been designated as an Environmentally Sensitive Area), enabling farmers to apply for grants to help them reduce their use of nitrogen fertilisers. In 2007 the Environment Agency reported that the chemical quality of water in English rivers was 'good' or 'very good' in 76% of cases – up from 55% in 1990. However, pollution of river water can still result from industrial leaks and spillages, as well as from inefficiencies in the sewage treatment plants that return treated water into rivers.

Spatial variability

In the British Isles, most of the rainfall occurs in the upland areas of the north and the west, but most of the people live in the south and the east. Water supply is often, therefore, obtained from reservoirs in areas of high rainfall and then transported by pipeline to areas of high population. Birmingham's water, for example, is supplied by pipeline from three reservoirs in the Elan and Claerwen valleys of mid-Wales (Figure 10), more than 100 km away.

Figure 10: The Craig Goch Dam in the Elan Valley, Wales, storing water for Birmingham

Seasonal variability

Places that have a lot of seasonal variation in their rainfall can experience supply problems at certain times of the year. On the Costa del Sol in Spain, for example, the rainfall in July and August is virtually nil (see table on page 143) – just at the time when the demand for water increases significantly because of the high numbers of holidaymakers. Swimming pools have to be supplied with water and the golf courses need to be watered (there are over 70 courses here), as well as the demand for normal domestic use.

Average monthly rainfall in Malaga, Costa del Sol (mm)

Jan	Feb	Mar	Apr	May	Jun	Jul	Aug	Sept	Oct	Nov	Dec
64	53	60	39	20	10	1	3	16	56	90	70

To meet this high demand the area has developed six major reservoirs, including La Concepción with a capacity of 65 million m^3. A desalination plant was built near Marbella in 2005 to produce 20 million m^3 per year of fresh drinking water from seawater. However, there are still problems sometimes with meeting the demand, and water pressure is often low. Restrictions are sometimes placed on the filling of swimming pools and water supply is sometimes cut during the night. Another desalination plant opened near Mijas in 2009.

Loss through broken pipes

The water distribution system in London was mainly constructed in the 1830s and 1840s, and many of the original pipes have been in continuous use since then. As demand has increased, the water pressure has had to be increased in order to move water quickly through the system. But many of the pipes have been unable to cope with this pressure – and have burst. These 'burst water mains' are major contributors to the loss by leaks of almost 20% of the water in the London system.

In the 1990s, a new ring main was built, 40 metres below the surface, consisting of 80 km of 2.5-metre-diameter pipeline. This £250m project has greatly reduced the leakage losses and has given London a much more reliable distribution system. The ring main links the treatment plants to the supply points and storage reservoirs and allows 1.3 billion litres per day to be supplied. North and south extensions to this scheme are currently under construction which will give the system even greater flexibility and storage potential.

Water supply problems in LICs

The problems faced by LICs in meeting the demand for water are rather different from – and often more serious than – those faced by HICs, including:

- Lack of available clean water
- **Water-borne disease**
- **Water pollution.**

Lack of available clean water

It is estimated that over 1 billion people in LICs lack access to safe water. Figure 11 on page 144 shows, country by country, the percentage of people with access to safe water – people who have a reasonable means of getting an adequate amount of water that is safe for drinking, washing, and essential household activities. This measure shows the country's capacity to collect, clean, and distribute water to consumers – and reflects the health of its people.

Skills Builder 4

Study the table showing average monthly rainfall in Malaga.

(a) Draw a bar chart to show the data.

(b) Which month has the highest rainfall?

(c) Which two months have the lowest rainfall?

(d) Describe the seasonal variability in rainfall.

(e) Suggest two reasons why this is a problem.

Skills Builder 5

Look at Figure 11.

(a) Which parts of the world have the highest percentage of people with access to safe water?

(b) Which have the lowest?

(c) Do these places have similar levels of development?

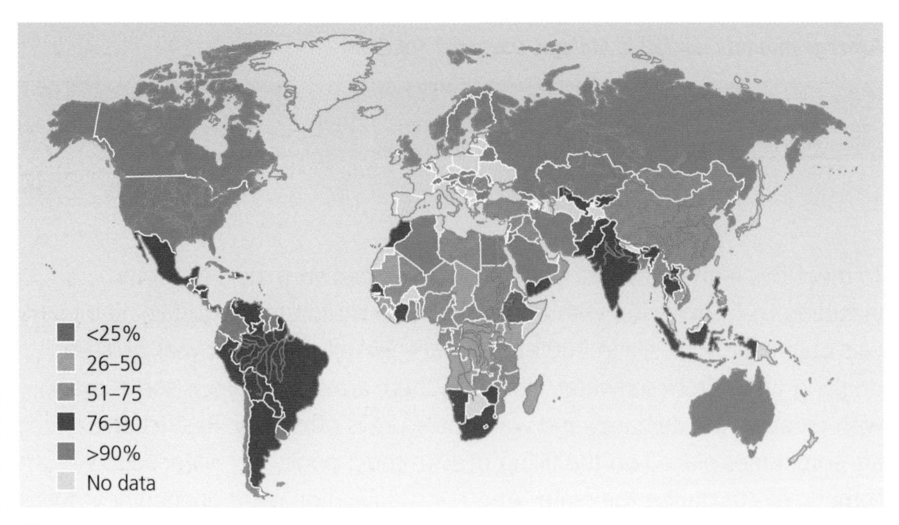

Figure 11: Access to safe water

The countries with the lowest percentage figures are Eritrea (7%), Cambodia (13%), Central African Republic (19%), Democratic Republic of Congo (27%), Papua New Guinea (28%) and Madagascar (29%).

Because of the lack of available water, many women and children in rural areas spend huge amounts of effort and time – in some cases as much as 6–8 hours per day – collecting and carrying water. Even in urban areas, relatively few people have water piped into their homes. Figure 12 shows the sources of water for the people of a Mozambique city. You will see that only about one-third of the population have water piped to their home or yard. Figure 13 shows small children in Ghana carrying water.

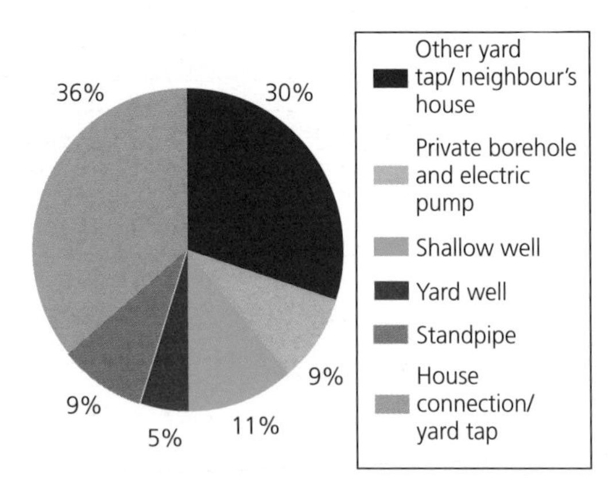

Figure 12: Sources of domestic water in a city in Mozambique

Water-borne disease

Without safe water, people cannot lead healthy, productive lives. An estimated 900 million people suffer – and approximately 2 million die – from water-related diarrhoeal illnesses each year. Most, but not all, of these people live in LICs, and those at greatest risk are children and the elderly. Millions more people worldwide suffer from other water-borne diseases, such as bilharzia, cholera and typhoid.

Bilharzia is a life-threatening parasitic disease caused by a worm that lives in a host snail. Humans can become infected when they come into contact with water in ponds and rivers where the snail lives. Over 300 million people in LICs are thought to have this disease.

Cholera is a disease caused by bacteria from faeces getting into drinking water. This is very common in LICs where sewerage systems are inadequate. In Zimbabwe, for example, 50,000 people were infected in 2008/2009 and over 3,000 died.

Figure 13: Collecting water in Ghana

Typhoid, also known as enteric fever, is a serious infection that is caused by the bacterium Salmonella typhi. The disease is transmitted from human to human and spread by eating food or drinking water that is contaminated with typhoid bacteria. During 2000 up to 15 people per week were dying of typhoid in Sierra Leone.

Water pollution

Surface water and groundwater can both be affected by pollution, although groundwater pollution is much less obvious and visible. If pollution comes from a single location, such as a discharge pipe from a factory, it is known as **point-source pollution**. Other examples of point-source pollution include an oil spill from a barge or someone pouring oil from their car down a drain.

A great deal of water pollution happens not from one single source but from many different scattered sources. This is called **non-point-source pollution**. This might be the case with run-off from fertilised fields adding nitrates to rivers. Sometimes pollution that enters the environment in one place has an effect hundreds or even thousands of kilometres away. This is known as transboundary pollution.

A major cause of water pollution in LICs is resource exploitation. Deforestation can result in the death of the root systems that previously held soil in place, leaving sediments free to run off into nearby streams, rivers, and lakes – seriously affecting fish and other aquatic life.

Poor farming practices that leave soil exposed to the elements also contribute to sediment pollution in water. If this water is used for drinking, the high sediment concentrations can affect people's digestive systems. If deforested areas are used for farming, the use of fertilisers and pesticides often results in water pollution. A recent study in Costa Rica found that deforested areas used for cattle ranging produced between three and eight times the level of contamination in rivers compared to areas left as tropical forest.

Mining of metallic minerals affects fresh water through the heavy use of water in processing ore, and through water pollution from discharged mine effluent and seepage from waste rock. It is estimated that extracting 1 tonne of copper produces 99 tonnes of waste. When large quantities of rock containing sulphide minerals are excavated from an open pit or opened up in an underground mine, they react with water and oxygen to create sulphuric acid, which causes serious pollution of rivers and streams. In 2006 a major pollution episode occurred in the copper mining industry in Zambia. The Environmental Council of Zambia directed Konkola Copper Mines to cease operations following the company's pollution of water in the Kafue River. The residents of Chingola in Copperbelt Province were faced with water shortages as a result of the pollution. A river polluted by copper mining can be seen in Figure 15.

Figure 15: A river polluted by copper mining

Skills Builder 6

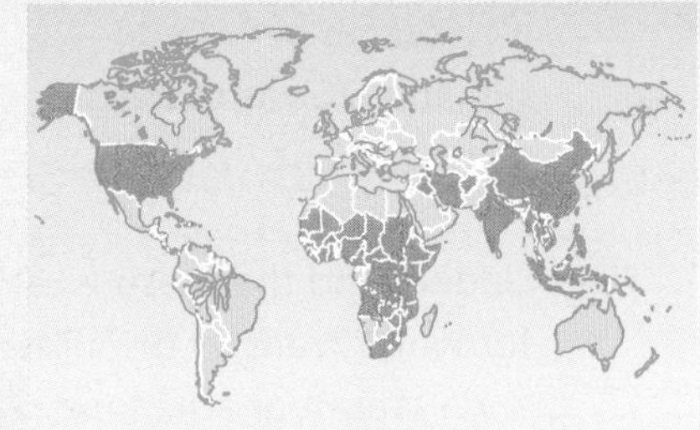

Figure 14: Countries reporting cholera cases, 2004–2007

Study Figure 14.

(a) Describe the global pattern of reported cases of cholera.

(b) Explain why cholera is more common in some countries than in others.

(c) Name two other water-borne diseases.

Activity 4

Use the internet to research the symptoms of bilharzia, cholera and typhoid.

The focus of this part of the topic is on water pollution. Do not get sidetracked in examination answers on to other impacts of deforestation, such as loss of habitat for wildlife.

Objectives

- Know the different ways of managing water usage.
- Be able to describe the differences between HIC and LIC water usage management.
- Understand the reasons for water transfer disputes and water management schemes.

Figure 16: A dual-flush toilet

Activity 5

Find out what water management methods are used in your home or school/ college.

Management of water usage and resources

The management of water usage in HICs

In HICs, domestic use of water is being managed in a number of ways. One of the most common is the use of water meters in homes. In the UK, about 30% of homes already have water meters, but this is set to increase because all new homes built now must have meters installed by the water supply companies. Householders with meters pay for the volume of water they use – measured in cubic metres – as recorded by the meter. The 70% of householders currently without meters pay a standard charge, irrespective of how much water they use. The **metering** system encourages people to be careful about how much water they use. In 2008 the cost was about £1.60 per cubic metre, and a typical family of four might use about 200 cubic metres per year.

Another approach is to devise more efficient means of using water. For example, most new homes now have dual-flush toilets fitted (Figure 16). An old-style single-flush toilet can use up to 13 litres of water in one flush. New, more water-efficient dual-flush toilets use only 6 litres for a full flush and 4 litres for a reduced flush.

The efficiency of old-style toilets can be improved by adding a cistern displacement device such as a 'Save a flush' bag or a 'Hippo'. These are available for free from most water companies in the UK. These devices are easy to install and are placed in the toilet cistern to displace – and therefore save – approximately one litre of water every time it is flushed.

Industrial use of water can be managed by using new manufacturing techniques and by recycling water. In the steel industry, for example, 11,000 litres of water are used in a traditional blast furnace-related process to produce 1 tonne of steel. However, modern electric arc systems and the use of recycled steel reduce this to about 2,300 litres per tonne. A lot of water is used for cooling in steel plants, and this can also be recycled so that the total actually consumed is reduced. Water can be condensed back into a liquid state after it has been evaporated by the heat produced.

In agriculture, use of water in irrigation systems is being carefully managed by the use of modern, more efficient systems. Drip-feed and sprinkler systems use much less water than irrigation channels and ditches. It is estimated that drip systems are 90% efficient, with only 10% of water being lost to evaporation. Sprinkler systems are about 70% efficient, whilst channels and ditches may be only 50% efficient.

The management of water usage in LICs

In LICs, the key to water use management is generally the application of **appropriate technology** to supply small communities. One particularly effective method has been the use of cost-effective boreholes (CEBs) in Africa. The 'Rural Water Supply Network' aims to provide cost-effective boreholes that are appropriate to meet the needs of the local communities. CEBs cost between $5,000 and $6,000 to install. Currently, 2,000 new CEBs are planned for Mozambique, to be installed at a rate of about 400 per year. This scheme will be part funded by the Dutch government. Figure 17 shows a borehole under construction.

In Kolkata, India, sewage water is being recycled for use in fish farming and agriculture. Sewage is piped to shallow lagoons which allow sunlight to reach the bottom to promote growth of algae and photosynthetic oxygen. Using aquatic plants like water hyacinth and duckweed, dirt and some metals are removed and it is also purified by exposure to sunlight and aeration. Fish, especially carp, flourish in this productive ecosystem and fishermen's co-operatives have been formed which are proving to be successful commercial operations. The stages in the scheme are shown in Figure 18. Another aspect of the scheme is the use of treated, solid waste as a natural fertiliser for the growth of vegetables.

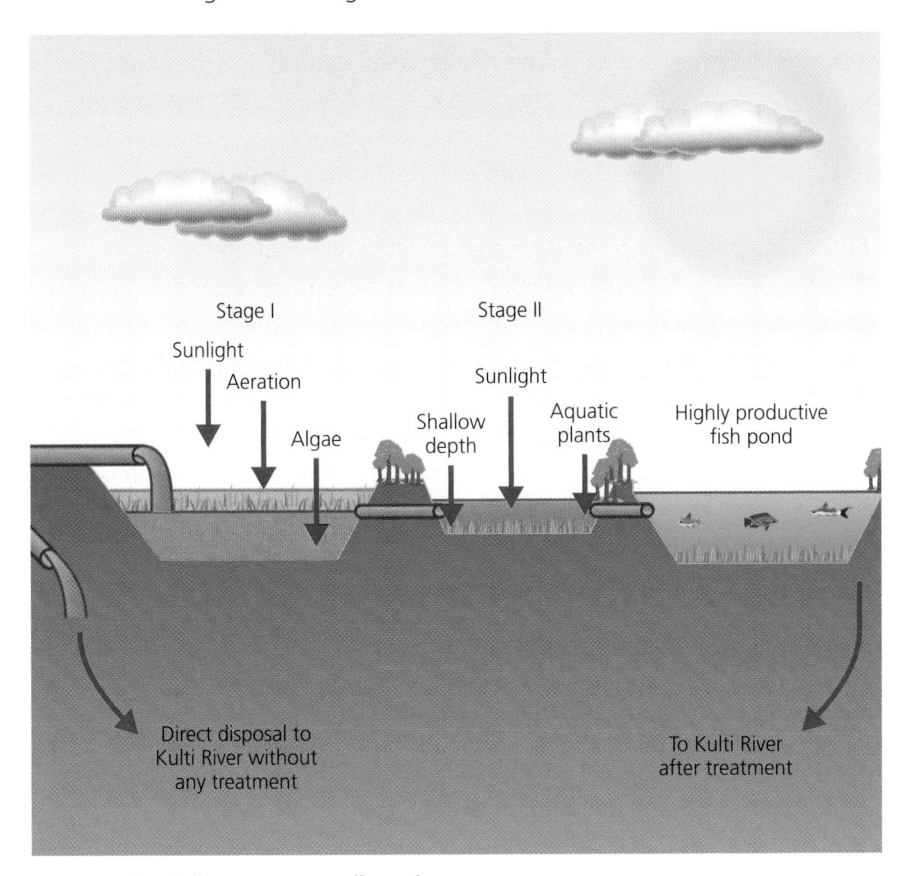

Figure 18: The Kolkata water recycling scheme

Figure 17: A borehole under construction in Mozambique

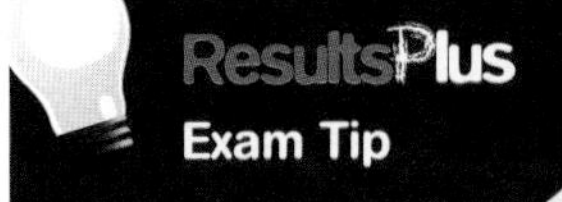

Exam Tip

Technology is appropriate if the local community is able to use it relatively easily and without much cost. If water pumps need expensive spare parts or are difficult to maintain, then the local community may not find them appropriate. You may need to explain this in an examination answer.

Activity 6

Describe the three stages of water recycling being used in the Kolkata system.

The management of water resources

A dispute over water transfer – a case study of the Colorado River basin

Figure 19 shows the Colorado River basin, which is 630,000 km^2 in area. The Colorado River has its source about 4,000 m high in the Rocky Mountains of Colorado and flows south-west for 2,300 km to the Gulf of California in Mexico. The Colorado and its tributaries drain south-western Wyoming and western Colorado, parts of Utah, Nevada, New Mexico and California, and almost all of Arizona.

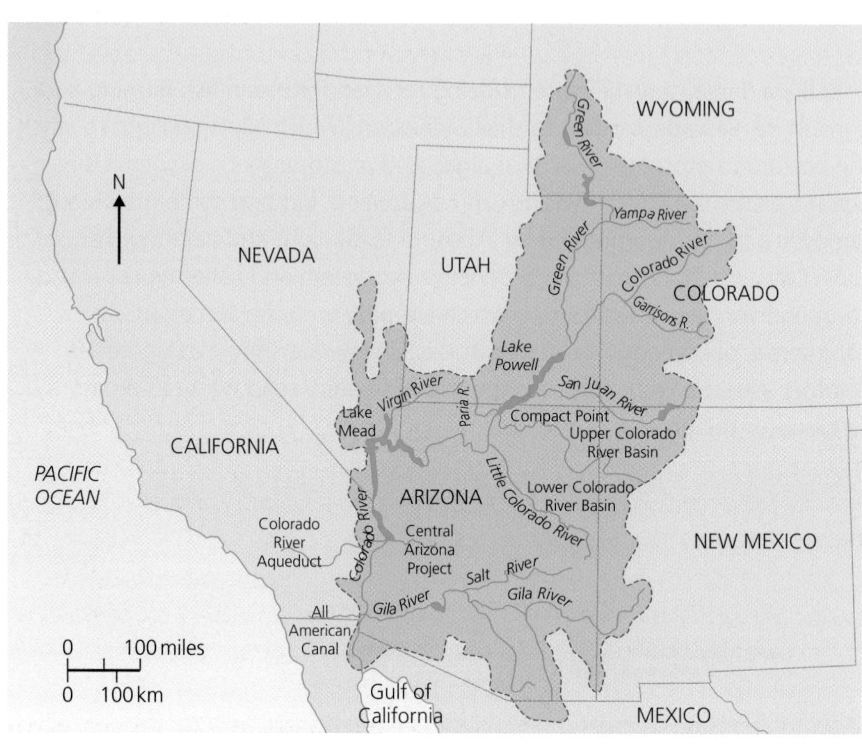

Figure 19: The Colorado River basin

Skills Builder 7

Study Figure 20.

Compare the allocations of water usage to the different parts of the Colorado basin.

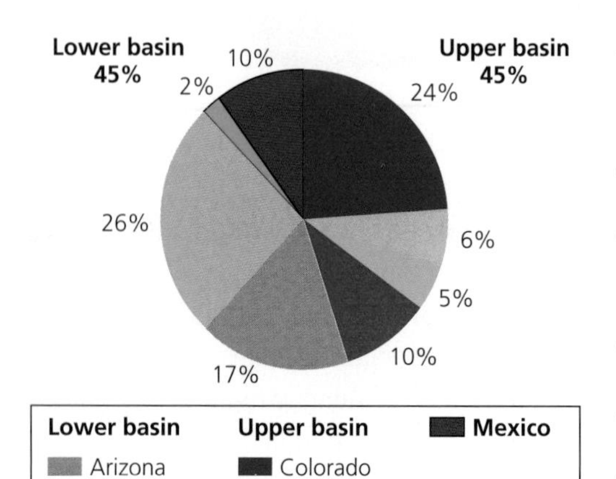

Figure 20: How the water was allocated in the Colorado River Compact, expressed as percentages (rounded)

The use of water from the Colorado is a controversial issue, and there is much conflict between the potential users. These include:

- Those who live in the upper part of the river basin.
- Those who live in the lower part of the river basin, including Arizona, whose population increased by 40% in the 1980s.
- Farmers, who want water for irrigation.
- The native Indian population who claim rights to the water.
- Neighbouring settlements – such as Las Vegas, Nevada – which are in dry areas but have a high demand for water.
- Settlements that need water to generate electricity by hydro-electric power.

After a number of treaties and agreements established by the government between 1922 and 1948, water was allocated to the various users in different parts of the basin, as shown in Figure 20.

The first major development began in 1928 when the government passed the Boulder Canyon Project Act, authorising the construction of the Boulder Dam (now called the Hoover Dam). Construction of this dam was considered one of the major engineering accomplishments of its time. Since its completion in 1936, the Hoover Dam and Lake Mead (which it created) have become major tourist attractions (Figure 21).

Figure 21: The Hoover Dam

Shortly after the completion of the Hoover Dam, planning and construction began downstream on the Parker Dam. From Lake Havasu, the reservoir created by the dam, water is transported some 400 km across California to supply a portion of the water needs of Los Angeles and most of the water supply for San Diego. The Davis, Imperial, Laguna and Morelos dams further regulate flows and diversions in the lower basin.

Many additional projects have been completed since then. In the mid-1960s, Glen Canyon Dam was completed, creating Lake Powell. This dam was controversial, and opposition to its construction helped to shape policy more towards concepts of water management and environmental protection.

In 1963 a decision of the US Supreme Court made explicit the amount of water apportioned among the lower-basin states, as well as the amounts that had been implicitly 'reserved' for native Indian tribes and public lands. This decision prompted funding of the Central Arizona Project, completed in the 1980s. The project comprises a mountain tunnel through which water from the southern end of Lake Havasu is pumped up and into an aqueduct that flows southward to the two cities of Phoenix and Tucson – although a lot of water is lost to evaporation because temperatures are high for much of the year.

More conflicts and disputes have occurred in recent years. In 2007 Arizona, which has to supply the great demands of Tucson and Phoenix, was in dispute with the government over the amount of water being extracted further upstream.

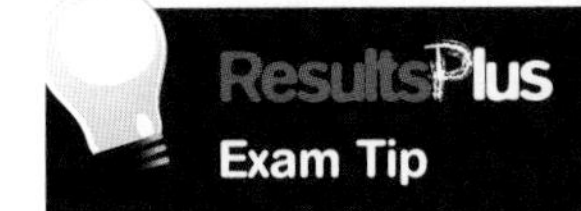

Exam Tip

Demand changes over time – so agreements over water use may need to be changed.

The biggest loser would seem to be Mexico, because so much water has been taken out of the river before it reaches the international border. Irrigation water extracted from the river is supposed to be put back in to maintain flow in Mexico. However, this water is very saline and unsuitable for Mexican farmers to use. About 20% of farmland is not being cultivated for this reason. The government has also had to build an expensive desalination plant near the border at Yuma. Environmentalists are unhappy, because the reduced flow at the mouth has led to a loss of habitats for birds (such as the egret) and fish (such as the totoaba).

Mexico has also objected to the lining of the All American Canal, which runs near its border and takes water to California. Previously, a lot of water leaked out of the canal and into the Mexical Aquifer. Lining the canal has stopped leakage and the aquifer is now holding less water.

The increasingly severe competition for whatever small quantities of water remain in the Colorado River keeps the basin tied up in legal disputes and controversy. Water projects must now undergo thorough environmental-impact studies in accordance with US environmental protection laws.

Case study quick notes (Colorado River basin):

- Water supply is a controversial issue in this area.
- Many different groups of people want and need water from the river.
- Laws have been passed to try to regulate water use.
- So much water is taken out that the river is now virtually dry at its mouth.

ResultsPlus
Build Better Answers

For a water management scheme you have studied, explain why the scheme was necessary. (6 marks)

■ **Basic answers** (Level 1)
Describe the scheme and what it involves without valid explanation.

● **Good answers** (Level 2)
Also provide some broad explanation in terms of the need for water.

▲ **Excellent answers** (Level 3)
Provide full explanation of the reasons why the scheme was set up, with specific reference to a named scheme, giving details, facts and figures, etc.

A water management scheme – a case study of Sydney Olympic Park

Sydney Olympic Park was developed to enable the city to host the Olympic games in 2000. As part of the legacy of that event, the 640-hectare park has been developed to provide a broad range of commercial, residential, recreational, leisure and public uses. These were designed to make use of the facilities and infrastructure left behind after the Olympics. The 'Water Reclamation and Management Scheme' was set up to provide:

- A water reclamation plant (Figure 22) that removes water from sewage at a rate of up to 3 million litres per day.
- A brickpit – a water reservoir holding up to 300 million litres.
- A water treatment plant to filter and disinfect water from the water reclamation plant and storage reservoirs at a rate of up to 7.5 million litres per day.
- A separate, dedicated supply system to pipe water from the treatment plant through Sydney Olympic Park and Newington (a new suburb, originally constructed to house the Olympic athletes).

Figure 22: Sydney Olympic Park water teatment plant

The overall aim of this scheme was to reduce pressure on the demand for drinking water supply in the Sydney area by providing recycled water for other uses. Figure 23 shows how the scheme works.

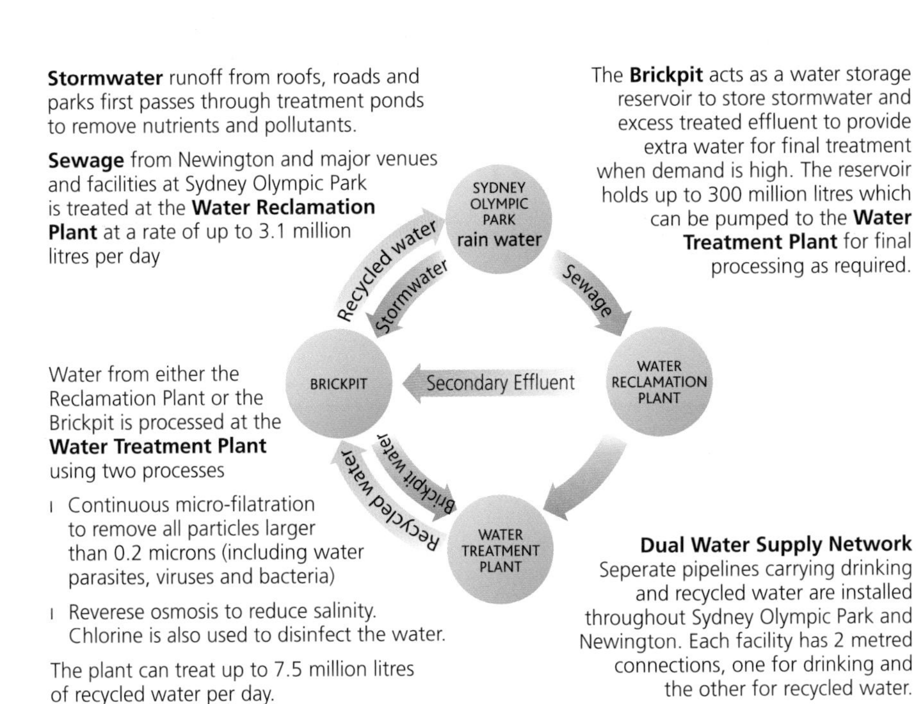

Figure 23: How the Sydney Olympic Park scheme works

The positive effects of the scheme have been that it:

- Saves approximately 850 million litres of drinking water each year by reducing drinking water consumption at Sydney Olympic Park and Newington by around 50%.
- Treats and reuses almost 100% of the sewage, contributing to a reduction in discharge of sewage effluent to waterways and the ocean.
- Contributes to developing greater public confidence in using recycled water.
- Develops an understanding of how water supply in urban areas can be managed sustainably.
- Provides water at a price that is 15 cents per litre lower than the price of drinking water.

The scheme has been very successful and is highly regarded by the community.

The recycled water is suitable for many important uses such as:

- Toilet flushing
- Car washing
- External wash down
- Construction
- Parklands irrigation
- Residential gardens
- Playing fields irrigation
- Clothes washing
- Ornamental water features
- Firefighting
- Cooling towers
- Swimming pool filter backwashing

Activity 7

Describe the stages of water reclamation used in the Olympic Park scheme.

Exam Tip

Read examination questions carefully. Sometimes they will ask for the positive impacts of schemes, sometimes the negative impacts – and sometimes both.

Quick notes (Sydney Olympic Park):

- Recycling of water is important to meet growing demand.
- Recycled water has a range of uses; however, it is not designed for drinking.

Considerations for recycled water include:

- Recycled water and the risk to public health.
- Perception of recycled water quality.
- Management of dual water systems (separation of supply).
- Cost of recycling versus the cost of drinking water.

Know Zone
Chapter 8 A watery world

The twenty-first century is likely to be dominated by more and more problems over resource shortages. Water is the most basic of resources and is also one that is likely to be most disputed.

You should know...

- ☐ How and why water consumption varies from country to country
- ☐ Who the main water users in a country are, including domestic, industrial and agricultural users
- ☐ That some uses have grown quickly in recent years, especially in leisure and tourist activities
- ☐ From which sources we draw water, including reservoirs, aquifers and rivers
- ☐ The global distribution of water deficit and water surplus areas
- ☐ The problems of water supply in HICs
- ☐ The problems of water supply in LICs, including the issue of water quality
- ☐ How water is managed in HICs
- ☐ How water is managed in LICs
- ☐ A case study of conflict over water resources
- ☐ A case study of a water management scheme that shows both the positive and negative effects on people and the environment

Key terms

Agriculture
Appropriate technology
Aquifers
Domestic
Irrigation
Metering
Non-point-source pollution
Point-source pollution
Reservoir
Seasonal variability
Water-borne diseases
Water consumption
Water deficit
Water pollution
Water surplus

Which key terms match the following definitions?

A Water-bearing rocks

B An artificial lake created as part of a water supply system

C Addition of water to farmland by artificial means

D A situation in which the usable water supply does not satisfy the demand

E The use of meters allowing a service such as water supply or electricity to be charged according to how much is used

F The amount of water used by a person or group of people

G The tendency for something to change according to the time of year

H Diseases passed on by microorganisms being present in water

To check your answers, look at the glossary on page 289.

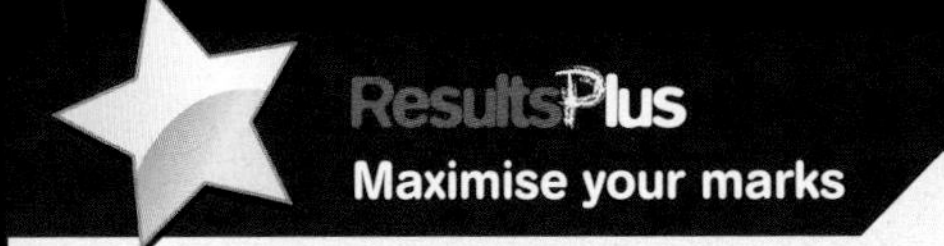

Foundation Question: Describe **two** ways in which water use can be reduced in HICs. (4 marks)

Student answer (awarded 2 marks)	Examiner comments	Build a better answer (awarded 4 marks)
Customers can be persuaded to use water more efficiently. *They can also be made to use less water in wasteful things, like stopping people watering their gardens*	• *Customers can be...* is not clear enough to be awarded a mark. It needs to describe the way in which people can use water more efficiently. • *They can also...* This is good and scores 2 marks. However, the method of *stopping* is not included in the answer.	The introduction of water metering in homes will reduce consumption by making people more aware of the costs. To stop a waste of water, hose-pipe bans can be used or new techniques such as short-flush toilets can be introduced.

Overall comment: Try to add examples to support the point you are trying to make in your answer.

Higher Question: Explain why water transfer can cause conflict between countries. Use a named example. (6 marks)

Student answer (awarded a Level 2)	Examiner comments	Build a better answer (awarded a Level 3)
There has been a lot of problems in the Mekong region and lots of quarrels between all the countries like India. *Dams built up the river have made the water levels fall in the delta region which has caused salt problems.* *The changing water levels of the river have affected the lake (Tonle Sap) in Cambodia, reducing the water level and the fish supply. It has also reduced food for the local population, although it does increase power for other countries.*	• *There has been...* does not gain any marks because the information is vague and the example location is inaccurate. • *Dams built up...* is good. However, the lack of location and explanation of the processes means it only receives 1 mark. • *The changing water levels...* scores 2 marks because there is a good level of detail about the relationship between the dams and the water level, and the description of the effect is clear.	The Mekong rises in Tibet and flows through China for most of its length. China is much more powerful than its neighbours and they use the river for HEP. Reducing water levels downstream in Vietnam has meant that the delta region has more sea water, which is very damaging for the huge rice crop in the region. The changing water levels of the river have affected the lake (Tonle Sap) in Cambodia, reducing the water level and the fish supply. It has also reduced food for the local population, although it does increase power for other countries.

Overall comment: Be as clear as you can in your answers – if you discuss conflict then try to explain what these conflicts are and how they have occurred.

Unit 3 The human environment

Your course

This unit investigates the human geography of the human world and the issues relating to the people living on our planet. There are two sections:

Section A will cover the human world and you will study one topic:

- Topic 1 (Chapter 9): Economic change
- Topic 2 (Chapter 10): Farming and the countryside
- Topic 3 (Chapter 11): Settlement change
- Topic 4 (Chapter 12): Population change

Section B will cover people issues and you will study one topic:

- Topic 5 (Chapter 13): A moving world
- Topic 6 (Chapter 14): A tourist's world

Your assessment

- You will sit a 1-hour written exam worth a total of 50 marks.
- There will be a variety of question types: short answer, graphical and extended answer, which you will practice throughout the chapter that you study. You will answer one question from Section A and one question from Section B.
- Section A contains four questions, one on each of the human world topics. You will have a resource booklet which contains any material that you will need to answer the exam questions.
- Section B contains two questions, one on each of the people issues topics. You might be asked to refer to material in the resource booklet to answer the questions.

Remember to answer the questions for the topics that you have studied in class!

Study the satellite photograph of Mumbai, India.

- Identify an area on the map that is occupied by:
  (a) an airport
  (b) a large shanty town
  (c) an area of better-quality housing.

Chapter 9 Economic change

Objectives

- Learn what economic sectors are, and how they change in importance.
- Explain the decline of the UK's primary and secondary sectors.
- Understand the significance of the growth of the tertiary sector.

Changes to different economic sectors

The economic sectors

The **economy** of a country is divided into three or four sectors. Each sector involves a different type of activity:

- The **primary sector** – working natural resources. The main activities are agriculture, fishing, forestry, mining and quarrying.
- The **secondary sector** – making things, either by manufacturing (e.g. a TV or a car) or construction (e.g. a house, road or new airport).
- The **tertiary sector** – providing services. These include services that are commercial (e.g. shops and banks), professional (e.g. solicitors and dentists), social (e.g. schools and hospitals), entertainment (e.g. restaurants and cinemas) and personal (e.g. hairdressers and fitness trainers).
- The **quaternary sector** – a rather new sector that is mainly found in high-income countries (HICs). It is concerned with information and communications technology (ICT) and research and development (R&D). Universities are part of it.

Because the numbers are still very small, in this chapter we will include the quaternary sector as part of the tertiary sector.

Figure 1 shows how the sectors change with time over three phases. The critical phase is the industrial one. This occurs when manufacturing (the secondary sector) becomes more important than the primary sector, and cottage industries are replaced by mechanised industries, housed in factories. This process of **industrialisation** is accompanied by important economic and social changes.

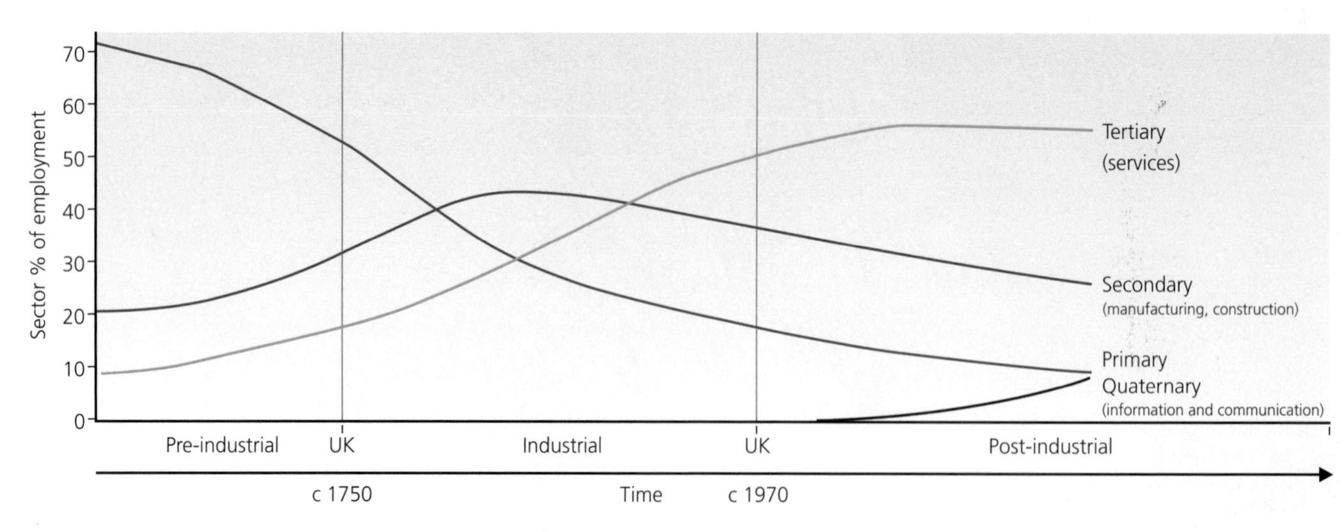

Figure 1: Economic sector shifts over time – the Clark-Fisher model

The three phases are:

- The pre-industrial phase – the primary sector leads the economy and may employ more than two-thirds of the working population. Agriculture is by far the most important activity.
- The industrial phase – the secondary and tertiary sectors increase in importance. As they do so, the primary sector declines. The secondary sector peaks during this phase, but rarely provides jobs for more than half of the workforce.
- The post-industrial phase – the tertiary sector establishes itself as the most important sector. The primary and secondary sectors continue their decline. The quaternary sector begins to appear.

How do we measure the relative strength of these economic sectors? Figure 1 does it according to their percentage share of the total workforce. It is also possible to do it according to how much they contribute to a country's economic output – their percentage of the **gross national income** (GNI).

What causes the relative importance of the sectors to change over time? This is a much more difficult question, and the best answer is that these shifts are part of the **development** process. As countries develop, they move along a broad pathway. The character of that pathway changes along its length. The **sector shifts** are one of those character changes. Others include a rise in the standard of living and a growth of towns and cities.

Today, the world's countries are strung out along this development pathway. Some have made little progress along it, while others have gone a long way. Some are moving quickly, others hardly at all.

Figure 2 classifies countries according to where they are on the development pathway.

Activity 1

1. Approximately when did the UK start the industrial phase?
2. Approximately when did the UK start the post-industrial phase?
3. Describe how the importance of each sector changes over time. Make sure you use numerical data (i.e. percentages) in your answer.

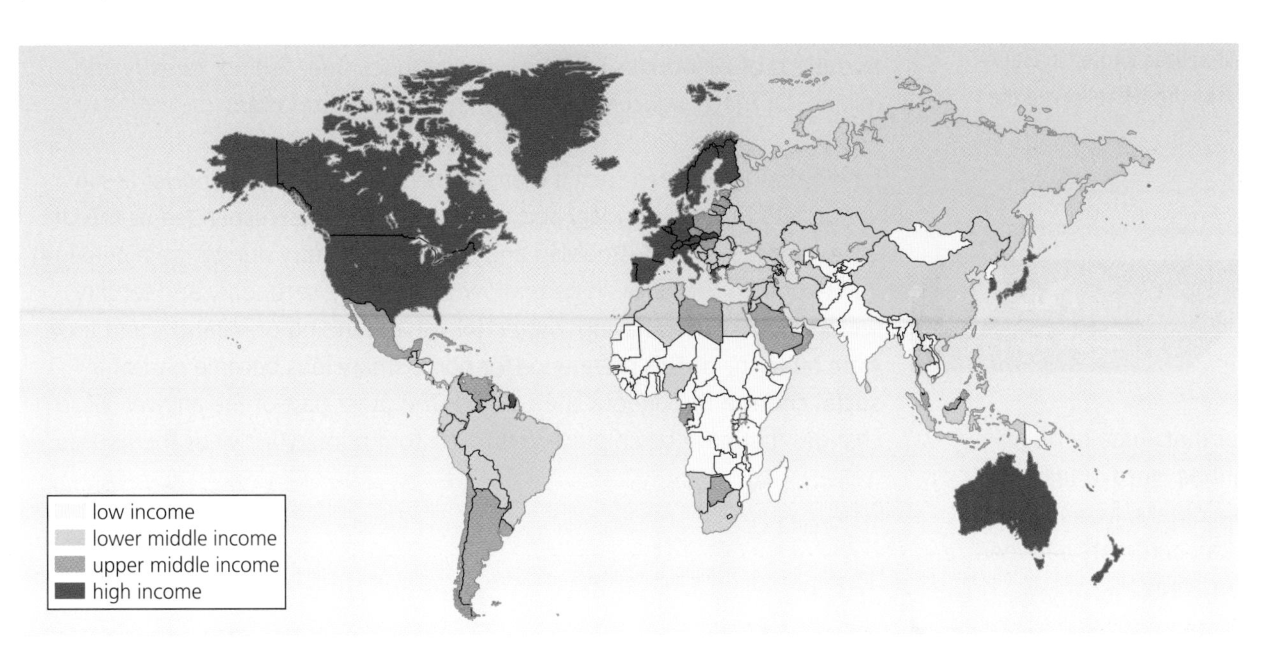

Figure 2: Global distribution of development

Figure 2 on page 157 divides countries into three broad groups, based on a measure of development known as **per capita GNI**:

- Low-income countries (LICs) – occur largely in Central Africa and in South and Southeast Asia
- Lower and upper middle-income countries (MICs) – most common in South America, North and South Africa, parts of the Middle East, Eastern Europe and Asia
- High-income countries (HICs) – found mainly in North America, Western Europe and Australasia.

In each of those three groupings, the balance of the three main sectors is different. The primary sector is the most important one in most LICs. The secondary sector is quite strong in most MICs. The tertiary sector is strongest in all HICs. To illustrate this point, Figure 3 shows the importance of the three economic sectors in Ethiopia (an LIC), China (an MIC) and the UK (an HIC).

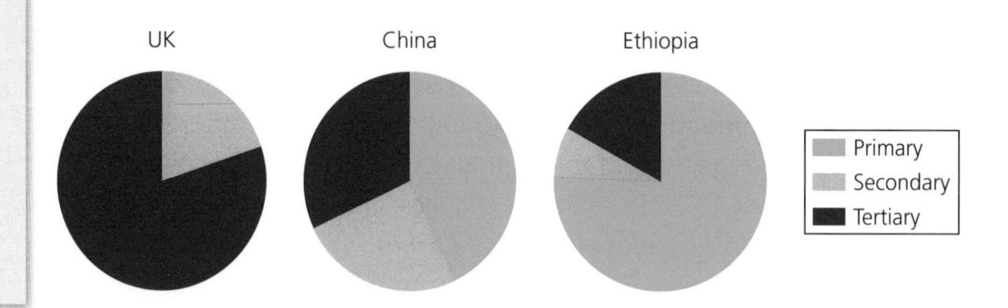

Figure 3: Balance of economic sectors in a typical LIC, MIC and HIC

Skills Builder 1

In Russia, the primary sector accounts for 11% of all jobs, the secondary sector 29% and the tertiary sector 60%.

(a) Draw a pie-chart to show these proportions.

(b) How does this pie-chart compare with those in Figure 3?

(c) Do you think Russia is ahead of or behind China along the development pathway? Give your reasons.

Activity 2

Read the first two paragraphs about the decline of the UK's primary sector again. How does the decline in this sector compare with what was shown in Figure 1 on page 156? Has the UK followed the model?

ResultsPlus Exam Tip

Remember that some parts of countries might be more or less developed than others. In many LICs, there are marked contrasts between rural and urban areas.

The decline of the UK's primary sector

The number of people working in the UK's primary sector has shrunk to less than 2% of the total workforce. The decline started about 250 years ago, at the beginning of the Industrial Revolution. At this time, around two-thirds of all workers made a living in agriculture, fishing, forestry and mining. Of these, agriculture was by far the most important.

So why has there been such a dramatic decline? Part of the answer lies in the growth of the secondary and tertiary sectors. These expanded as the UK developed into an industrialised and urbanised country during the nineteenth and twentieth centuries. In general, workers in the secondary and tertiary sectors were seen as having 'better' jobs than farming or mining – and they were better paid. This preference for non-primary jobs became part of a **social change** throughout the country. The other part of the answer lies in changes that have taken place within the four primary activities themselves.

Agriculture

The number of farm workers in the UK today is very small. There are three main reasons for this:

- The UK used to produce nearly all of its own food, although the limitations of climate have always meant that certain foods had to be imported. But today it produces less than 60% of its food, because – even with foods that can be grown here – it is now much cheaper to import some types from abroad.
- Agriculture has become highly mechanised – a single machine, such as a combine harvester, can do in a day what it used to take tens of men to do in a week (Figure 4). Livestock, particularly poultry, are reared by factory-like methods.
- Much of our farming has been taken over by **agribusiness**. Everything has been scaled up. Small family farms have been put together to make much larger and more commercial farms. Hedgerows have been removed to create larger fields that are more suitable for farm machinery. Much money is being invested in the latest machinery and in the latest technology – better seeds, pesticides, herbicides and drugs, as well as the **genetic modification** of crops.

Figure 4: A modern agricultural landscape in eastern England

Activity 3

Why is so much of the UK's food supply imported?

Fishing

Not so long ago, most settlements around the UK's coast were involved in fishing. Fish were an important part of the nation's food supply. Today, however, most of the fishing fleets have gone. Most of the total UK catch is now landed at just three Scottish ports – Peterhead, Lochinver and Fraserburgh. The decline of the fishing industry has been quite marked. It is the result of:

- Over-fishing – the depletion of fish stocks caused mostly by the **mechanisation** of fishing. Larger vessels have been using modern electronic equipment to help locate fish shoals and then trawling larger nets to catch them (Figure 5).
- Tight international controls on the amount of fish that can be caught today in order to conserve the remaining fish stocks.

Figure 5: A modern trawler in action

Mining and quarrying

Most of the mining in the UK has been for coal. Coal provided the fuel for the Industrial Revolution, and at one time there were hundreds of coal mines (Figure 6 on page 160). Today there are only five deep mines and seven surface ('opencast') workings. So why the decline?

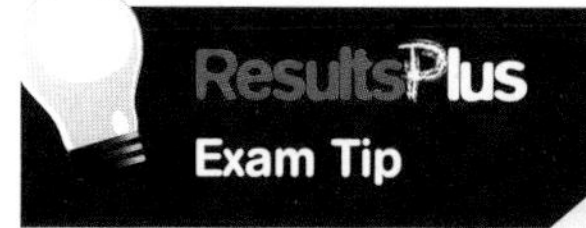

Exam Tip

A decline in employment is not the same thing as a decline in output. Industries such as agriculture are still important in terms of output, whereas coal mining is not.

Figure 6: A derelict mining landscape

ResultsPlus
Watch out!

The UK has not run out of coal. There are still large reserves left, and perhaps they will be worked one day.

Activity 4

What is to be gained from planting new forests?

Suggest reasons for the decline of primary industry in the UK. (4 marks)

Basic answers (0–1 marks)
Identify one reason, usually to do with resources running out, but the reason given is not illustrated with an example or developed.

Good answers (2 marks)
Outline a number of reasons for the decline, such as cheaper imports and mechanisation reducing employment, but do not contain examples or added detail.

Excellent answers (3–4 marks)
Offer at least two reasons in some detail with examples, figures and data added to the overall idea.

The following points help to explain the situation:

- It is cheaper to import coal from other countries, particularly from Poland and Australia.
- The collapse of large-scale traditional industries, such as iron and steel, has cut the demand for coal.
- Modern electricity generating stations are powered by natural gas rather than coal.
- It became clear by the 1950s that burning coal was not good for the environment, particularly for air quality – and now we know that it contributes to global warming.

Quarrying still takes place. Much of it produces building materials, such as sand, gravel and cement, and hard rocks which are crushed to make surfacing for our roads.

Forestry

There has been little or no decline in forestry, but in terms of the numbers employed it is the least important of the primary activities. The main employer is the Forestry Commission, which is responsible for planting new public forests and managing existing ones.

The decline of the UK's secondary sector

It is amazing to think that the UK was the world's first industrial nation. It led the Industrial Revolution. Fifty years ago, manufacturing produced 40% of the country's economic wealth and employed a third of the workforce. Today, however, it only produces about 25% of the wealth and employs less than 20% of the workforce.

Factories have closed in the UK because the things they were making can now be made more cheaply elsewhere (Figure 7). The main factors encouraging this global shift of manufacturing from its old strong holds in HICs to new locations in MICs and LICs are:

- Cheaper land and labour in the poorer parts of the world, with fewer regulations about working conditions and less concern about environmental impacts.
- Fast, efficient and cheap transport to move goods from the new factories to the UK and other major markets.
- The **global superhighway** created by modern **communication networks**. The fast transfer of information by the internet and mobile phones means that the new factories can keep in close touch with market trends in the UK. Changes in fashions and styles – as well as new orders – can be instantly relayed to the factory, whether it is in Mumbai or Shanghai.

Two other developments have also played a part. The first has been the growth of large organisations known as transnational companies (TNCs). These companies operate on a truly global scale and have a range of business interests in many countries. So they can set up a factory in China, assemble the necessary raw materials, say from Africa and Australia, and then transport the manufactured goods to the main markets in Europe and North America.

The second development is the process of **globalisation**. Most countries are becoming tied into one huge global economy. Each country has a part to play – which might be as a supplier of raw materials or cheap labour, as a wealthy consumer market or as a location inventing new technologies.

The UK has now lost most of its traditional industries, such as iron and steel, shipbuilding, textiles and pottery. The numbers employed in the secondary sector have declined, not just because of this **deindustrialisation**. Modern production methods and **automation** have meant that those goods still produced in the UK are made by machines rather than human hands.

Although the UK is deindustrialising, it is still ranked sixth in the global league table of manufacturing countries (see the table). By 2020, however, it is expected to drop two places, and by then China will head the table.

Figure 7a: Redundant factories in the UK

Figure 7b: Industrial plants in India

The world's top 15 manufacturing nations

Rank	2002	Prediction 2020
1	US	China
2	Japan	USA
3	China	Japan
4	Germany	Germany
5	France	South Korea
6	UK	France
7	Italy	India
8	Korea	UK
9	Canada	Italy
10	Mexico	Brazil
11	Spain	Russia
12	Brazil	Indonesia
13	Chinese Taipei	Mexico
14	India	Taiwan
15	Russian Federation	Canada

Source: OECD

Skills Builder 2

Look at the table of the world's top 15 manufacturing nations.

(a) Which country is predicted to move up the rankings most between 2002 and 2020?

(b) Which country is predicted to move down the rankings most?

The rise of new industrial countries

Many of the manufacturing industries that the UK has lost are now found in other parts of the world – in particular, in newly industrialised countries, like South Korea and Taiwan, and in recently industrialising countries, like China and India. New factories are being set up in these countries because the costs of producing goods are lower, which means bigger sales and profits.

Case study: Made in China

Take a look at the labels on your clothes and shoes. You will probably find that a number of them say 'Made in China'. The same is likely to be true for your TV set, mobile phone and photocopier. As we have seen, China is currently ranked as the third most important manufacturing country in the world and is famous for its **consumer industries**. For example, China produces:

- Half of all the world's clothes
- Two-thirds of the shoes
- One-third of the mobile phones
- Two-thirds of the photocopiers
- Half of the microwave ovens.

But China has many other industries too – from iron and steel to chemicals and fertilisers, from ships to aircraft, and from military equipment to space satellites. Thirty years ago, there were far fewer industries in China and 'made in China' products were scarce outside the country. So what has caused this spectacular 'industrial explosion'? Figure 8 identifies six major factors. One of these has been particularly important – the political shift from **communism** towards **capitalism**.

Thanks to its industries, China now has the third largest economy in the world. Perhaps this is not too surprising because it is a vast country with a huge population. Figure 9 shows the rate at which its economy grew between 1995 and 2005. Clearly, the annual rate of growth was generally higher than in the other four rival countries.

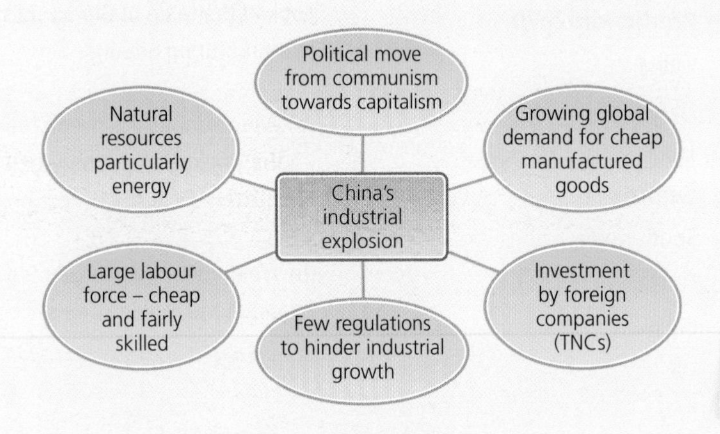

Figure 8: The factors behind China's industrial explosion

Case study quick notes:

- Industrialisation can take place very rapidly, particularly when governments and businesses work together.
- Industrialisation can happen quickly if a country has plenty of natural resources and cheap labour.
- Industrial success brings both benefits and costs.

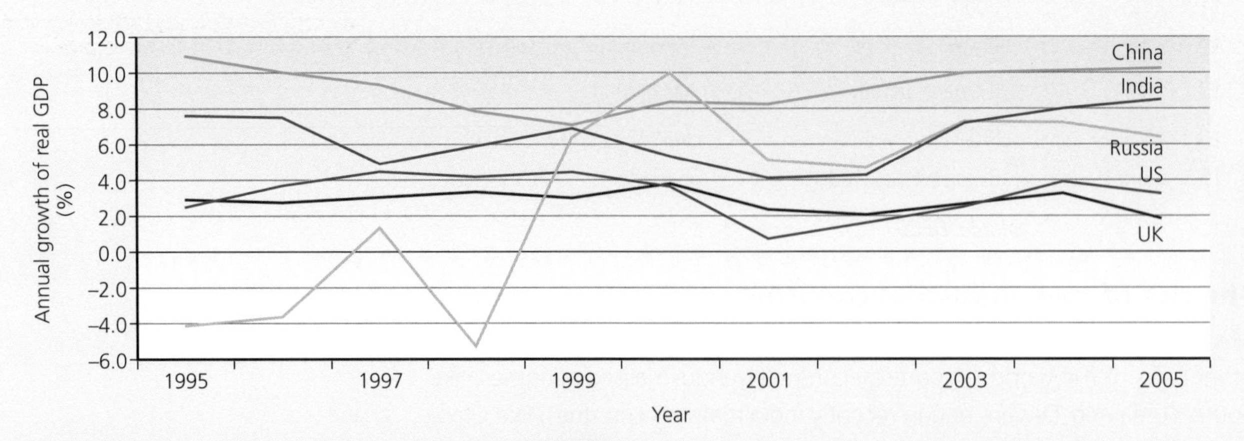

Figure 9: Rates of economic growth in five countries, 1995–2005

Becoming an industrial nation brings many benefits. These include:

- Rising incomes and a higher standard of living for its people (at least those living in towns and cities)
- Improved working conditions for some, but not all
- A healthy economy with exports exceeding imports
- Better housing in urban areas
- More influence in global affairs.

But industrialisation also has its downside:

- Damage to the environment caused by the working of natural resources, such as coal, oil and iron ore
- Having to find new sources of raw materials as reserves begin to run out
- Much pollution of air, water and land by factories and power stations
- Massive rural-to-urban migration as people leave the countryside for factory jobs in towns and cities
- A widening gap between the rich and the poor.

Growth of the tertiary sector

The tertiary sector, which provides a wide range of services, is by far the most important part of the economy in all HICs, accounting for more than two-thirds of the jobs. Why have these services become so important? The short answer is 'development'. As a country moves along the development pathway several things happen:

- It is able to afford more and better social services, such as schools, medical centres, hospitals and libraries.
- People earn more and have money to spend in the shops on 'basic' things, such as food and clothing.
- After they have bought the 'basics', people have more money left (**disposable income**) to spend on 'luxuries', such as entertainment, holidays, eating out and recreation.
- Technology creates or makes possible new services. Think of all those new services connected with ICT – broadband service providers, website designers and the servicing of PCs and laptops.

Skills Builder 3

Study Figure 9 on page 162.

(a) In which year was China's rate of economic growth highest?

(b) Which country recorded a higher rate of growth than China?

(c) In which years?

(d) In what ways are the graphs for China, India and the UK similar?

Activity 5

Why has there been concern about the safety of goods made in China?

You may come across the term 'service industries'. The name might suggest that they are part of the secondary sector. They are not. They are simply services and they are part of the tertiary sector.

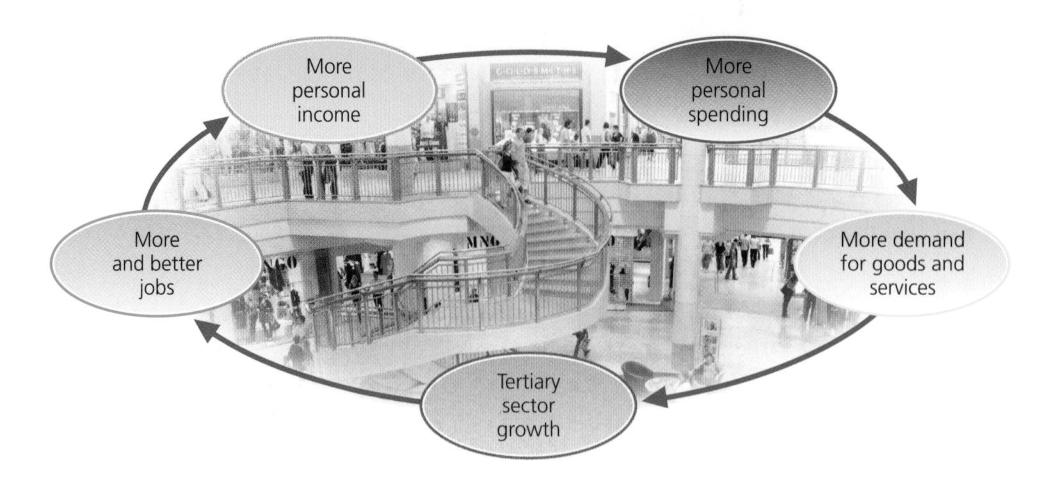

Figure 10: The cycle of growth in the tertiary sector

Figure 10 shows the cycle of growth within the tertiary sector. With development comes an increased demand for a variety of goods and services. This leads to the tertiary sector growing and the creation of more and better paid jobs. That, in turn, means more income and more personal spending on services. This increased demand encourages the provision of even more goods and services. And so another round of growth is started.

There is another reason for the 'rise' of the tertiary sector, particularly in the UK. Its percentage importance in the country's economy has increased even further as the primary and secondary sectors have declined.

Over the last fifty years, as the tertiary sector has grown, its services have changed:

- There are more of them, as the population and disposable income have increased. Think of the many new supermarkets that have sprung up in and around towns and cities.
- Advances in technology have led to new services – as illustrated by ICT.
- People's tastes have changed – cinemas have closed because many people prefer to watch DVDs at home.
- The age structure of the population has changed as people live longer. This has created a bigger demand for services for elderly people – care homes, day centres and Saga holidays.

Explain why tertiary employment has grown in most HICs in recent years. (6 marks)

■ **Basic answers** (Level 1)
Identify one reason why tertiary employment has grown and usually give the 'attractiveness' of these jobs as the reason.

● **Good answers** (Level 2)
Explain one reason for the growth and offer some evidence to support the idea.

▲ **Excellent answers** (Level 3)
Provide at least two reasons with good detail supporting them. The reasons might cover the growth in ICT, disposable income and growth of holidays, with useful data to support them.

The 'grey pound'

The population of the UK is becoming 'greyer'. Fifteen per cent of the population are over 64 years old, compared with 5% in 1911. The rate of spending amongst retired people is expected to rise faster than in any other age group. Far from staying at home with a cardigan and slippers, the 'Ski-ers' (so-called because they are Spending the Kids' Inheritance) are keen shoppers and travellers. There are now signs that businesses are beginning to wake up to the growing power of the 'grey pound':

Saga	Saga is now a multi-million pound business (Figure 11). It started as a retirement holiday business, offering cut-price winter breaks to seaside towns, but now it earns most of its money from car, holiday and pet insurance, sold to people aged over 50.
B&Q	DIY giant B&Q has recently started selling a new range of products for the over-50s that includes Stannah Stairlifts, easy-grip electric plugs, special garden tools and remote controls for a range of electrical goods. It is the first time a major UK retailer has offered such a wide range of products for older or infirm shoppers.
L'Oréal	L'Oréal, the beauty products manufacturer, unveiled Jane Fonda, at the age of 70, as the new face of its television advertising, while Marks & Spencer uses the famous 1960s model Twiggy – 60 in 2009 – to sell its clothes.

Just remember, however, that not all pensioners are 'Woofs' (well-off older folks). Inadequate state pensions mean that there is also much poverty. This has encouraged the growth of charities concerned with older people, such as Age Concern, Help the Aged and Age Care. They too are part of the tertiary sector.

Figure 11: Passengers about to board a Saga cruise

Activity 6

Read the 'grey pound' section.

(a) Create a table with two columns headed 'Goods' and 'Services'. Record the goods and services that you think are specifically for the elderly.

(b) Which two goods and which two services do you think would help most to improve the lives of pensioners with little money. Give your reasons.

Exam Tip

Remember that there are positive as well as negative impacts of an ageing population. Try to cover both in your answers.

Quick notes (the 'grey pound'):

- Changes in the age structure of a population can alter the demand for goods and services.
- The 'grey pound' is becoming increasingly strong.
- Not all pensioners are well off and able to take advantage of this increasing number of goods and services.

Objectives

- Learn the main factors affecting the location of economic activities.
- Explain why economic locations change over time.
- Understand the costs and benefits of deindustrialisation in rural areas.

Skills Builder 4

Study Figure 12.

(a) Can you identify one locational advantage of the power station?

(b) Suggest one locational disadvantage.

Figure 12: The gas turbine power station in Shoreham, Sussex

Activity 7

How might the location factors of a nuclear power station differ from those of a thermal power station?

Economic locations

Location factors

Nearly every economic activity – whether it is a quarry, a factory or a shop – is found in a particular place for good reasons, not by chance. Those reasons are called **location factors**. They are, if you like, the 'needs' of the activity. Those needs, the location factors, vary according to the type of economic activity. Let us look at two examples, one from the secondary sector and the other from the tertiary sector – a power station and a supermarket.

Case study: Locating a thermal power station

A thermal power station converts fuels, such as natural gas and coal, into electricity. Fuels are burnt to convert water into steam, which is then used to turn the generators that create the electricity. These basic facts draw our attention to two location needs – fuel and water. Power stations are huge buildings and need much space.

Not so long ago, nearly all power stations were burning coal. Because coal is bulky and costly to transport, power stations tended to be located on or close to coalfields. More recently, power stations were also burning oil. This was imported from either the North Sea or overseas, so it made good sense to locate on or near the coast. A coastal or estuary location also met these older stations, need for cooling water. Once steam has passed through a station's generators, it has to be cooled back into water so that it can be reheated to make more steam. Water taken from the sea or an estuary is used for this purpose but the cooling process needs special structures and therefore space.

Since the 1990s, there has been a 'dash for gas'. Gas-fired stations are now generating around 40% of the country's electricity. These stations are relatively small and cheap to build, and they need less storage space and emit less carbon dioxide than coal or oil. Shoreham power station, on the Sussex coast, was built on the site of an old coal-fired station that was closed in the late 1970s (Figure 12). It is supplied with gas from a nearby port depot. The water it uses for steam is cooled by an air condenser which covers an area the size of one and a half football pitches. Much of the water is recycled, and the nearby River Adur can provide the necessary 'top-up' water.

Most of the electricity generated at power stations is used in our towns and cities. But because electricity is easily transported by power lines over long distances, there is no need for power stations to be located close to where it is eventually used.

Quick notes (locating a thermal power station):

- There are two types of location factor – physical (e.g. land, water) and economic (e.g. cost of land, fuel efficiency, markets, etc.).
- The location factors of an economic activity can change over time, particularly as a result of new technology.

Case study: Locating a supermarket

New supermarkets are still springing up around the edges of our towns and cities. Often they are part of big retail parks or shopping complexes (Figure 13). But why are they being built there and not in the centres of towns and cities – where most big shops used to locate? One important reason is that supermarkets require a lot of space – space inside to display the thousands of different goods on open shelves, and space outside to provide hundreds of car-parking spaces. This all takes up a lot of land, and land is much cheaper on the edges of urban areas than in their centres.

Another reason for this location is that supermarkets want to be reached by as many customers as possible. Being on the edge of town will often mean that the supermarket can be fairly easily reached by people from nearby towns. Urban centres suffer from traffic congestion and car-parking charges, and that tends to put would-be shoppers off. At one time, if you didn't have a car, you couldn't use these out-of-town or edge-of-town supermarkets. Nowadays, however, bus services are provided to many of them.

Figure 13: An aerial view of The Trafford Centre – an edge-of-town shopping centre

Deindustrialisation in rural areas

We perhaps think that industry is only found in towns and cities. Two hundred years ago, a number of different industries could be seen in the countryside. They included the making of woollen cloth, flour-milling, lime-making, brick-making and brewing. Many small mines were scattered around the countryside. They were working small deposits of minerals, such as iron, lead, tin, copper, silver and even gold. The first coal was mined in rural areas. Because of their dangerous nature, some factories – making gunpowder, fireworks and guns, for example – were deliberately located in the countryside, away from people.

Activity 8

1. Read the case studies again and make a grid like this. Identify the factors in each case study and write them in the correct place in your grid.

	Thermal power station	Supermarket
Physical factors		
Economic factors		

2. What evidence is there that the locational needs of both activities have changed over time?

3. Draw a diagram to show the factors which account for the location of your school. Consider factors such as site, transport, cost of land, catchment areas of pupils, etc.

Quick notes (locating a supermarket):

- The location of much retailing has changed, moving from urban centre to urban fringe.
- The reasons for this shift are linked to accessibility, the rise of big supermarkets and lifestyle changes.

Activity 9

Investigate a named rural area near to where you live using old and new photographs, OS maps and land-use maps. What evidence can you find that there used to be industrial activities in the area?

Skills Builder 5

Study Figure 15 on page 169.

What is the map evidence for saying that the old gravel workings are located on a flood plain?

Activity 10

Here is a chance to see whether you can use what you learnt earlier in this chapter.

1. Add a third column to the location factors grid that you produced for the Activity on page 167, and analyse the factors affecting the location of gravel working.

2. Make a new grid with two columns, labelled 'Costs' and 'Benefits', and two rows, labelled 'During the working of gravel' and 'After working has finished'. Fill in the grid with your notes. At each of the two stages (during and after), which do you think are greater – the costs or the benefits?

Very few of these traditional industries are still going. Mineral deposits are either exhausted or no longer economic to work. Farmers today use chemical fertilisers rather than lime. Flour-milling and brewing are now concentrated in a few large units, mostly in or near urban areas. The countryside is always changing. As part of these changes, old industrial buildings are either demolished or put to new uses. The table below lists some of the things that you might look out for in the countryside as reminders of these former industries.

Landscape feature	Former industrial activity
Water mill	Power for grinding cereals into flour; power for the woollen industry
Lime kiln	Burning limestone to make agricultural fertiliser
Tip or spoil heap	Working of a localised mineral
Dry pits	Extraction of sand (for building) and clay (for brick-making)
Windmill	Power for grinding cereals into flour
Brick kiln	Brick-making

Deindustrialisation, therefore, is not something that only occurs in towns and cities. In the countryside, it has been a mixed blessing. There have been costs and benefits.

Costs	The loss of jobs in rural areas.
	The break-up of rural communities, as people move to towns and cities to find work.
	Derelict industrial buildings and disused quarries scar the landscape.
	The need to clean up old industrial sites – demolishing old buildings, filling in old pits and removing toxic waste.
Benefits	Less environmental pollution.
	Old industrial buildings that can be made into tourist attractions.
	The opportunity to remove ugly industrial buildings from the landscape.
	The chance to return land to farming (reagriculturalisation) or forestry – or to create new wildlife habitats.
	The opportunity to use **brownfield sites** for new housing.

Despite deindustrialisation, there is still industry to be found in rural areas. But it is not traditional rural industry. Today it is more likely to be a power station, a cement works, or a water supply reservoir, servicing the needs of urban areas.

Old gravel pits

Gravel is one of the many materials (like bricks, cement and sand) used in vast quantities to build our towns and cities. Deposits of gravel are widespread in rural areas, particularly on flood plains. They are extracted from open pits, which are often waterlogged. The gravel is literally dredged from beneath the water table. Generally speaking, the deposits nearest to expanding urban areas were the first to be worked. Since the deposits are usually fairly thin, they quickly become exhausted. So as one pit is finished, extraction moves on to another (Figure 14).

Figure 14: A recently abandoned gravel pit

But what should happen to the abandoned pit? There are three options:

- Just leave the pit as it is, and let it become a wildlife area.
- Make the pit into a water sports area (boating, fishing, etc.) for use by local people.
- Fill it up with rubbish and rubble, and convert it into usable land – perhaps for housing.

Quick notes (old gravel pits):

- There are costs and benefits during the working of gravel.
- There are costs and benefits after the gravel has been worked out.

The last option used to be a popular choice. Disposing of the huge amounts of household rubbish we create every day is a real challenge. Using it to fill old pits was a chance not to be missed. However, we now know that tipping waste into a water-filled hole can easily cause serious pollution, because chemicals seep out into the local water table. The decay of household rubbish in pits also creates methane which is a hazardous gas. For this reason, today many pits remain unfilled, and are managed as either wildlife reserves or water parks. In the Cotswold Water Park, a network of 140 old gravel pits has been converted into an important recreation area (Figure 15). Activities include angling, birdwatching, camping, canoeing, paddling, sailing and swimming.

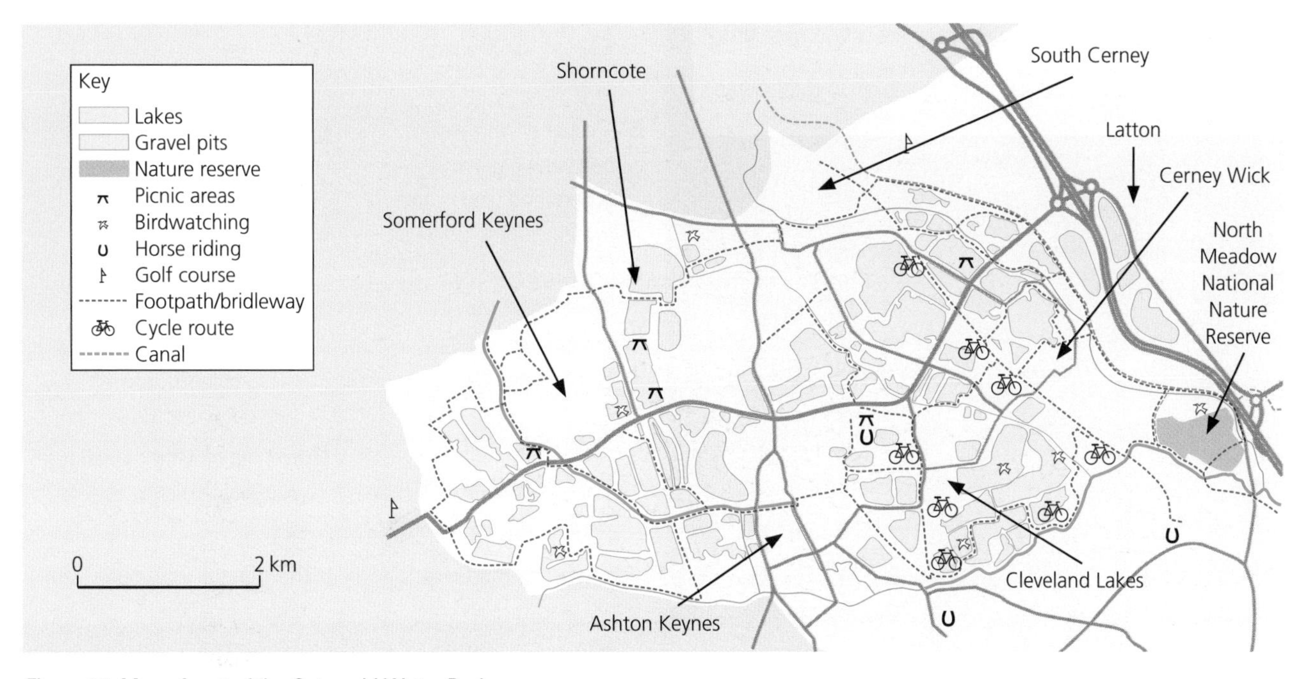

Figure 15: Map of part of the Cotswold Water Park

Know Zone Chapter 9 Economic change

The rate of economic change is quite startling. More people work in Indian restaurants in the UK than in the shipbuilding, coal and steel industries put together. These changes impact on all of our lives.

You should know...

- ☐ The meanings of primary, secondary and tertiary activity
- ☐ The relative importance of these sectors in countries at different stages of development and how they have changed
- ☐ Why the primary sector has declined in the UK, including changing technology and the role of imports
- ☐ Why the secondary sector has declined in the UK, including globalisation and improved global communications
- ☐ A case study of the growth of manufacturing industry in **one** LIC or **one** MIC
- ☐ The reasons why the tertiary sector has grown so fast in the UK
- ☐ Why location factors vary from economic activity to economic activity
- ☐ Why economic activities may shift their location over time
- ☐ The costs and benefits of deindustrialisation in rural areas

Key terms

Agribusiness
Automation
Brownfield site
Capitalism
Communication networks
Communism
Consumer industries
Deindustrialisation
Development
Disposable income
Economy
Genetic modification
Global superhighway
Globalisation
Gross national income (GNI)
Industrialisation
Location factors
Mechanisation
Primary sector
Quaternary sector
Secondary sector
Sector shifts
Social change
Tertiary sector

Which key terms match the following definitions?

A The process, led by transnational companies, whereby the world's countries are all becoming part of one vast global economy

B A piece of land that has been used and abandoned, and is now awaiting some new use

C Any large company involved in food production, including farming, seed supply, agrochemicals, farm machinery, food processing, and food marketing and retailing

D The manipulation of the genetic material of a plant or animal to produce desired traits, such as nutritional value or resistance to herbicides

E The replacement of human (or animal) labour with machines

F The move from an economy dominated by the primary sector to one dominated by manufacturing (the secondary sector)

G Something which affects where a particular activity is found

To check your answers, look at the glossary on page 289

Foundation Question: Study Figure 1 on page 156.
i. Describe the changes in primary employment shown. (2 marks)
ii. Give two reasons for the changes shown. (4 marks)

Student answer (awarded 1 mark + 2 marks)	**Examiner comments**	**Build a better answer** (awarded 2 marks + 4 marks)
i. It goes down all the time to less than 10%. *ii. The mines have run out and we have had to import stuff instead.* *In some industries like farming tractors have replaced people.*	*i. It goes down...* This is true but it goes down at different rates. It is sensible to offer some data and time periods in order to get high marks. This scores 1 mark only. *ii. The mines have...* scores 1 mark for the idea of imports. This statement is true of some mines but not others. The student should state which mines they mean, such as coal, tin or gold. *In some industries...* receives only 1 mark. *Like farming* does not help because tractors are a very specific example of mechanisation, which is the point the candidate is illustrating.	i. Primary employment starts at over 70% in the pre-industrial period. It then falls steeply in the early industrial period before flattening out a little but still declining to less than 10% in modern times. ii. Some resources such as iron ore and coal are expensive to mine in the UK. It is now cheaper to import. In industries such as agriculture, labour has been replaced by machinery (such as tractors). So employment has fallen as a result of mechanisation.

Overall comment: It is important to follow command words in exam questions – students often make mistakes with command words such as *describe* and *explain*.

Higher Question: Study Figure 1 on page 156.
i. Define the term tertiary sector. (1 mark)
ii. Suggest reasons why tertiary employment increases as countries become richer and more developed. (4 marks)

Student answer (awarded 1 mark + 1 mark)	**Examiner comments**	**Build a better answer** (awarded 1 mark + 4 marks)
i. The tertiary sector is made up of services for people. *ii. As countries become richer people want to do cleaner jobs.* *They are more educated as well and can do jobs like accountants and lawyers rather than farming which is dirty and tiring.*	*i. The tertiary sector is...* This is correct although the student could have provided a clearer answer. 1 mark is awarded. *ii. As countries become...* This may be true of a small number of people but many do not have this choice as manufacturing and primary jobs have disappeared. *They are more...* scores only 1 mark. These examples are highly paid and quite unusual jobs in the tertiary sector. The point here is about 'education' but the student returns to the theme of *dirty* jobs.	i Tertiary activities provide services, unlike secondary and primary activities which produce goods. ii As people have got richer they have more to spend their money on. This creates jobs in things like shops, bars and the tourist business. They also have more time now for things like holidays and outings. There has been a real growth in financial services like banking. These jobs create other tertiary jobs such as taxi drivers, shop workers, etc.

Overall comment: The student could have improved their answer by considering the limits to people's choices as other industries have declined.

Chapter 10 Farming and the countryside

Objectives

- Learn the more important changes and their consequences.
- Explain the reasons for these changes and their consequences.
- Understand that the changes and their consequences vary from place to place and over time.

Changes to the UK countryside

Change is everywhere – and that includes the UK's countryside. 'Change' is a big umbrella idea, under which many things happen. The nature of countryside change varies from place to place. The most important factor is distance from a major city. Because of this, it is possible to recognise four different types of countryside beyond the urban area, as shown in Figure 1:

- The **urban fringe** – This type of countryside is being quickly lost to urban growth, and will not be discussed much in this chapter.
- The **commuter belt** – This is countryside, but the settlements within it are used as dormitories by urban-based workers and their families.
- The **accessible countryside** – This is beyond the commuter belt, but within day-trip reach. Still very much a rural area.
- The **remote countryside** – This takes the best part of a day to reach from a city. Almost totally rural.

Figure 2 shows the distribution of these four types of countryside in England and Wales.

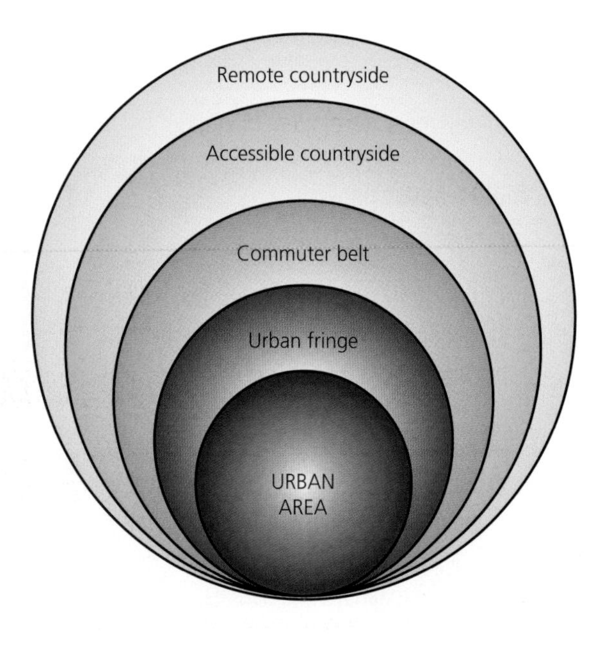

Figure 1: The four types of countryside outside the urban area

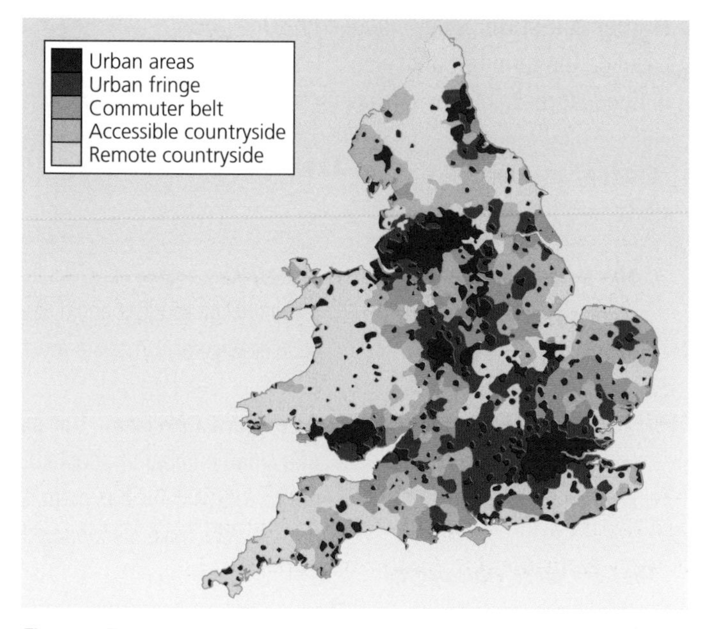

Figure 2: The distribution of the four types of countryside in England and Wales

Change within each type of countryside falls under four main headings:

- Population – growth and decline, migration, changing age structures
- Economic – decline in primary sector, farm **diversification**, growth of tourism, urbanisation and **counterurbanisation**
- Social – more leisure time for recreation and tourism, longer commuting, growth of second homes, retirement, counterurbanisation
- Environmental – loss of countryside to the urban built-up area, conservation and protection, pollution.

The decline in primary employment

The most important activity in this sector is agriculture. But Figure 3 shows that it has declined as an employer. Mechanisation has cut the number of jobs. Although the amount of land being farmed has decreased, what it produces has increased. Agriculture has also declined in terms of the part it plays in the UK's economy, for two reasons:

- The UK is now importing more than 40% of its food from other countries.
- The goods produced by the secondary and tertiary sectors are more costly than food products and therefore they contribute more to the economy.

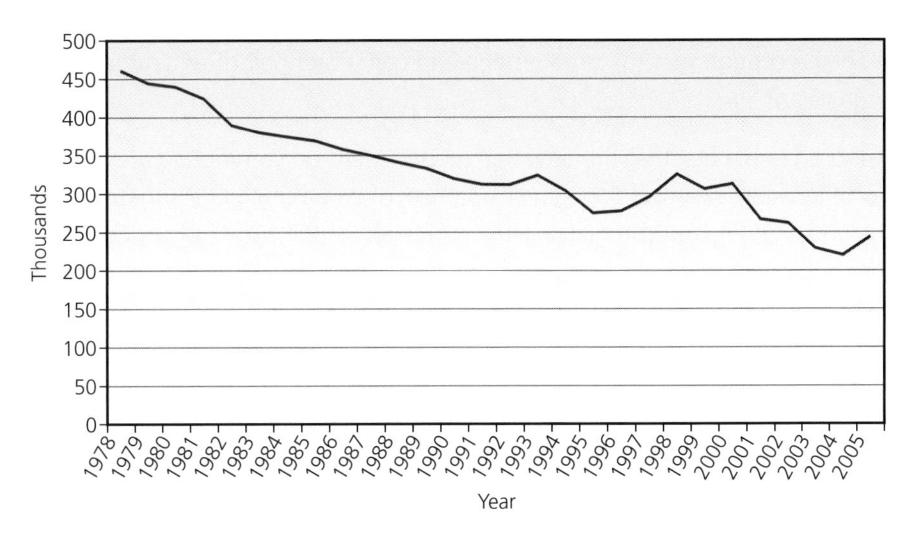

Figure 3: Employment in agriculture in the UK, 1978–2005

Two of the other activities in the primary sector – fishing and mining – have also declined. Only forestry has not changed very much. (The decline in primary emplyment is discussed in Chapter 9.)

Rural–urban movement

The decline in primary employment – especially in agriculture – is bound to impact on the countryside. In addition there is the deindustrialisation of the countryside (see Chapter 9). With the number of jobs falling, many people have chosen to leave the countryside, and have found work in towns and cities.

Skills Builder 1

Study Figure 2 and refer to an atlas.

(a) Where are the main areas of remote countryside?

(b) What are the physical features of these remote areas?

ResultsPlus
Build Better Answers

Study Figure 3, which shows the decline in UK employment in agriculture (1978–2005). Explain the decline in employment in UK agriculture. (4 marks)

■ **Basic answers** (0–1 marks)
Describe the changes but do not offer any explanation.

● **Good answers** (2 marks)
Offer one explanation, probably mechanisation, but do not provide any evidence or detail.

▲ **Excellent answers** (3–4 marks)
Provide at least two ideas, possibly including mechanisation and cheaper imported food, with some detail and evidence to support the ideas.

Skills Builder 2

Study Figure 3.

(a) How many people were employed in agriculture in 1978 and in 2005?

(b) What was the percentage change in agricultural employment between 1978 and 2005?

ResultsPlus
Exam Tip

Good quality examination answers recognise that migration is a two-way process, with people both arriving and leaving areas.

ResultsPlus
Watch out!

Remember that not all the people setting up home in the urban fringe and commuter belt have 'come out' from the city. Significant numbers are 'coming in' from accessible and remote areas of the countryside – to move nearer to the city, but not into it.

Skills Builder 3

Calculate the straight-line distances from London of the furthest commuting points shown in Figure 4.

This **rural–urban migration** has also been encouraged by:

- Higher wages and more opportunities in urban areas
- The availability of more and better services
- The perception that towns and cities offer a better quality of life.

As a result, the remote and accessible types of countryside have lost population – particularly people of working age.

The spread of commuting

For a long time now, many people have been moving from the countryside into urban areas. But now there are also movements in the opposite direction. People are leaving the cities and towns and buying homes in the countryside. But they continue to work in the city or town they have just left – they become **commuters**. This is happening in the urban fringe and in the commuter belt. The reasons for this outward movement include:

- The attraction of cheaper and more spacious housing.
- The availability of fast transport to the place of work. Most jobs are either in the city centre or in new industrial estates and business parks in the urban fringe.
- Fears about personal security in large urban areas.
- The perception that the urban fringe and commuter belt offer a better quality of life.

In short, people feel that the time and money spent on commuting are worth it. Figure 4 shows the complex pattern of commuting in south-east England. From it, you can gain a clear impression of the extent of London's commuter belt.

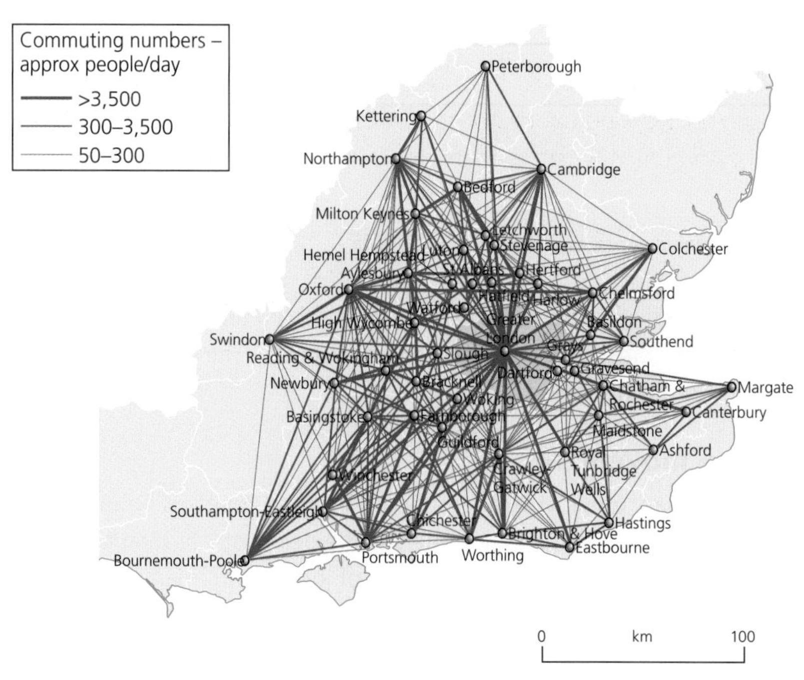

Figure 4: Commuting in south-east England

Retirement moves

It is common for people, when they retire, to move home. They do this for a number of reasons:

- It is no longer necessary for them to live close to what was their place of work.
- To downsize into a smaller home.
- To sell their home for something cheaper and use the difference in price as some sort of pension.
- To move into a quieter, calmer and more attractive environment.

The last point helps to explain why many retired people move away from cities. The dream for many is to retire to a small cottage in the countryside or to settle in some picturesque village. In reality, that often does not work out (see Chapter 13).

There are other movements of people involving the countryside. These are part of counterurbanisation (see Chapter 11). However, it is important to remember that movements of population are usually two-way. For example, while people may be moving out of the remote countryside and heading for towns or cities, there will be others heading in the opposite direction. This is true too for the other three types of countryside.

Leisure, recreation and tourism

One of the features of twenty-first-century living is that many people in the UK have spare time and **disposable income**, and both are being 'spent' on **recreation** and **tourism**. Recreational needs are being met in the urban fringe, commuter belt and accessible countryside. In the urban fringe and commuter belt, you will find sports centres, playing-fields, golf courses and country parks. In the accessible countryside, the facilities may be somewhat different. In order to reach them, city people will have to think in terms of a day-trip rather than just a short drive in the car. Maybe it will be a farm visit, a fun day at a theme park, some birdwatching around a nature reserve or an active day at a water park. Tourism is a booming business nowadays, thanks to shorter working hours, holidays with pay and modern transport (see Chapter 14). And this is where the remote countryside comes into its own.

The majority of our spare time is spent at home – leisure-time use of the countryside is not as great as you might think. People in Britain enjoy various indoor and outdoor activities. A recent survey found that the British spend about 45% of their free time watching television, 24% socialising with friends and family, 22% on sport and hobbies, and 9% on other activities. Other popular leisure activities are listening to the radio and music, reading, DIY, gardening, eating out and going to the cinema.

It is widely thought that most retirement moves are from cities and towns to the countryside. In fact, most retired people end up elsewhere (see Chapter 13).

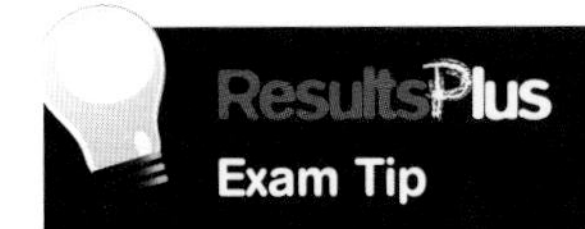

Good exam answers recognise that the decisions people make about moving involve both push and pull factors.

Skills Builder 4

Using the data in the final paragraph of this page, draw a pie-chart to show how the British use their leisure time.

The consequences of the changes in the countryside

In this section, we will look at the consequences of the changes just described. In doing this, it is important to remember the four types of countryside shown in Figure 1 (page 172).

Suburbanisation

The obvious outcome of the rural–urban movement of people is the building of new homes. They are most likely to be built in two locations – the urban fringe and the commuter belt. In the former, housing is mainly added to the outer edge of the city's built-up area. In the commuter belt, existing towns and villages become encircled by new housing estates. But the occupiers of all these new homes will need more than just a roof over their heads. They will require services, such as shops, schools and medical centres. So the building of new homes leads to a further round of building to provide these and other services.

In the urban fringe, the demand for both housing and services is increased by people moving out from the city. As Figure 5 shows, the urban fringe offers a number of benefits that attract people, businesses, shopping centres and recreational uses.

Around London and other major cities, growth in the urban fringe has been discouraged by the creation of 'green belts'. Within such belts, protection of the countryside has been the priority and planners have tried to keep **suburbanisation** to a minimum. Instead, population movements have been encouraged to new towns and expanded towns located further out in the commuter belt.

Activity 1

Which of the urban fringe benefits shown in Figure 5 do you think is most important to:

(a) a major retailer

(b) a local resident.

Give your reasons.

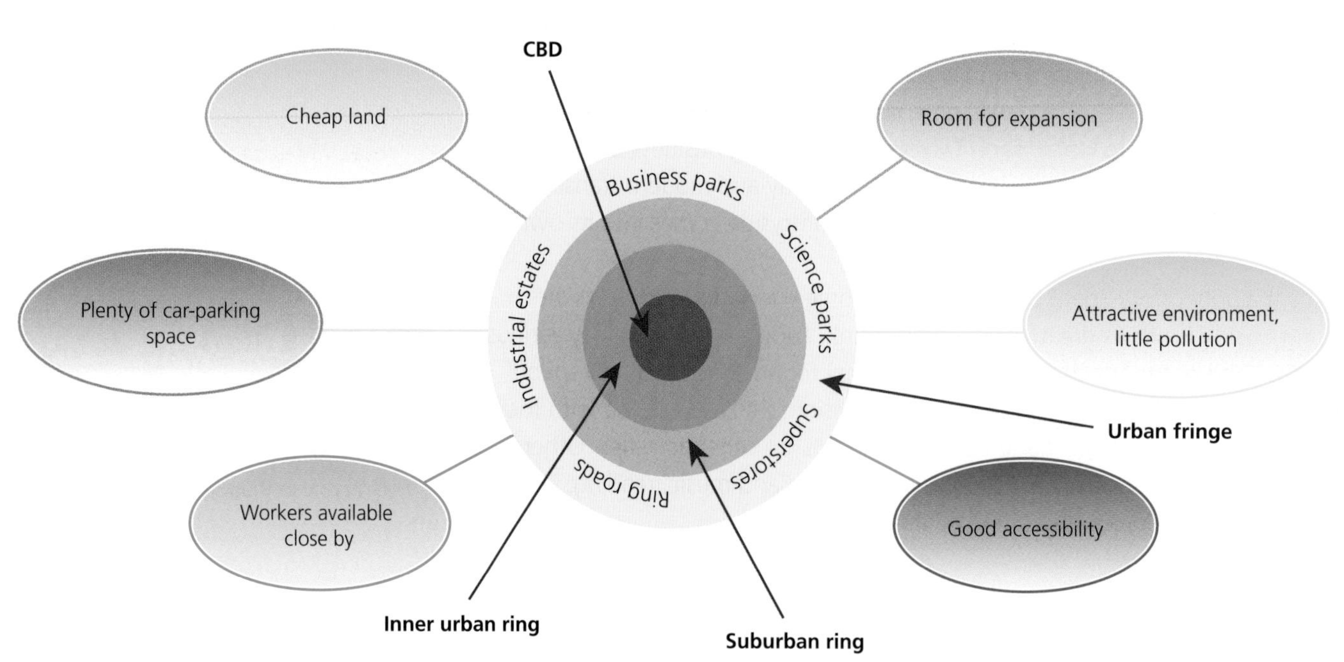

Figure 5: The benefits of the urban fringe

Changing population structures

The movement of people after they have retired from work has important consequences. This is especially true if those movements converge on particular places, such as 'honeypot' villages (discussed below).

Figure 6 shows that retired people account for more than 18% of the population over quite large areas of Great Britain. The age structures of these popular retirement areas become unbalanced (see Chapter 12), with the age pyramids tending to bulge towards the top end. Elderly people have particular needs. For housing, they require small bungalows and flats, sheltered accommodation and care homes. For social services, they require day centres, clinics, occupational therapy and so on. It is the local authority's job to make sure that the right sort of housing and services are made available.

In the urban fringe and commuter belt, the age structure is rather different. The influx of people of working age, often with young families, means that the population pyramids are broad at the base and in the middle age range.

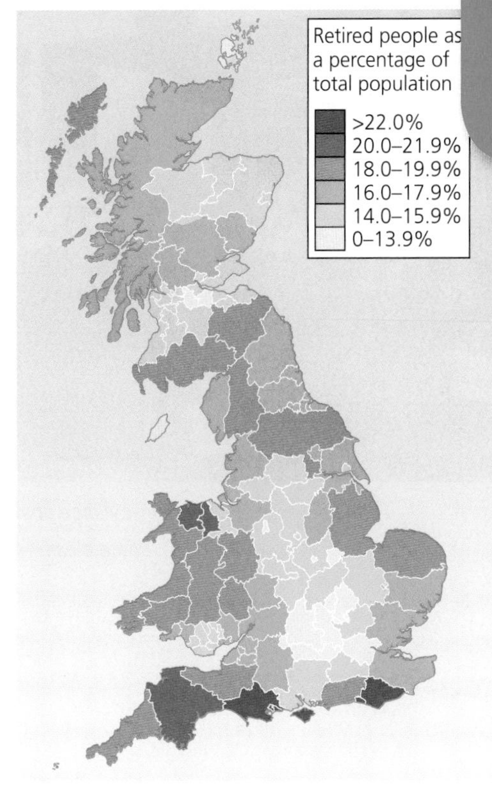

Figure 6: The distribution of retired people in Great Britain (2001)

The spiral of decline

There are two ends to every population movement. So far, we have looked mainly at the consequences at the destination end. Let us now look at the starting end, and in particular at what the loss of population means to remote countryside areas. It is usually those of working age who leave first. When they and their families are gone, the demand for local services (shops and schools) falls and they close. The loss of services makes the area even less attractive. More people then choose to leave. As Figure 7 shows, there is a chain reaction which leads the rural area into even faster decline.

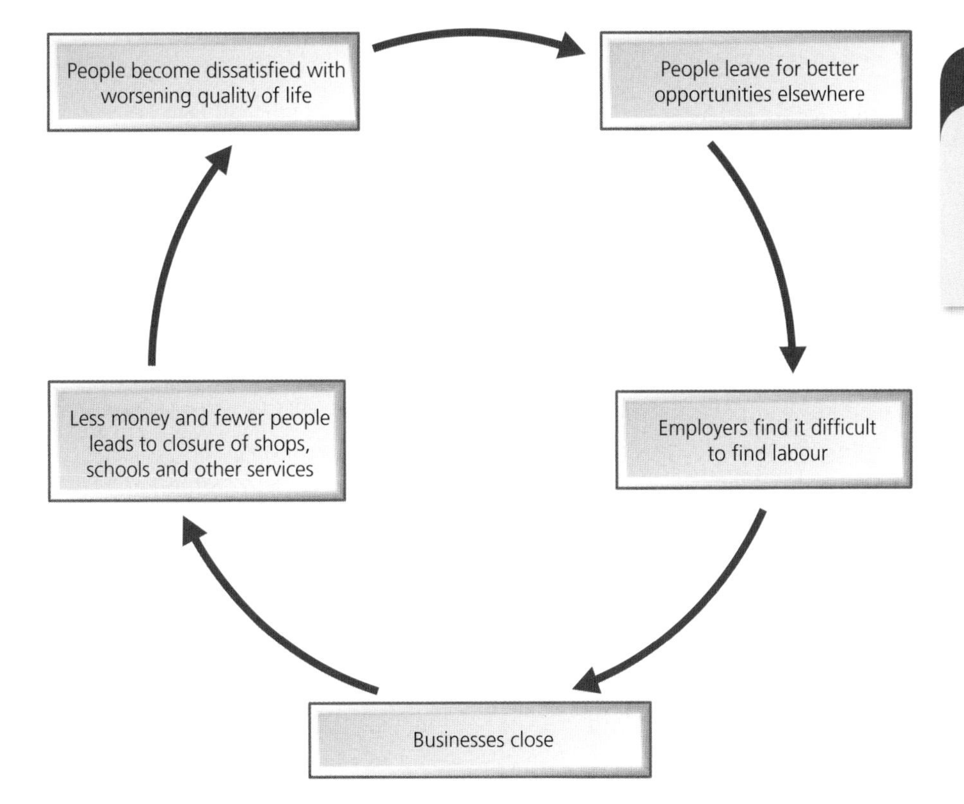

Figure 7: The spiral of decline in rural areas

Skills Builder 5

Study Figure 6 and, using an atlas, identify the main areas of rural Britain where less than 18% of the population are retired people.

Figure 8: A campaign against village post office closures

Village shop closures

A survey in 2000 found that 3,000 British villages, with a combined population of 4.5 million, had no shop and no post office. One-third of British villages were dependent on neighbouring towns for even basic goods. At that time, village post offices were closing at the rate of 400 a year. The Post Office announced in 2008 the closure of a further 2,500 of their branches – many of them in rural areas. About half of these double as local shops (Figure 8).

It is now extremely difficult for a specialist shop, such as a butcher or baker, to survive. Shops must sell a wide range of goods if they are to compete with the supermarkets and larger chains of stores in neighbouring towns. Even in some expanding commuter villages shops are closing because they are not being supported by local residents. Those with cars often prefer to drive to edge-of-town supermarkets. Goods may be cheaper there, but the costs of driving there and back are often forgotten.

The closure of village shops is an important factor in persuading village residents to move home. This is particularly true for those – particularly the elderly and not so well off – who do not own a car and have to rely on poor public transport. In a number of villages, residents have stepped in and taken over the threatened post office. In most cases, they have received grants from the Department for the Environment, Food and Rural Affairs (DEFRA).

Quick notes
(village shop closures):

- The closure of village shops and post offices is happening at a fast rate.
- The closures help persuade people to move out of villages altogether.

Honeypot villages

The growth of tourism in the countryside is not evenly spread. It tends to be focused on what are called **honeypots** – places which offer something that attracts large numbers of tourists. It might be because of:

- The immediate surroundings – e.g. Castleton in Derbyshire, with its limestone caves.
- The picturesqueness of the place – e.g. Finchingfield in Essex, often described as a 'chocolate-box' village.
- Its historic associations – e.g. Tintagel in Cornwall, with its legendary King Arthur's castle (Figure 9).
- Its use as the setting for some TV series – e.g. Holmfirth in West Yorkshire, the setting for the long-running *Last of the Summer Wine* series.

Becoming a honeypot village is fine for those making a living out of tourism, but there is a downside – constant crowds and traffic congestion. It is not so much fun if you are a resident and place a value on peace and quiet. It is not much fun either if all the food shops have become tea-rooms and souvenir shops.

Figure 9: The annual May Day fair at Finchingfield – a honeypot village

Diversification of farming

Farming today can no longer support the large numbers of families it once did. Many farmers are finding that they can scarcely make a profit from their traditional food production alone. Supermarkets are now paying UK farmers very low prices, and even cheaper food is being imported from overseas. So, if they want to stay where they are, hard-pressed farmers have no choice but to diversify – by doing one of two things:

- Finding other ways of making money out of the farm – while continuing to farm.
- Turning their farms into completely different businesses.

The first option is by far the most common form of diversification (see the case study on page 180). With the second option – changing to a different business – a move into leisure, recreation and tourism is a common choice. In other instances, the land has been sold off and farm buildings have been turned over to micro-businesses or cottage industries, such as making greetings cards, knitwear and beauty preparations. With the extension of broadband services into the remote countryside, some farmhouses have become telecentres or premises for telecottaging (see Chapter 11). Others have been renovated and turned into second homes. The table below gives some idea of the different ways of diversifying a farm.

New products	New outlets	Tourism	Leisure & recreation	Development	Energy
• Organic crops • Herbs • Bees • Goats • Ducks • Ostriches • Red deer • Llamas • Cheese • Bottled water	• PYO • Farm shop • Farmers' market	• B&B • Caravan or camping site • Café or restaurant	• Shooting • Off-road driving • Mountain biking	• Converting barns into housing • Industrial units • Telecentres	• Wind turbines

There are concerns about the impacts of diversifying into non-farming activities. What will happen to the landscape when there is no growing of crops, no grazing of livestock and no maintenance of field boundaries? And what will happen to local rural communities with few or no farming families?

Activity 2

Research a farm that you know or can visit and find out what the farm was like thirty years ago. How has the farm and what it does changed?

A UK farm that has diversified – a case study of Hazel Brow Farm

Hazel Brow Farm (Figure 10) is located in picturesque Swaledale in Yorkshire and covers an area of 80 hectares. It has been owned by the Calvert family for a number of generations. Lead mining was once quite important in this area, providing work for men and boys. Miners' families would farm a smallholding – keeping a few cows, sheep and perhaps a pig, and growing some vegetables. The smallholding usually consisted of a few productive fields on the lower land, some higher rough pasture and grazing rights on the moor.

Figure 10: Hazel Brow Farm, which has diversified as a tourist attraction

As the lead mines closed in the early years of the twentieth century, many families left Swaledale to find work. Some emigrated to the USA. The land they left behind was added to other farms, and during the last forty years the farms in the area have become still larger. This has been necessary in order to provide a basic living for a family. But for the Calvert family, increasing the size of the farm has not been enough. As the prices of farm products have fallen because of cheap food imports, so they have had to look to other ways of making money out of their farmland.

The farm is located in accessible countryside, and the Calverts' solution has been to keep it as a farm, but to exploit it as a tourist attraction. So the farm now has a visitor centre where people can learn about farming. There is a chance to handle farm animals. There are nature trails through the farm, a heritage project and a discovery room. There is also a café and a shop, where farm produce is on sale. In order to increase the appeal of this produce, the Calverts have converted to **organic farming**.

The Calverts now feel more secure about their future – and they continue to run a working farm. This is the case with the majority of farms that have diversified. Only a minority have completely diversified into activities unrelated to farming.

ResultsPlus
Exam Question Report

Choose a farm or farming system you have studied. Describe and explain the changes that have occurred.
(5 marks, June 2008)

How students answered

Some students answered this question poorly. They offered general answers that failed to identify either farms or farming systems. One or two changes were listed only, rather than explained.

14.9% (0–1 marks)

Most students answered this question reasonably well. They described at least one change, usually some aspect of diversification, but the explanation was limited to an assertion that change was needed.

44% (2–3 marks)

Many students answered this question well. They described at least two changes, often changing farming technology and diversification, and explained *why* these changes have taken place.

41.1% (4–5 marks)

Activity 3

Can you think of any additional ways that the Calverts might diversify more into tourism?

Twenty-first-century changes to farming

Farming, like industry, is constantly changing in order to increase production, reduce costs and increase profits. In the past, this has led to serious damage to the environment. Today, however, we can point to some changes that promise rather better things.

Moves to more environmentally friendly farming

The idea of 'environmentally friendly' farming is to work with nature, not against it. It is about making farming 'greener' by reducing its adverse impacts on the environment.

Making farming greener

There is a range of different actions that can be taken to make farming greener – that is, kinder to the environment. A selection is shown in the table below.

Drip irrigation	This is the best way to get water, fertiliser and pesticides to the roots of a crop. Computerised control systems deliver the right amount of water, fertiliser and pesticide at the right time, by means of a buried tape or pipe. Waste is minimised – but the system is costly to install.
Arable rotation	Arable rotation, e.g. rotating vegetables with legumes like peas and beans which fix nitrogen, helps to reduce the amount of fertiliser needed. It also helps to break disease and insect pest cycles.
Hedgerows	These play an important role on farms. They help to prevent soil erosion and water run-off, and they provide shelter, control livestock and protect crops from the wind. They also provide an important habitat for wildlife and are often seen as a defining character of the English landscape. During the 1960s and 1970s, vast numbers of hedgerows were ripped out in order to create larger fields. Fortunately, some are now being restored.

Activity 4

Find out more about one of the methods mentioned in the table. What are its advantages? What are its disadvantages?

The rise of organic farming

Organic farming is a form of agriculture that relies on crop rotation, green manure, compost and biological pest control to maintain soil productivity and control pests. It does not use chemical fertilisers, herbicides and pesticides, livestock feed additives or genetically modified organisms. It is an environmentally friendly method, but the yields from organic farming are on average 20% smaller than those from conventional agriculture. There has been a large rise in the demand for organic food, as people have become more aware of the link between good food and health. Retail sales are now estimated to be worth £1.2 billion a year. Figure 11 makes an important point – much of the organic food we buy and eat is imported rather than grown in the UK countryside.

Quick notes (making farming greener):

- There are both costly and cheap ways to greener farming.

Skills Builder 6

Study Figure 11.

(a) In which category of organic food did imports account for the smallest percentage of total UK retail sales?

(b) Suggest reasons for this.

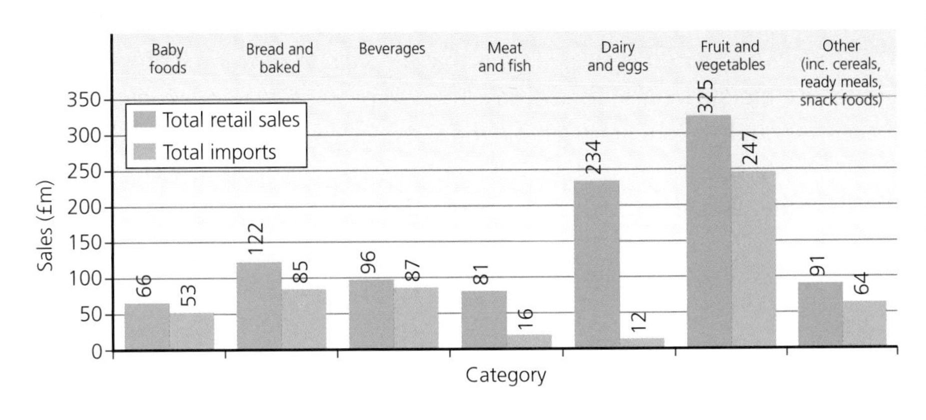

Figure 11: Organic food in the UK, total sales and imports

Figure 12: A farmers' market

Skills Builder 7

Study Figure 13.

(a) Which source of biofuels has the greatest potential in Switzerland?

(b) Which source of biofuels is most used?

Activity 5

Which one biofuel source in Figure 13 would you recommend as being best? Give your reasons.

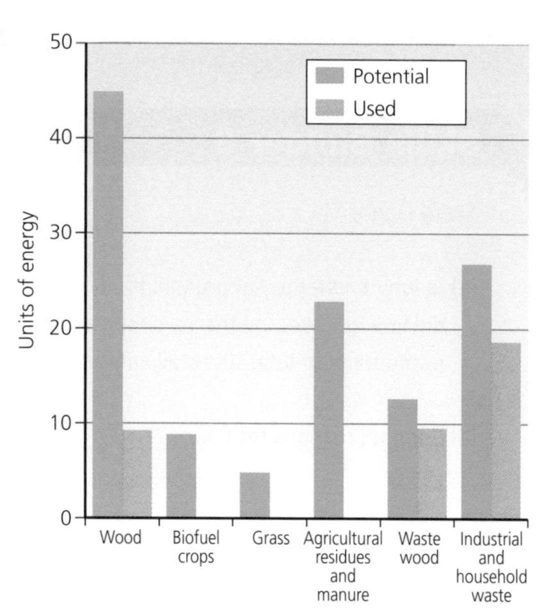

Figure 13: Sources of biofuels in Switzerland

Retailing of locally sourced products

The production and distribution of much of our food is now in the hands of a few major supermarket chains, such as Tesco, Asda and Sainsbury's. They are always on the look-out for sources of cheaper food, and they have now turned to low-income countries, which are being persuaded to grow cash crops for export instead of food for local consumption. This is having a disastrous effect on poorer families and the environment in these countries. Here in the UK, our agriculture is also suffering – and so too is our environment.

However, with rising fuel prices and growing concern about global warming, '**food miles**' have become an issue. The supermarkets have realised the public concern about flying in food – especially foods that could have been grown here – from other continents. The supermarkets have also noticed the rising popularity and success of farmers' markets. There are currently over 2,000 farmers' markets in the UK, where farmers sell a variety of their own produce direct to the public (Figure 12). By doing so, they keep food miles to the barest minimum. Because the supply chain is a short one, the food prices are competitive. Organic farmers seem to be doing particularly well at these markets.

Growing biofuels

Today, the term **biofuel** usually refers to the ethanol and diesel made from processing crops of corn, sugarcane or rapeseed. Interest in growing biofuels has increased as:

- oil and gas prices continue to rise
- stocks of both oil and gas go down
- people become more concerned about global warming.

It is the target of the European Union that by the year 2020 there should be a 10% mix of biofuels in all vehicle fuels.

Biofuels are a way of reducing greenhouse gas emissions. Emissions are lower compared with fossil fuels (coal, oil and natural gas). Although burning biofuels does release carbon dioxide, growing biofuel plants absorbs the same amount of the gas from the atmosphere. However, we need to understand that energy is used in farming and processing the crops. This can make biofuels as polluting as fossil fuels.

There are two other downsides. One is that a switch to growing biofuels will reduce the growing of food crops. This will push up food prices and make the country even more dependent on imported food. The other is that a countryside full of fields all growing the same crop is likely to reduce **biodiversity**. Wildlife would be badly affected.

Finally, we need to understand that crops are not the only source of biofuel. Figure 13 shows there are at least five others. The important point is that four of these exist in the countryside.

Management of the UK countryside

Objectives

- Learn why some areas of the countryside are protected.
- Recognise that there are different ways of managing the countryside.
- Understand the reasons for the different ways of managing the countryside.

Reasons for the different ways of managing

The management of the UK countryside is necessary because:

- There are conflicts between different land uses, as for example between agriculture and new housing, between recreation and nature conservation.
- Development threatens the natural and cultural heritage of the countryside. It is also likely to spoil stretches of attractive landscape and fine scenery.

Figure 14 shows some of the ways in which development threatens the countryside.

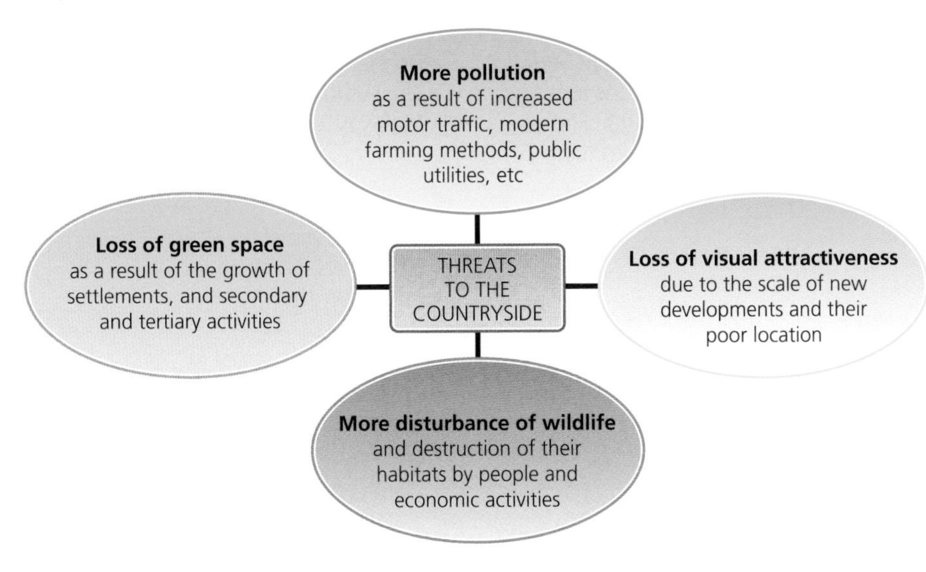

Figure 14: Development threats to the UK countryside

In general, it is planners who manage the countryside. It is their job to sort out the conflicts and to control the impacts of development. They may be working for a national government department, such as DEFRA, or for county or local government. But there are parts of the countryside that qualify for special management because they are thought to be especially valuable in terms of their heritage. Their management needs to be tailored to suit:

- The exact nature of what is thought to be special about the area
- The main aim for the area – whether it is to **protect**, **preserve**, **conserve** or **enhance**
- The size of the area.

The table on page 184 shows a range of protected and managed areas in England. (Some of these designations – AONBs, ESAs and NNRs, for example – are not made in other parts of the UK.) The areas that are designated are thought to be precious because of one or more of the following factors – their landscape and scenery, their wildlife, their scientific interest or their cultural heritage. Each designation involves a slightly different way of managing the area. The table omits a huge number of small nature or wildlife reserves. Some of these are owned and run by county councils and local authorities. Others belong to county and local wildlife trusts. They are mainly to protect wildlife and natural habitats.

Good answers about management usually recognise that different groups of people have different interests that are sometimes in conflict.

Types of protected areas in England

Designation	Number	Purpose	Examples
World Heritage Site	24	To conserve natural and cultural sites of outstanding importance in terms of global heritage.	• Jurassic coast • Hadrian's Wall • Bath
National Park	9 (8% of England)	To protect areas of landscape beauty, their wildlife and places of interest. To encourage open-air enjoyment of them.	• Lake District • Norfolk Broads • New Forest
National Nature Reserve (NNR)	222 (<1% of England)	To protect the most important areas of wildlife habitat and geological formations as places for scientific research. Places where wildlife comes first.	• Dungeness, Kent • The Lizard, Cornwall • Malham Tarn, N. Yorks.
Country Park	270+	To provide the perfect setting for people of all ages to experience nature, heritage and the great outdoors.	• Box Hill, Surrey • Lickey Hills, Birmingham • Rutland Water
Area of Outstanding Natural Beauty (AONB)	36 (15% of England)	To protect areas that are outstanding in terms of their flora, fauna, historical and cultural associations as well as scenic views.	• Cotswolds • Lincolnshire Wolds • Malvern Hills
Environmentally Sensitive Area (ESA) now replaced by Environmental Stewardship Scheme (ESS)	22 (10% of agricultural land in England)	To work in cooperation with farmers to enhance the conservation, landscape and historical value of the key environmental features of an area. Where possible, to improve public access.	• Brecklands, East Anglia • North Kent marshes • Somerset Levels
Site of Special Scientific Interest (SSSI)	4,000+ (7% of England)	To preserve the country's very best wildlife and geological sites.	• Bamburgh Dunes, Northumberland • Epping Forest, Essex • Ribble estuary, Lancs.
Local nature reserve	1,280	To ensure the conservation of nature and/or the maintenance of special opportunities for study, research or enjoyment of nature.	• Little Paxton pits, Cambridge • Kirtlington Quarry, Oxon.

Activity 6

Which of the designations in the table can be found near to your school?

Activity 7

Using examples from the New Forest, explain the difference between a **pressure** and a **conflict**.

Managing the pressures and conflicts in the New Forest National Park

The New Forest National Park is one of fifteen in Great Britain. Each National Park is unique in terms of its heritage and landscape. But, despite this, all of them are faced with the same general pressures and conflicts.

Case study: New Forest National Park

This is one of the most recent parts of Britain to be made a National Park. It was designated in 2005 and covers an area of 570 km² to the west of Southampton and immediately north of the Solent. It was once a royal hunting ground. Today it is a mix of heaths, bogs and ancient woodlands. It is renowned for its wildlife, including rare and endangered species, as well as for its ponies, cattle and pigs – which roam freely through much of the Forest. (These are not wild – they are owned by local farmers known as 'Commoners'.)

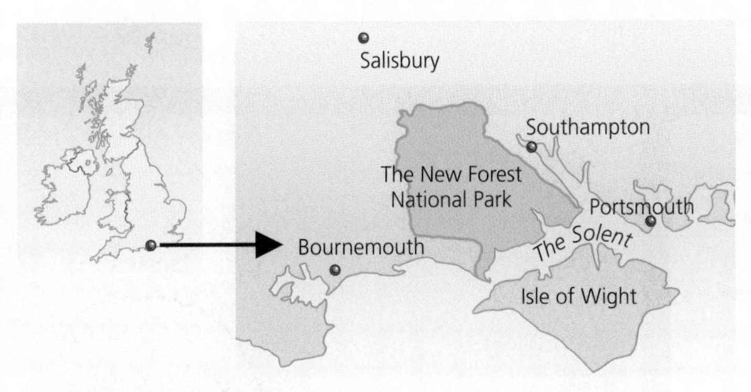

Figure 15: The location o

Figure 16: A New Forest scene

Like all National Parks, the New Forest is managed by a National Park Authority (NPA). The main pressures and conflicts that it has to deal with are:

Pressures

- The Forest is accessible to millions of people living in London, south-east England and the Midlands. Too many visitors mean traffic congestion, over-full car parks, disturbance of wildlife, footpath erosion, etc.
- Local people need to make a living, and that is not easy when there are all sorts of restrictions on what can and cannot be done inside a national park.

Conflicts

- There are conflicts between the Forest's function as a recreational area for the enjoyment of people and the need to protect its wildlife and habitats.
- And there are conflicts between incomers and local people. Housing is being bought by well-paid London-bound commuters and by wealthy second-homers. It is very difficult for local young people to find affordable housing.

The task of the NPA is not an easy one. It needs to steer a middle course between rival claims and different interest groups. It is very easy to upset everyone. Successful management lies in the 'three Cs' – **consultation** (with the public, special interest groups, local government), **compromise** (getting people to give ground) and **control** (ensuring that the main aims are achieved).

Case study quick notes:

- National Parks are unique in terms of heritage and landscape.
- However, all of them are faced with a number of pressures and conflicts.
- These pressures and conflicts are the same for all National Parks.

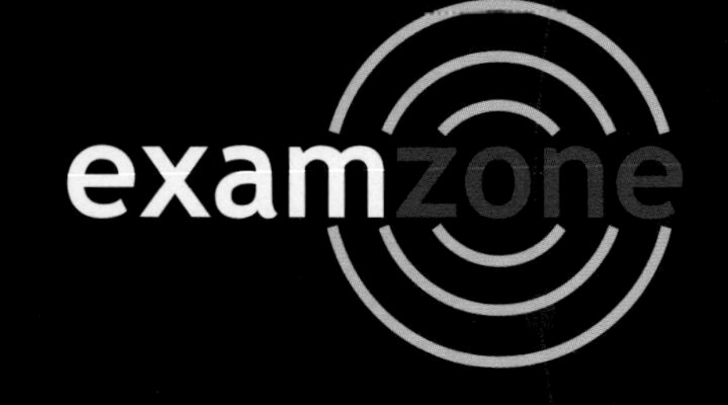

Know Zone
Chapter 10 Farming and the countryside

There is something of a crisis in UK farming and the changes over the last fifty years have been very marked. Most rural communities have very few resident farmers or farm workers left and services have also declined very rapidly.

You should know...

- ☐ Why primary activities have declined in the UK's countryside
- ☐ Why people have left rural areas
- ☐ The reasons for the inward migration of other groups, including commuters
- ☐ The destinations chosen by people who move when they retire
- ☐ How important leisure and tourism is in many rural areas
- ☐ How these changes impact on the countryside by using up rural land
- ☐ About the changes in rural population structure
- ☐ How services have declined in rural areas
- ☐ Why some villages have turned into 'honeypots' and the impact this has had
- ☐ How and why some farms have diversified into other profit-making activities
- ☐ A detailed case study of a UK farm that has diversified
- ☐ The changes that have taken place in UK farming practices, including organic farming, biofuels and local sourcing
- ☐ How the UK countryside is managed and who the various 'players' who take part in that process are
- ☐ How the pressures and conflicts in a UK National Park are being managed

Key terms

Accessible countryside
Biodiversity
Biofuel
Commuter
Commuter belt
Counterurbanisation
Disposable income
Diversification
Food miles
Honeypots
Organic farming
Recreation
Remote countryside
Rural–urban migration
Suburbanisation
Tourism
Urban fringe

Which key terms match the following definitions?

A An environmentally friendly form of agriculture that relies on methods such as crop rotation, green manure, compost and biological pest control

B A place of special interest or appeal that attracts large numbers of visitors and tends to become overcrowded at peak times

C The movement of people from the countryside into towns and cities

D A countryside area with settlements that are used as dormitories by urban-based workers and their families

E Spreading business risks by adding new activities and removing complete dependence on the one original activity

F Countryside that is being quickly lost to urban growth

G Fuel derived from biological material, such as palm oil

H The amount of money which a person has available to spend on non-essential items, after they have paid for their food, clothing and household running costs

To check your answers, look at the glossary on page 289.

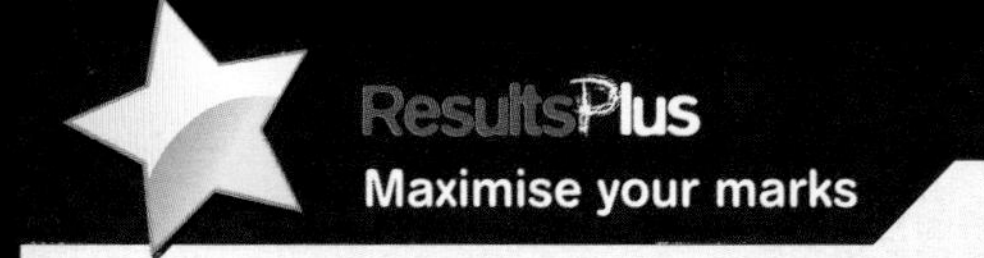

Foundation Question: Study Figure 6 on page 177, which shows the distribution of retired people in Great Britain.
i. Identify a coastline or coastal region that has a high number of retired people in the population. (1 mark)
ii Outline **two** impacts on an area that has a high percentage of retired people. (4 marks)

Student answer (awarded 1 mark + 1 mark)	**Examiner comments**	**Build a better answer** (awarded 1 mark + 4 marks)
i. South coast. *ii. It is much nicer for people to live in quiet areas.* *The places that they go to may be honeypots and get very busy with vistors.*	• The *South coast* is the correct answer and is awarded a mark. • *It is much nicer...* This is a reason and not an impact. The quality of the language is also basic. • *The places that...* The idea of *honeypots* scores 1 mark but the answer does not focus on retired people as the question asks.	i. South coast. ii. The population structure will change with far more old people and a smaller proportion of young people. Areas with larger numbers of elderly people develop resources for that age group. These include medical services and home visitors, as well as retail facilities.

Overall comment: Confusing reasons and impacts is a common examination error. Make sure that you know the difference.

Higher Question: Study the table on page 179, which shows different types of farm diversification.
i. Define the term farm diversification. (1 mark)
ii. Outline why rural diversification schemes, such as these, have taken place. (4 marks)

Student answer (awarded 1 mark + 1 mark)	**Examiner comments**	**Build a better answer** (awarded 1 mark + 4 marks)
i. When farmers look for other means of earning money. For example by letting out cottages. *ii. It keeps farms going and keeps people in the countryside.* *Some of this is really annoying like shooting and off-roading and it will disturb other people who like rural peace and quiet.*	i. *When farmers look...* is a good answer and scores 1 mark. However, it does not need an example as this is not asked for in the question. ii. *It keeps farms...* is rather brief and the phrase *keep farms going* is not precise. The second idea is better and gets 1 mark. *Some of this...* is an impact of rural diversification and not a reason so cannot be credited.	i. When farmers look for other means of earning money to supplement their core activity. ii. The main reason is falling income from agriculture and fewer subsidies available. This is because of competition from abroad and changed EU rules. If farmers have other income sources they may be able to continue to operate. This means farmers can stay in local communities which do not decline as fast, keeping services such as shops and schools open.

Overall comment: Make sure that you leave a little time at the end of the examination to re-read your answers. Re-reading this answer might have helped the student spot the lack of precision in the answers and the confusion between reasons and impacts.

Chapter 11 Settlement change

Objectives

- Learn the meaning of 'site' and 'situation' and the processes affecting settlements in remote rural areas.
- Recognise different settlement patterns and the consequences of processes at work in remote rural areas.
- Understand the factors affecting settlement characteristics and the processes of change in rural areas.

Factors affecting settlements

Factors affecting the site, situation, growth and shape of settlements

A settlement can range in size from a single farmstead to a vast mega-city, such as Shanghai. No matter what their size, all settlements have two features:

- A **site** – the ground on which the settlement stands, described in terms of its physical characteristics.
- A **situation** – the location of the settlement relative to its surroundings, described in relation to other settlements, rivers, relief features, transport lines, etc.

An interesting question to ask about any settlement is – why is it where it is? Why does it occupy that particular site, in that particular situation? In prehistoric times, the density of lowland vegetation in the UK meant that early settlements occupied the higher ground, where the vegetation was thinner and more easily cleared. Sites on higher ground were dry, accessible and easy to defend. Over time, as agricultural techniques improved, lowland areas were cleared of their vegetation and cultivated. Lowland valleys offered shelter, good soils and a supply of water. Later, as trade developed between settlements, sites at crossroads, river crossings and the heads of estuaries were settled for commercial reasons.

ResultsPlus
Watch out!

Try not to confuse site and situation. Many examination questions use these terms so be clear what they mean. Revisit the definitions of these terms on page 20.

Figure 1 shows the three main types of settlement shape (sometimes called settlement 'form') that have evolved over time:

- **nucleated**, where dwellings and buildings are packed closely together – at a crossroads, for example, or at a river crossing point where routeways converge.
- **dispersed**, where individual dwellings are spread out – for example, over a flat lowland that is fairly uniform in terms of soils and vegetation.
- **linear**, where dwellings and buildings are strung out along a road, a river valley, a ridge or a stretch of coastline.

Nucleated or clustered villages form at route centres. They may originally reflect the need for defence, or the place where farming was carried out communally.

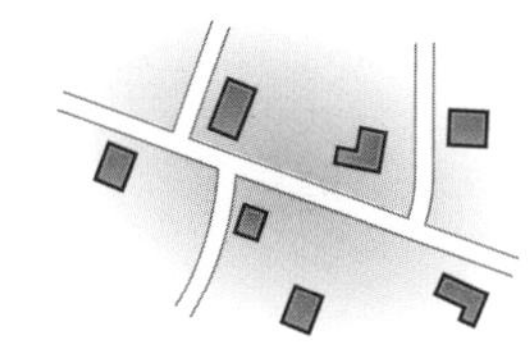

Dispersed villages have no original nucleus. They probably developed where areas of woodland were gradually cleared for agriculture.

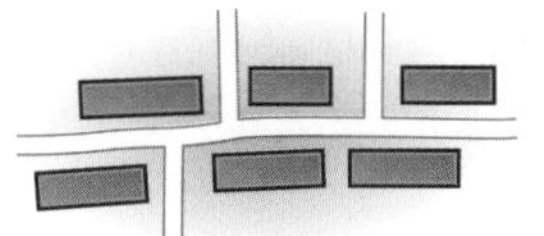

Linear villages are strung out along roads, river valleys, ridges or the coastline

Figure 1: Three types of settlement shape

Settlements grow most readily in the direction of fewest physical obstacles. Steep slopes and land liable to flooding are likely to be avoided. On the other hand, route ways, such as roads, provide lines along which settlements can easily spread. This is obviously the case with linear settlements. But in the case of a settlement situated where route ways converge, tentacles of growth will reach out along each of those routes. As these tentacles grow longer, the space between them will become filled in by more housing, helping the settlement to keep its nucleated form.

Activity 1

(a) Use an atlas to help you name a British town or city situated at the head of an estuary.

(b) What are the advantages of such a situation?

(c) What shape is a settlement at the head of an estuary likely to take? Give your reasons.

The South Wales valleys

The valleys of South Wales are deep and run roughly parallel to each other. They rise in the Brecon Beacons and drain in a south-easterly direction towards the Bristol Channel. From the late eighteenth century to the second half of the twentieth century, this was an important area of coal mining and iron working. The coal and iron were moved along the canals, railways and roads which ran along the valley floors. Because the valleys were so narrow, there was little space in which the settlements could expand. Much of the housing for the miners, iron workers and their families was arranged in parallel terraces up the valley sides (Figure 2). Individual settlements spread up and down the valley, along the transport routes. Eventually, settlements became joined together by this growth to form ribbons of continuous urban development. Thus the area acquired its characteristic linear settlement pattern, which is well illustrated by the most famous of the valleys – the Rhondda (Figure 3).

Figure 2: Terraced housing in a South Wales valley

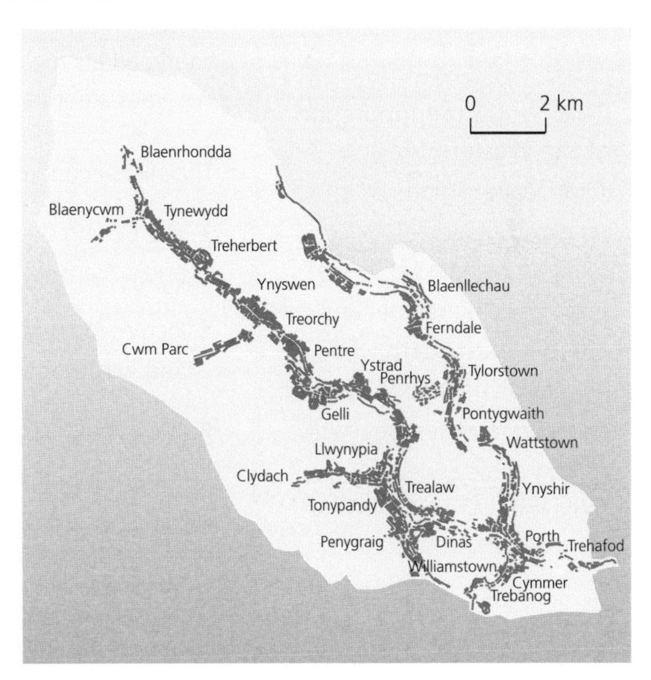

Figure 3: Settlements in the Rhondda valley

Changes in rural communities

There are two main processes causing change in the rural communities located in areas of remote countryside – **depopulation** and **counterurbanisation**. Depopulation is simply the loss of population. In the UK, the Western Isles of Scotland illustrate well the common causes and consequences of depopulation.

Quick notes (the South Wales valleys):
A linear settlement pattern is strongly influenced by relief and drainage.

Depopulation of the Western Isles

During the twentieth century, the population of the Western Isles of Scotland declined by almost a half to around 26,500 (Figure 4). Between 1991 and 2001 alone, the population shrank by 10%. The factors encouraging this depopulation include:

- A harsh physical environment
- A sense of remoteness from the centre of things
- Difficulties of transport and access to cities and services
- Difficulties of making a living, particularly in agriculture and fishing
- Limited job opportunities and low wages
- The impact of the media, making islanders aware of the 'bright life' to be found in other parts of the country
- The ambitions of young adults.

It is common for 18-year-old islanders to go to a mainland college or university, perhaps in Edinburgh or Glasgow. After graduation, they find that there are few good jobs back home – and they never return to the islands, other than to visit relatives and friends. The loss of young adults in particular starts a spiral of decline. (See Chapter 10, Figure 7 on page 177.) The loss of these islanders has two immediate effects on the population structure. Fewer young adults mean fewer young children, and the population as a whole becomes older in structure.

The socio-economic consequences of depopulation include:

- The closing down of services, such as schools, shops, post offices and public transport
- Increased costs per head of providing public utilities, such as water, energy and waste treatment
- Abandoned crofts and empty villages.

As for the environment, depopulation is a mixed blessing. It might be good for wildlife, but the signs of neglect in the landscape can reduce its scenic appeal.

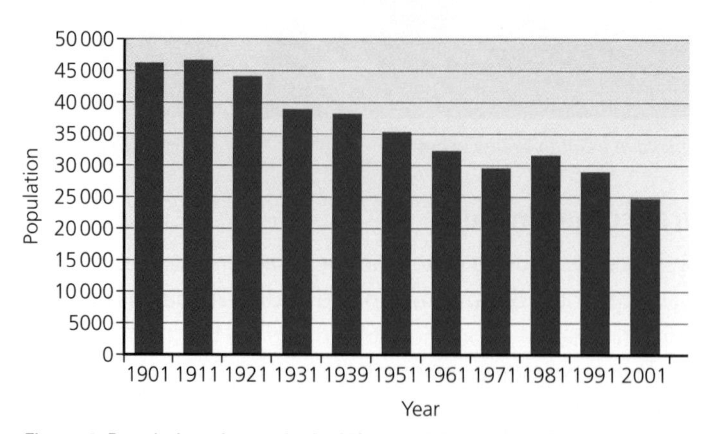

Figure 4: Population change in the Western Isles, 1901 to 2001

Skills Builder 1

(a) Make a copy of Figure 4, but add plot points for the years 2011 and 2021. Extend the trend of the graph to 2021. What does your graph suggest the population of the Western Isles will be in 2011 and in 2021?

(b) Look at a map of Scotland and note down the names of the larger Western Isles.

Quick notes (depopulation of the Western Isles):

- Rural depopulation has a variety of causes.
- Being an island adds to the sense of being remote and cut off from the modern world,
- The effects of depopulation on rural communities are mainly demographic and socio-economic.

Counterurbanisation involves two types of movement. One is of people and activities from large cities to smaller cities and towns. The other type – which we are concerned with here – is from towns and cities to rural areas. The main cause of counterurbanisation is that people become discontented with the way of life in larger cities. They become unhappy about things like the cost of housing, the pollution, their personal security and the general quality of life. The urban–rural movement of population in the UK has been most obvious in the commuter belt and the accessible countryside. But the outward ripple of urban people is now beginning to reach parts of the remote countryside. Perhaps the increase in holiday homes in the countryside is an early sign of counterurbanisation (Figure 5).

The impact of counterurbanisation on rural areas is the opposite to depopulation – it increases population. In some areas, however, both processes are at work, in which case the important question is – which process is stronger? Is the number of people moving into the countryside as a result of counterurbanisation greater or smaller than the number leaving it? If it is greater, then the rural population will increase, and there will be what is known as a rural turnaround. If the number moving in is smaller, then it will slow the rate of depopulation – but it will not stop it.

Skills Builder 2

Study Figure 5. Describe the distribution pattern of second homes. What types of countryside seem to be most popular?

■ Counterurbanisation is a misused term. The movement of people (and activities) from the inner to the outer parts of the urban area is *not* counterurbanisation. That is 'suburbanisation'. The term 'counterurbanisation' only applies when people become detached from their original city in terms of either where they work or where they reside. The new location must be either in a smaller town or city, or in the countryside.

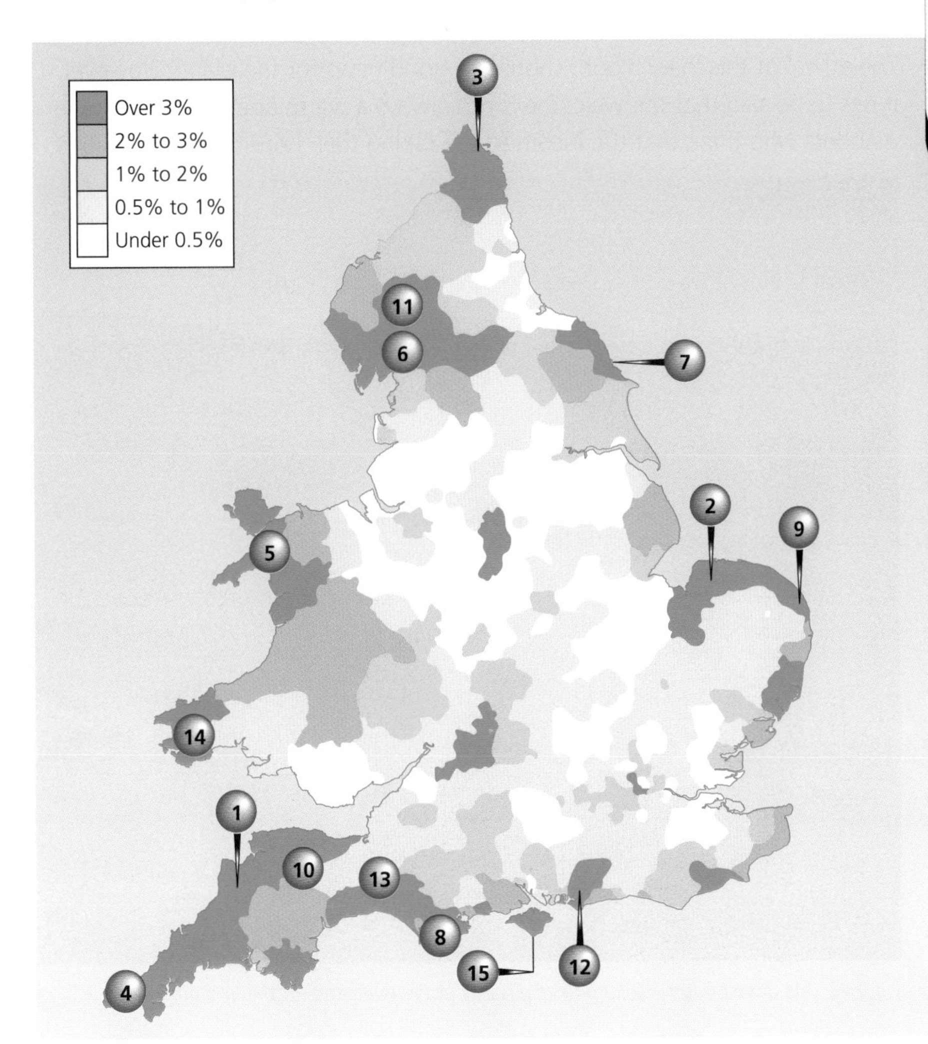

	District	% second homes
1	North Cornwall	9.71
2	North Norfolk	9.54
3	Berwick	9.15
4	Penwith	7.96
5	Gwynedd	7.41
6	South Lakeland	7.36
7	Scarborough	6.86
8	Purbeck	6.85
9	Great Yarmouth	6.16
10	West Somerset	5.70
11	Eden	5.36
12	Chichester	5.32
13	West Dorset	5.30
14	Pembrokeshire	5.20
15	Isle of Wight	5.06

Source: Savills Research 2007

Figure 5: Distribution of second homes in England and Wales

Skye bucks the trend

Skye is one of the largest Western Isles. It has an area of 1,656 km^2 and a population of just over 9,000. Population numbers have been rising for some time as a result of net in-migration. But why should Skye be different from the rest of the Western Isles? The answer is that Skye is part of the Inner Hebrides and lies much closer to the mainland than the other islands. And in 1995 a bridge was opened which gave Skye a direct link to the mainland for the first time (Figure 6). It is this bridge, plus two good ferry links, which has improved the accessibility and appeal of the island. The removal of tolls on the bridge in 2004 also provided a boost.

Skye is principally an agricultural island, but tourism is the main industry. There is also some fishing, forestry work and distilling. The incomers to the island fall into three main categories:

- People involved in tourism, either working as hotel staff or setting up their own B&Bs and guest houses
- People moving to the island to retire
- People taking advantage of modern ICT and telecottaging.

A survey has shown that 30% of the incomers come from outside Scotland. The arrival of this 'new blood' should be good news for the island. However, it has to be said that the welcome is not always a warm one. There are some islanders who think that the incomers are taking their housing and their jobs.

Activity 2

Can you think of any telecottaging businesses that could be run from Skye?

Quick notes (Skye bucks the trend):

- Overcoming the remoteness of rural areas is important if the cycle of decline is to be stopped.
- Tensions easily arise between local people and incomers.

Figure 6: The Skye bridge – linking Skye directly to the mainland for the first time

Changing land use in urban areas

Urban areas are constantly changing. Two of the most important drivers of change in the towns and cities of the UK today are: the need for more housing and the need to adjust to deindustrialisation.

It is easier if we now take each of these drivers of change separately and look at their causes and consequences together.

Objectives

- Learn the ways in which urban areas are changing.
- Recognise the reasons for these changes.
- Understand the relative merits of using brownfield and greenfield sites.

More housing

Causes

The need for more housing in the UK is the result of three changes:

- An increase in the national population. Between 2001 and 2007, for example, it increased from 58.8 to 60.9 million.
- An increase in the number of households. More people are living alone (through choice, divorce or death of a partner) – and more people are living longer.
- Increased personal wealth has resulted in more people wanting to own a home of their own.

Each household – whether it is made up of one person or a family of five – needs a separate dwelling, if possible. In 2001, there were 21.6 million households in the UK. Of these, 30% were single-person households.

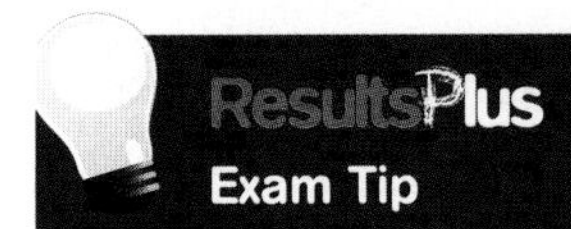

ResultsPlus
Exam Tip

Remember that the demand for housing is not just a question of population changes: it also involves social changes.

Consequences

It is estimated that, by 2020, around 3 million more dwellings will need to be built. So the question is – where to build all this new housing? The four main possible locations are shown in Figure 7. We will consider them in rank order – in terms of the amount of new housing that can be provided by each.

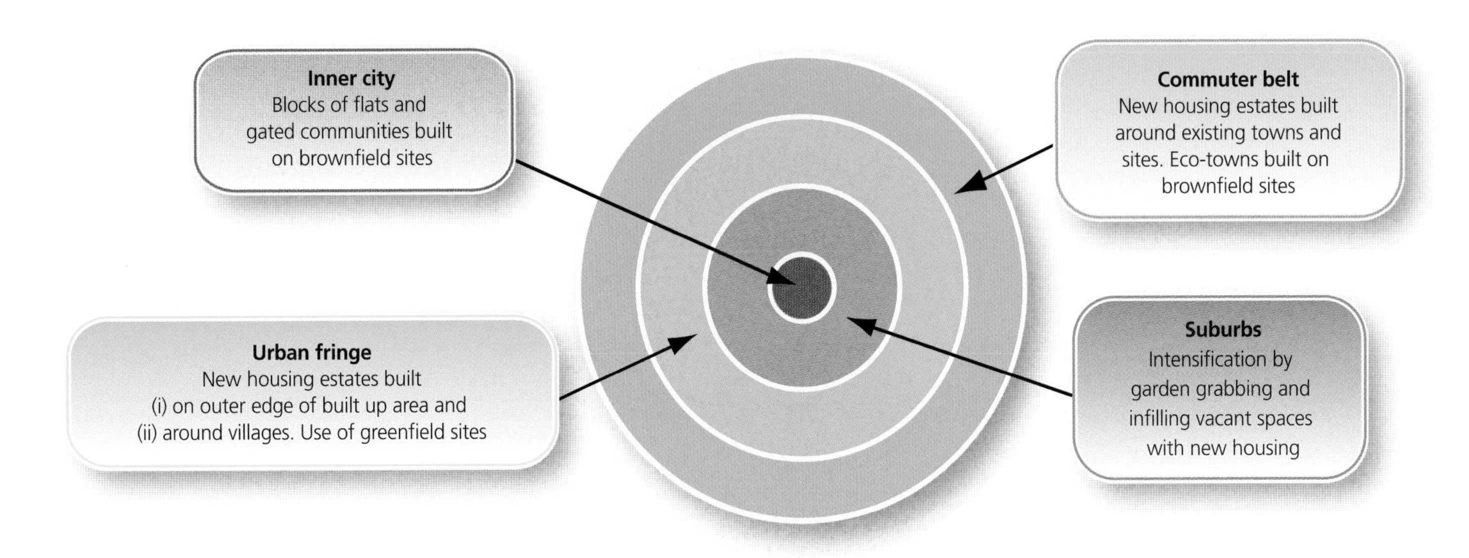

Figure 7: The four main possible locations for new housing

Figure 8: A new suburban development in the urban fringe

The first possible location for new housing is on the inside of the urban fringe – at the outer edge of the built-up area. New housing estates built here are near to the open countryside, but not too far from the town or city centre (Figure 8). Remember that these centres are where many people still work and where many of the best shops are still located. Recently, however, offices and factories have been moving to the outer suburbs – shortening the journey to work for many employees. Also the building of large edge-of-town retail complexes has brought shops and leisure facilities closer to their customers. Much of this new building will be in the form of compact estates built on **greenfield** sites – which means a loss of countryside.

Suburban sprawl

Unless there is tight planning control, it is easy for suburban housing to sprawl over large areas of countryside. This happened in the UK during the inter-war period between 1918 and 1939, particularly around London. Figure 9 shows the main factors encouraging this spread of low-density sprawl.

A house and garden in the suburbs is the dream of many. But when tens of thousands of dreams come true, it means:

- Vast areas of low-density housing
- A wasteful use of greenfield sites
- A high dependence on the car to move around the sprawl – whether it is for shopping, taking the children to school or getting to the railway station in order to catch the commuter train
- A uniformity (monotony) of architecture, layout and socio-economic class.

Thankfully, the days of allowing yet more suburban sprawl in the UK are over. Planners now recognise the need to keep future suburbs compact, with plenty of services and a good social mix.

Quick notes (suburban sprawl):
Remember that suburban sprawl is not a good form of settlement – socially or environmentally.

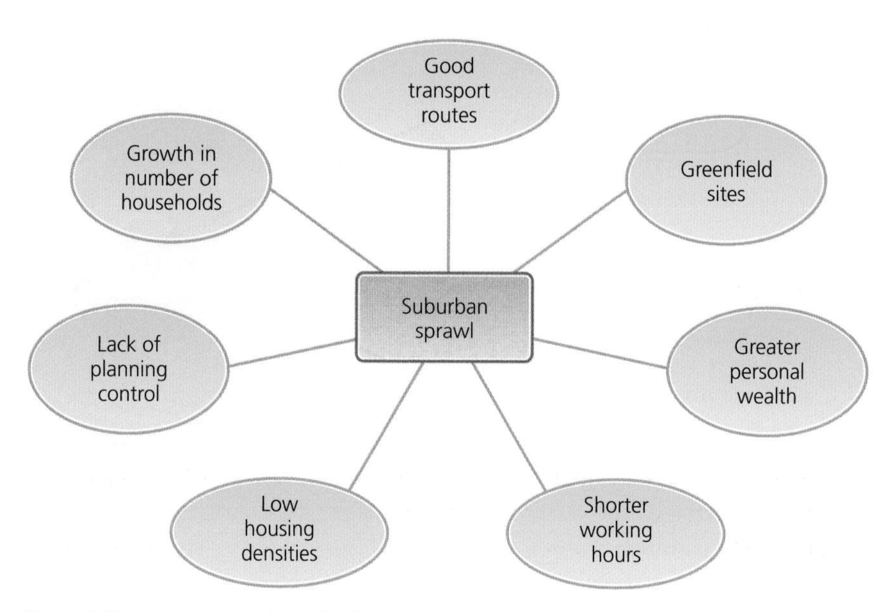

Figure 9: Factors encouraging suburban sprawl

The second possible location for new housing is within the urban fringe and the commuter belt. Again, housing is most likely to be in the form of compact estates built on greenfield sites around towns and villages. A possible disadvantage of living in such housing is the journey to work, which may well be located some distance away in a city. A new way of providing housing in this second location has recently been proposed.

Eco-towns

Ten new towns – known as '**eco-towns**' – are to be built in England over the next ten years or so (Figure 10). The largest of them will provide up to 20,000 new homes. Over 30% of the housing will be for less well-off households. The towns are expected to be environmentally friendly – carbon emissions will be kept low and full use will be made of recycled materials. They will be largely car-free.

Most of the proposed sites are **brownfield** ones – including an old airfield, an army depot, a colliery and a china-clay quarry. One early snag is that little has been said about providing employment. So it looks as if these eco-towns may be little more than commuter dormitories – which is hardly environmentally friendly.

Describe the choices to be made when trying to satisfy the demand for more houses. (4 marks)

■ **Basic answers** (0–1 marks)
State one possibility, such as the need to build more houses, in broad terms but fail to identify where this might take place.

● **Good answers** (2 marks)
Identify that there are various possibilities about where to build, but lack development.

▲ **Excellent answers** (3–4 marks)
Offer good detail of the distinction between greenfield and brownfield sites. Eco-towns might also be used to illustrate the greenfield option.

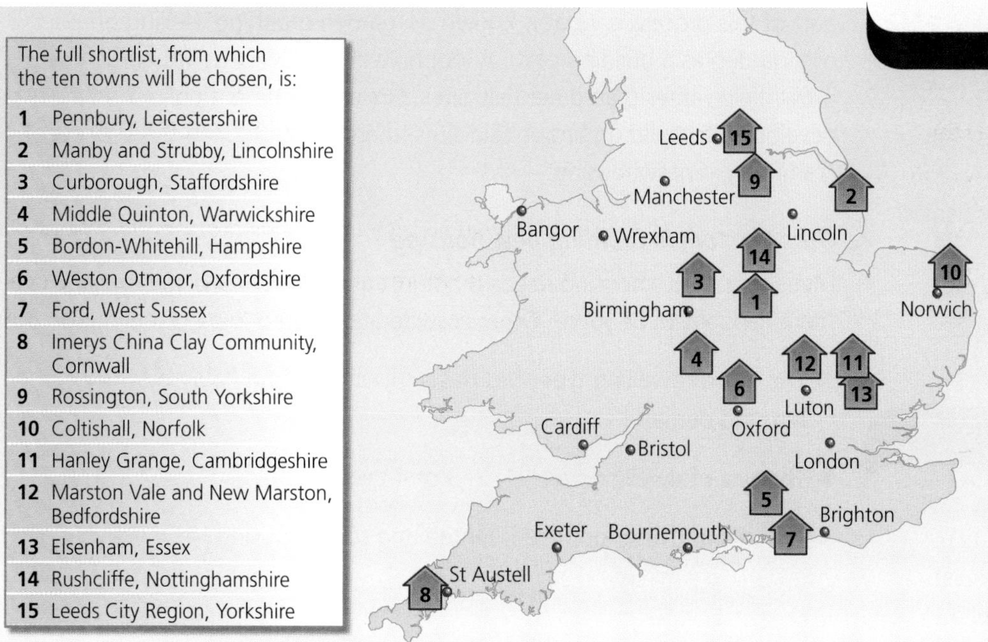

Figure 10: The proposed eco-town sites

The third possible location for new housing involves the redevelopment of inner-city brownfield sites. In this case, housing is located in the central and inner parts of the towns and cities – the parts that were once occupied by factories or slums. Public opinion about living in such areas is changing. At one time, everyone wanted to leave for the suburbs. Now people are being attracted by the advantages of living in a new home close to the bright lights of the town or city centre. Concern about personal security in these areas is being met by developers building some new housing as **gated communities.**

Quick notes (eco-towns):
- There is a growing wish to reduce building on greenfield sites.
- New residential developments need to be made environmentally friendly.

Figure 11: A gated community

Gated communities

A gated community is a group of homes shut away in a high-security compound, with CCTV cameras, electronic gates and sometimes even private security guards (Figure 11). Entry is restricted to residents and vetted visitors. Some of these developments are to be found in the new suburbs, but they are more common in brownfield developments in older and more rundown parts of urban areas. They are designed to offer a safe environment that provides personal security and protects private property. They keep out unwanted visitors and shut out areas of deprived housing that may be nearby. Gated developments are ideal for families. They allow parents a greater degree of freedom when it comes to letting their children play outside.

Some people believe that gated developments separate the 'haves' from the 'have nots'. Homebuyers have to pay a high price to live in one, so it is only an option for the more affluent. For this reason, gated communities can become the targets of crime. Criminals assume that those living behind the gates have something valuable to protect. No community can be made totally secure.

Quick notes (gated communities):

- There is a growing concern about personal security and the protection of property.
- Gated communities increase social segregation.

The fourth possible location for new housing is a relatively new one. Suburban intensification occurs in the older suburbs and involves building on vacant plots and public open spaces (e.g. parks and playing fields). One part of this process is what is known as 'garden grabbing' (selling off part of a garden as a building plot). A loophole in the law classifies gardens as brownfield rather than greenfield sites. Since government policy prefers developers to build on brownfield sites, there is little to stop this process of cramming in new housing.

Activity 3

Which of the four locations for new housing shown in Figure 7 on page 193 would you prefer to live in? Give at least three reasons for your choice and one reason why you rejected each of the other locations.

Other factors in building new housing

Meeting the UK's housing needs is not just a matter of building homes where there happens to be space. Other considerations need to be taken into account:

- The mix of dwelling types (houses, maisonettes and flats) to fit the housing demand
- The size of dwelling units (1, 2, 3 or 4 bedrooms)
- The mix of owner-occupied, rented and social housing
- The availability of work in an area
- The availability of public utilities (water, gas, electricity, etc.)
- Access to services (shops, schools, pubs, etc.)
- The environmental impact of the new housing.

The last of these – the environmental impact – will lead us into a big debate that is going on at the moment. It is between those who think that most of the new housing should be built on brownfield sites, and those who say that there is plenty of unused countryside that can provide greenfield sites.

Deindustrialisation

Much of the manufacturing that once took place in the UK has now moved to other parts of the world, particularly Asia. It is the deindustrialisation in urban areas that has had the most severe consequences, bringing unemployment which affects whole settlements in many parts of the country. The most obvious signs of deindustrialisation are derelict buildings and vacant, unused land – in short, there are plenty of brownfield sites. So a town or city experiencing deindustrialisation looks distinctly rundown. (See Chapter 9.)

Figure 12: Dereliction caused by deindustrialisation

Redevelopment and renewal

Those towns and cities that were once important centres of manufacturing are faced with two challenges:

- To find new economic activities and jobs to compensate for those lost as a result of deindustrialisation
- To find new uses for empty factory buildings or to clear them and their waste heaps and so make way for some new land uses (Figure 12).

Meeting these two challenges can be combined into one **redevelopment and renewal** strategy. Actions are taken to improve or change the popular image of the town or city and, where necessary, give parts of the urban area a facelift. Improving the image can do much to attract much-needed new investment and employment. The following examples show what can be done to recover from deindustrialisation.

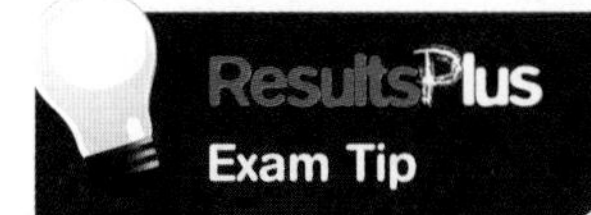

Exam Tip

Remember that answers to examination questions asking you to describe the impact of deindustrialisation should not include explanations of why that deindustrialisation took place. Take care over the command words.

Changing Bradford and its image

Bradford in West Yorkshire, with a population of over 450,000, is one of the ten largest cities in the UK. Its growth was based on the textile industry, particularly wool (Figure 13 on page 198), but this industry collapsed during the second half of the twentieth century. Faced with deindustrialisation, Bradford has been forced to do three related things:

- Find a new range of economic activities to support the city
- Find new uses for the land once occupied by the woollen industry
- Shake off its old image as a rundown city of closed woollen mills – and create a new image for the city.

Bradford has a large immigrant population drawn from Pakistan and Bangladesh. They came to Bradford to fill the thousands of jobs in the textile industry and now make up more than a fifth of the city's population. Since the mills closed, there have been tensions between the different ethnic groups, and as part of its re-imaging Bradford has had to find ways of achieving greater social harmony.

Figure 13: Bradford at the height of the woollen industry

Figure 14: Putting old mills to new uses

Activity 4

How are the two challenges – improving the quality of housing and creating more harmony between ethnic groups – linked?

Bradford has done well in terms of building up a new economy. It now has some modern engineering, chemicals and ICT industries. It has also developed a leisure and tourism industry by turning its industrial heritage into tourist attractions. Many of the mills still stand, and their exteriors have been smartened up. Inside, they have become museums, craft centres and galleries, or been converted into small business units or flats (Figure 14). Other old buildings associated with the woollen industry have been demolished and the resulting brownfield sites filled with blocks of offices or flats, shopping centres or premises for Bradford's new industries.

So from this redevelopment and renewal, a new version of Bradford has emerged. New life has been breathed into the city, giving it a new image. But two challenges remain – to improve the quality of housing and to create more harmony between the ethnic groups.

In Bradford, the need to meet the challenges of deindustrialisation has really involved the whole city. The next case study illustrates what might be done with a single brownfield site.

Figure 15: The cleared site awaiting redevelopment

Vaux: a new urban quarter in Sunderland

Vaux is one of north-east England's most significant brownfield regeneration sites (Figure 15). It provides an exciting opportunity for the development of a major city-centre, mixed land-use scheme. The site is just over 10 hectares, and was occupied by an old brewery which closed in 1999. It is situated between the present centre of Sunderland and the River Wear, and on top of a natural cliff overlooking the river. The brewery buildings have now been cleared. A public-interest private company, known as Sunderland arc (area regeneration company), has acquired the site and drawn up a master plan. The main aims are to: diversify the city's economy by creating new offices and 3,000 jobs; provide new shops, particularly cafés, to complement the shops in the present city centre; include a large hotel; and promote city-centre living by building up to 1,000 new homes.

Green space will be provided along the riverside, and the hope is that the architecture and design will be of the highest quality. The aim is to create a new and distinctive urban quarter that will be a good example of renewal and help the image of Sunderland.

Quick notes (Bradford and Vaux):

- The redevelopment of a single industrial brownfield site can serve a number of useful purposes.
- Towns and cities can survive deindustrialisation, because it creates opportunities for them to redevelop and renew themselves.
- An industrial past can be put to good use in tourism.

Brownfield versus greenfield

The question of where to build – on greenfield sites at the edge of the built-up area or on brownfield sites well inside the built-up area – arises not just with new housing. It also arises with the growth of other urban land uses, such as shops, new factories and offices. With all land uses, there are arguments for and against each type of site because, as the table shows, each has its advantages and disadvantages.

Activity 5

Make a list of examples of redevelopment and renewal in a town or city you know.

	Advantages	Disadvantages
Brownfield site	• Reduces the loss of countryside and land that might be put to agricultural or recreational use. • Helps to revive old and disused urban areas. • Services such as water, electricity, gas and sewerage already in place. • Located nearer to main areas of employment, so commuting reduced.	• Often more expensive because old buildings have to be cleared and land made free of pollution. • Often surrounded by rundown areas so does not appeal to more wealthy people as residential locations. • Higher levels of pollution; less healthy. • May not have good access in terms of modern roads.
Greenfield site	• Relatively cheap and rates of house building faster. • The layout is not hampered by previous development so can easily be made efficient and pleasant. • Healthier environment.	• Valuable farm or recreational land lost. • Attractive scenery lost. • Wildlife and their habitats lost or disturbed. • Development causes noise and light pollution in the surrounding countryside. • Encourages suburban sprawl.

There is no clear winner in this particular debate. It all depends on:

- The particular land use. Housing is fairly flexible in terms of where it might be built, but shops, offices and industries are more 'choosy' about location.
- The circumstances of the particular town or city. Is the green space really valuable? Are there serious problems and high costs involved in reusing the brown space?
- Your own set of values. Do you think that the countryside should be protected at all costs? Or do you think that more should be released for urban growth?

Activity 6

(a) Which do you think is the strongest argument for using brownfield sites for new housing? Give your reasons.

(b) Which do you think is the strongest argument against using them? Give your reasons.

Objectives

- Learn the reasons for the rapid growth of urban areas
- Recognise the major problems facing Dhaka
- Understand the reasons for those problems.

ResultsPlus
Exam Question Report

Urban areas in LICs are undergoing rapid growth. For a chosen urban area describe the reasons for, and the results of, this rapid growth and explain how growth is being managed.
(8 marks, June 2007)

How students answered

Many students answered this question poorly. They failed to focus on the theme and sometimes confused LICs with HICs, as well as concentrating on size rather than growth.

33% (0–3 marks)

Most students answered this question reasonably well. They offered some detail about the growth of urban areas, but had little on management other than occasional comments about the 'need' to control the situation.

59% (4–6 marks)

Very few students coped with the different angles of this question. Those who scored well dealt with the management issues by commenting on the management throughout. So the growth is explained by rural–urban migration, the results are often slums and the management is the top-down and bottom-up schemes to deal with this.

8% (7–8 marks)

Rapid growth of urban areas in low-income countries

Figure 16 compares the world's low-income countries (LICs) and high-income countries (HICs) in terms of their urban populations. There are two important features to notice:

- Since 1950 the LIC urban population has been growing much faster and at an accelerating rate (note how the curve continues to rise)
- Since 1970 the LIC urban population has become larger than the HIC urban population. At present it is nearly three times larger, and by 2025 it is expected to be five times so.

We might describe this remarkable growth as an 'urban explosion'.

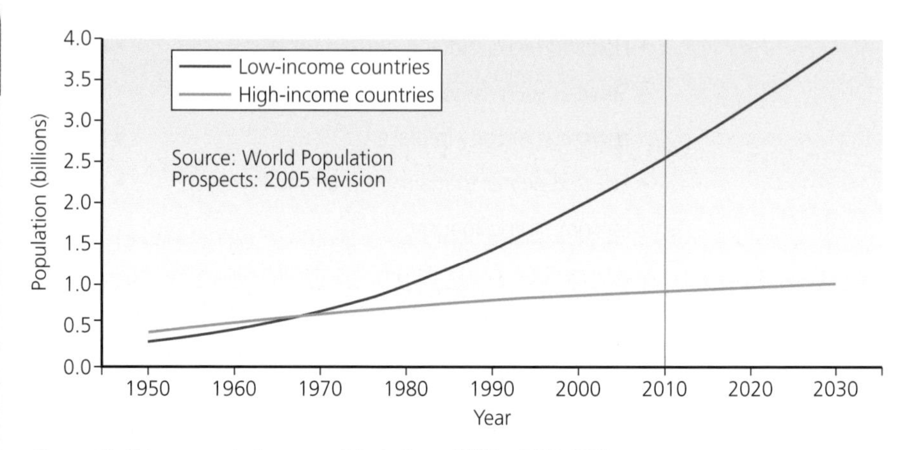

Figure 16: Urban population growth in LICs and HICs, 1950–2030

Reasons for the rapid growth of urban areas

Figure 17 shows the two main factors causing the urban explosion in LICs. First, a high **birth rate** and a falling **death rate** mean that there is a high rate of **natural increase**. Secondly, many people are migrating from the countryside to the towns and cities. Remember that those migrants, once they are settled in urban areas, also contribute to the high rate of natural increase – especially as many of them are young adults.

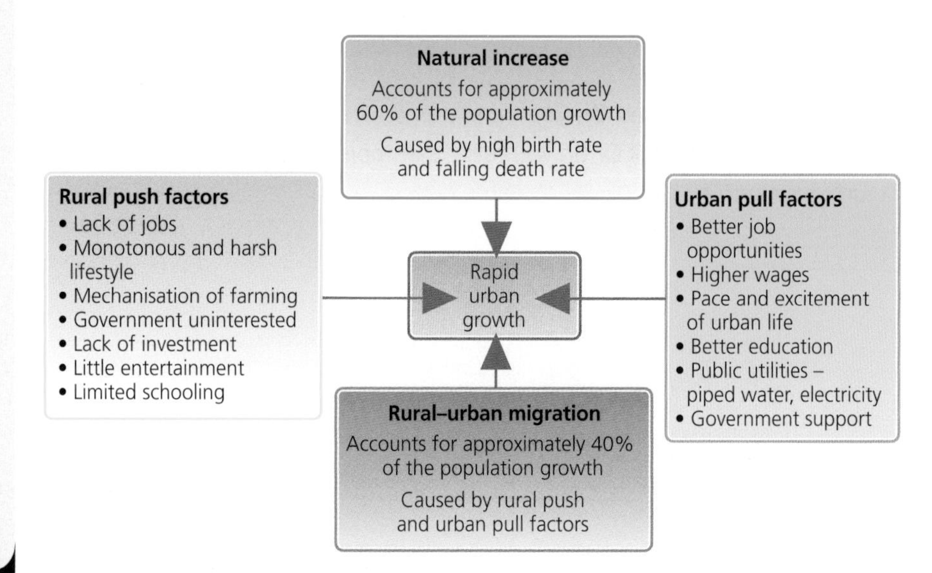

Figure 17: The main causes of rapid urban growth in LICs

Reasons for migration in both directions are given below:

Economic	Urban areas offer more jobs, better job prospects and higher wages. Rural employment is largely limited to farming.
Social	Many see urban areas as having better schools and medical services, more entertainment and excitement. Life in the countryside can seem rather harsh and dull.
Political	LIC governments are keen to help towns and cities because they create more economic growth and wealth. Rural areas receive little help.
Environmental	Urban areas suffer from higher levels of air and water pollution. Urban housing conditions are congested and can be very poor.

On balance, the factors stack up in favour of rural–urban migration.

Activity 7

(a) What are the four worst problems facing Dhaka?

(b) Rank the problems from 1 to 4.

(c) What could be done to reduce each of these problems?

Case study: An LIC urban area – Dhaka

Dhaka is the capital city of Bangladesh, one of the poorest countries in the world. The city's population is currently estimated as 12 million, and is forecast to increase to 21 million by 2025.

Dhaka is hemmed in by large river channels and lesser channels thread their way through the site. Its low-lying areas suffer widespread flooding during the monsoon season, June to November. It is on such areas that we find the 'bostis' – the illegal shanty or squatter settlements that have been built mainly by rural–urban migrants. They live there because they cannot afford to live elsewhere and still be close to job opportunities. They also live there because the city authorities have provided little housing to accommodate the huge increase in the city's population. With so many people living in such overcrowded and unhealthy conditions, disease is a major problem. Water pollution and rotting piles of waste dumped beside the bostis are a real health hazard. So too is the smoke from domestic fires and the air pollution from the factories and the traffic on the congested roads. The most common diseases are diarrhoea, tuberculosis, measles, malaria and dengue fever.

As in most LIC cities, the rate of population growth is outstripping the rate at which new jobs are being created. Although willing to work long hours for very low wages, most people are forced to find their own ways of making a living. This may involve selling in the street, shoe-shining, rubbish collecting or scavenging bottles and other types of waste for recycling. Begging, petty crime and prostitution are other, less legal ways of scratching a living. These activities make up what is known as the **informal economy**. In Dhaka, the driving of rickshaws (by pulling, pedal power or motor) is the most common informal activity. With little by way of public transport, rickshaws are an important factor in keeping the city moving. But they also add to the general congestion on busy and inadequate roads. Half a million children are estimated to be involved in informal activities. Most of them work from dawn to dusk, earning on average the equivalent of about 12p per day to help support their families. Their jobs vary from begging and scavenging to domestic service and collecting the fares on minibuses. By working like this, these children are constantly exposed to hazards such as traffic accidents, toxic fumes, street crime, violence and other forms of abuse.

Case study quick notes:

This account of Dhaka has focused on the worst of its problems. You may wonder why people in their thousands choose to move to Dhaka. Is it that life is so much worse in rural areas? Or is it that most people are optimists and believe that they will have their lucky break. Unfortunately, the sad truth is that relatively few poor people ever escape from urban poverty.

Know Zone
Chapter 11 Settlement change

Cities, towns and villages are changing every day. Many are growing fast, especially in the south of the UK. Elsewhere, some cities are having to adjust to very different economic circumstances. The effects of these changes are complex and not always positive.

You should know…

- ☐ How to define the site, situation and the shape of settlements
- ☐ How to recognise different types of settlements
- ☐ How to use maps to correctly identify the site and form of settlements
- ☐ The reasons for changes in rural communities caused by counterurbanisation and depopulation
- ☐ How and why land use is changing in the UK's urban areas
- ☐ The reasons for the increasing demand for new housing in the UK, including social, economic and political changes
- ☐ The consequences of deindustrialisation, including the redevelopment of 'brownfield' sites
- ☐ The advantages and disadvantages of 'brownfield' sites

Key terms

Birth rate
Brownfield
Counterurbanisation
Death rate
Depopulation
Dispersed settlement
Eco-towns
Gated community
Greenfield
Informal economy
Linear settlement
Natural increase
Nucleated settlement
Redevelopment
Renewal
Site
Situation

Which key terms match the following definitions?

A Where individual dwellings are spread out

B Forms of employment that are not officially recognised, e.g. people working for themselves on the streets of LIC cities

C The number of births per 1,000 people in a year

D Where dwellings and buildings are packed closely together – e.g. around a crossroads

E The ground occupied by a settlement

F An area of wealthy private housing with a secure perimeter and a controlled entrance for residents and visitors

G Where dwellings and buildings are arranged along a road, a river valley, a ridge or a stretch of coastline

H Proposed new towns that are designed to be much more sustainable than traditional settlements

To check your answers, look at the glossary on page 289.

ResultsPlus
Maximise your marks

Foundation Question:
i. Define the term 'brownfield' site. (1 mark)
ii. Describe **two** changes that have taken place in the industrial cities of the UK. (4 marks)

Student answer (awarded 1 mark + 1 mark)	**Examiner comments**	**Build a better answer** (awarded 1 mark + 4 marks)
i. A brownfield site is a place that is available for development in a city. *ii. These cities have lost employment and people have had to find different jobs such as working in call centres and in shops.* *These cities have got smaller, although some of them have succeeded in rebranding*	*i. A brownfield site...* is not correct. *ii. These cities have...* is worth 2 marks. It could be improved by including the type of jobs lost. *These cities have...* is not precise. Two ideas are mixed together and the first one is more or less the same as the reason given above. There is not enough detail here.	i. Brownfield sites are areas in towns and cities that are either abandoned or occupied by buildings that have no current use. ii. Factories have closed, such as textiles in Bradford and automobiles in Coventry. This has led to major unemployment. The cities have become rather rundown as a result of factory closures, with large areas of abandoned land, such as Salford Quays in Manchester. Many of these areas have now been redeveloped.

Overall comment: This answer lacks detail. Try to build up 'fact files' of the examples you have studied and use data to add detail to your answers.

Higher Question: Describe the attempts to redevelop urban areas that have suffered deindustrialisation. Use an example or examples in your answer. (4 marks)

Student answer (awarded 2 marks)	**Examiner comments**	**Build a better answer** (awarded 4 marks)
The first thing to do is to find a new image for the city, such as Sunderland tried in Vaux. *After that it needs new jobs to get the city going. This can involve lots of projects at different scales.* *Not all of these projects are successful and many of them fail because they are not set up right.*	• *The first thing to do...* This is a good start, which locates an appropriate example and offers an example. This scores 1 mark. • *After that it needs...* The language is not very precise here. The answer could be made stronger with some examples, as asked for in the question, but it is enough to gain 1 mark. • *Not all of these...* This final sentence does not add anything to the answer. The student needs to be clear what ***set up right*** actually means.	The first thing to do is to find a new image for the city, such as Sunderland tried in Vaux. Sunderland tried to attract new investors, such as the redevelopment of the Vaux brewery site which aims to create 3,000 new jobs. Of course it is difficult to reverse long-term decline. Although nearly £40m has been spent, unemployment is still high and much remains to be done.

Overall comment: The student could improve their answer by focusing more closely on the question. There is very little descriptive comment and the location, Sunderland, is hardly visible.

Chapter 12 Population change

Objectives

- Recognise the main patterns of global population distribution.
- Explain the distribution of population in terms of both physical and human factors.
- Understand how and why population growth rates vary from place to place and from time to time.

ResultsPlus
Watch out!

Remember that although the rate of global population growth is slowing down, the total population is still increasing, but not as quickly as it was a few years ago.

Population growth and distribution

Global population growth

During 2008, at least another 65 million people were added to the world's population. This annual increase was less than during the 1980s and 1990s. **Population growth rates** have fallen from 2.1% per year to around 1.95%. Besides showing the upward curve of global population since 1800, Figure 1 shows how long it has taken for the world population to increase by 1 billion. It took 118 years from 1804. Since then the length of time has shortened considerably to a mere 12 years between 1987 and 1999. What the graph also shows is that the length of time is just beginning to increase, from 12 to 14 years. In short, the rate of population growth is predicted to slow down.

The interesting question is – given this slowing down, when exactly will the point of **zero population growth** be reached? There is much disagreement about the likely date. Some have suggested it may be as early as the 2020s. Others say it will not be before 2060. Forecasting future population is a tricky business. The problem is that so often there are surprise changes. For example, population growth in the low-income countries (LICs) was lower than expected during the 1990s. In the UK, **birth rates** have increased since 2000 and this has caused the population to surge past the 60 million mark. Despite these difficulties, however, most forecasters expect a peak global population of just over 9 billion. This compares with 6.5 billion today.

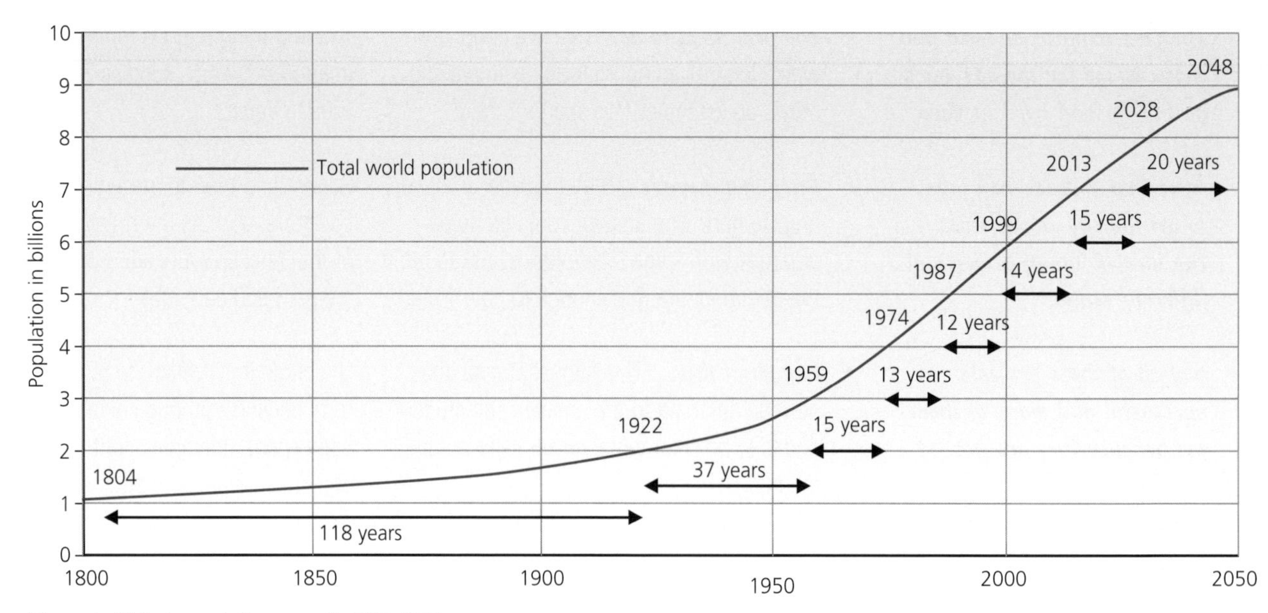

Figure 1: Global population growth, 1800–2050

Global population distribution

'**Population distribution**' means where people are located. On a map, it is most simply shown by using dots or symbols to represent a certain number of people. This will show us where people are and give an impression of how the numbers of people vary from place to place. In Figure 2, one dot represents 100,000 people.

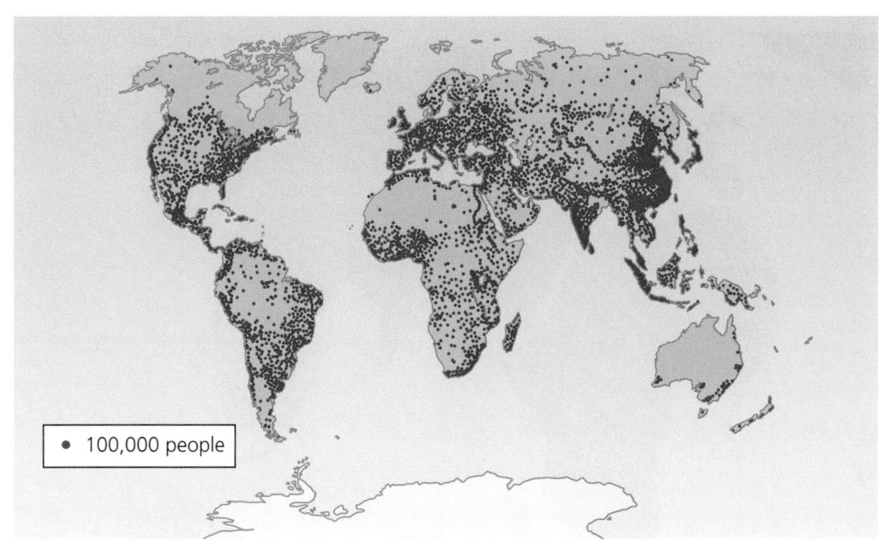

Figure 2: The global distribution of population, 2000

One of the most common ways of showing population distribution, however, is to relate population numbers to the space they occupy. This is **population density** – the number of people per unit of area – per km² or per hectare, for example.

Figure 3 shows the distribution of the world's population in 2006. Individual countries have been coloured according to their average population density. In that year, the average density of population in the world was 48 persons per km². Two areas stand out on the map as containing many people – Asia and Europe. They show high population densities of over 75 persons per km². In some countries, densities exceed 300 persons per km². At the other extreme, the areas of least population (less than 10 persons per km²) are Canada, Greenland, Russia, Australia and parts of Africa.

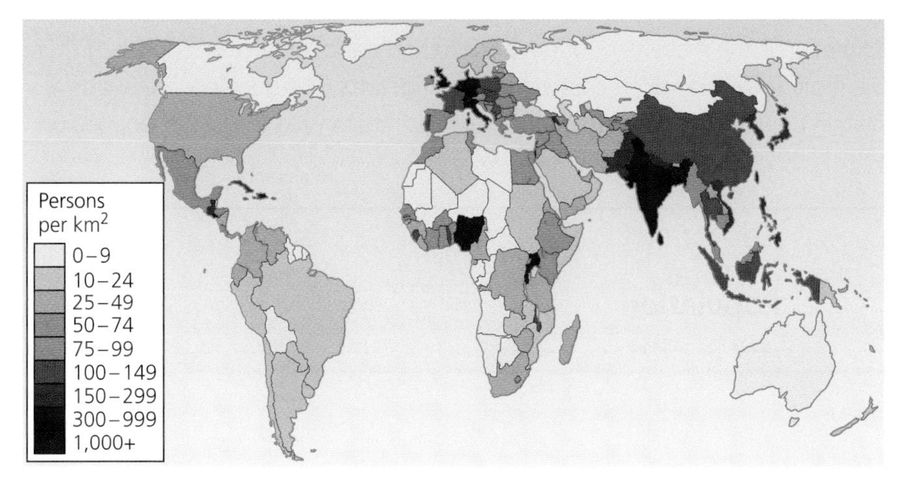

Figure 3: The global distribution of population, 2006

Skills Builder 1

Which do you think gives a better picture of global population – Figure 2 or Figure 3? Give your reasons.

ResultsPlus
Watch out!

Be careful not to confuse population distribution with population density. Distribution is where people are. Density is how closely together people are living.

But is the global distribution of population changing? Figure 4 gives us a snapshot of change between 1990 and 2000. It shows some strong contrasts, as for example between the high rates of growth in Africa and the Middle East and the little or no gain in North America, Greenland, Europe and Russia. We will look later at the factors that help explain these differences in population growth rates.

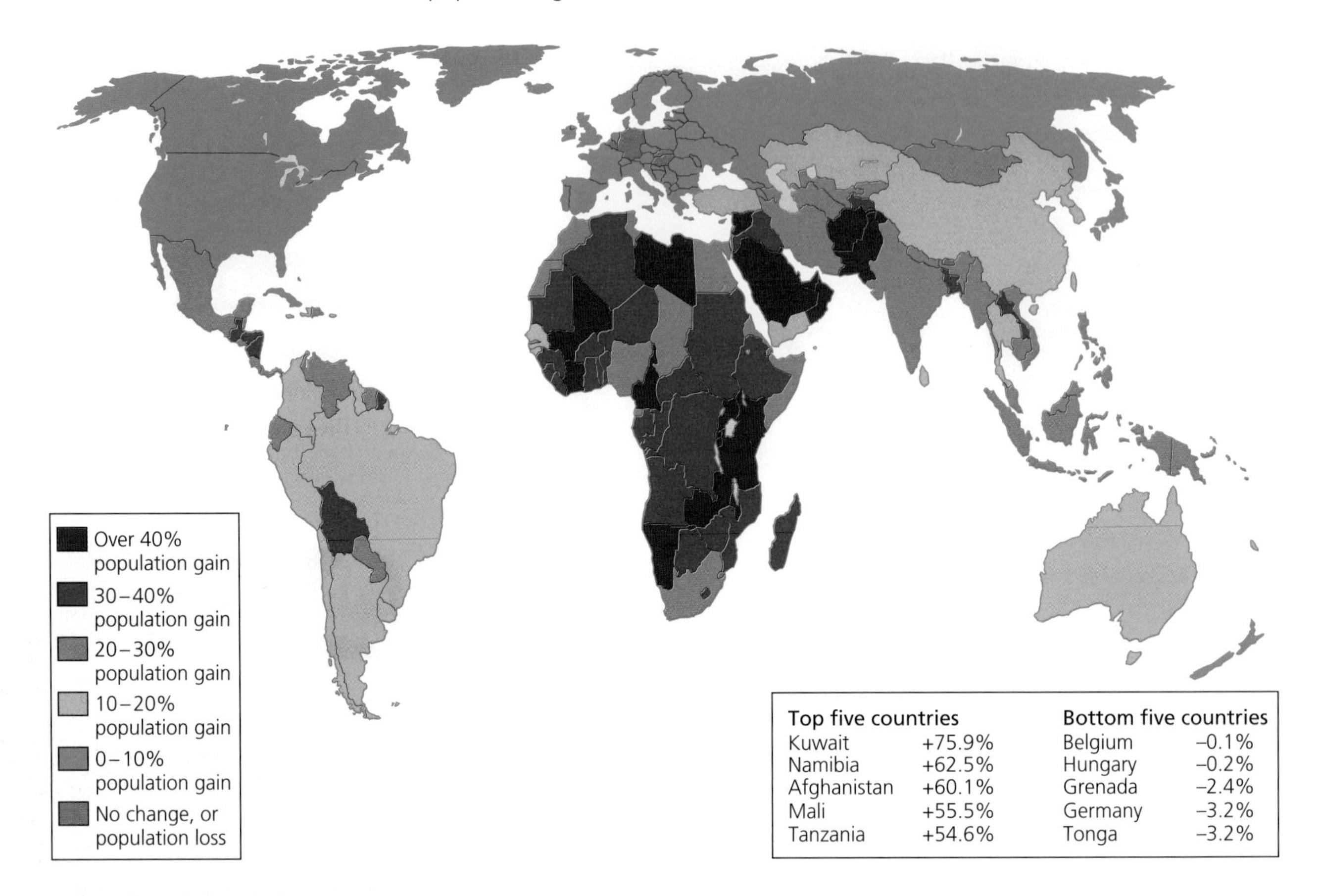

Top five countries		Bottom five countries	
Kuwait	+75.9%	Belgium	–0.1%
Namibia	+62.5%	Hungary	–0.2%
Afghanistan	+60.1%	Grenada	–2.4%
Mali	+55.5%	Germany	–3.2%
Tanzania	+54.6%	Tonga	–3.2%

Figure 4: The global distribution of population change, 1990–2000

Activity 1

Using Figure 4, make a note of the top four countries showing the greatest population gains, and the four showing the greatest loss. How does the UK compare with these countries?

Changing birth and death rates

Population change is produced by two processes – natural change and migration (Figure 5). Natural change depends on the balance between birth rates and **death rates**. If there are more births than deaths, population will increase. If there are more deaths than births, population will decrease. Migration is the movement of people into and out of an area or country. If there are more **immigrants** (in-comers) than **emigrants** (out-goers), there will be a gain in population. If the situation is reversed, there will be a loss of population.

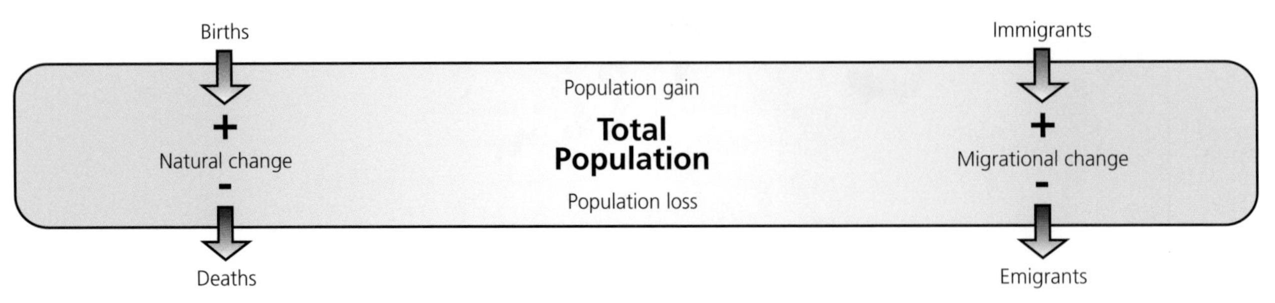

Figure 5: Elements of population change

So the overall rate of population change will depend on the natural change combined with the change as a result of migration (see Chapter 13). If both are positive, then rates of population growth will be high. If they are both negative, then there will be a high rate of population loss. If one is positive and the other negative, then it is possible that they might balance each other out – resulting in little or no change.

As countries develop, their birth and death rates change and, as a result, so too their rate of natural change. These changes underlie a generalisation known as the '**demographic transition model**', which is shown in Figure 6.

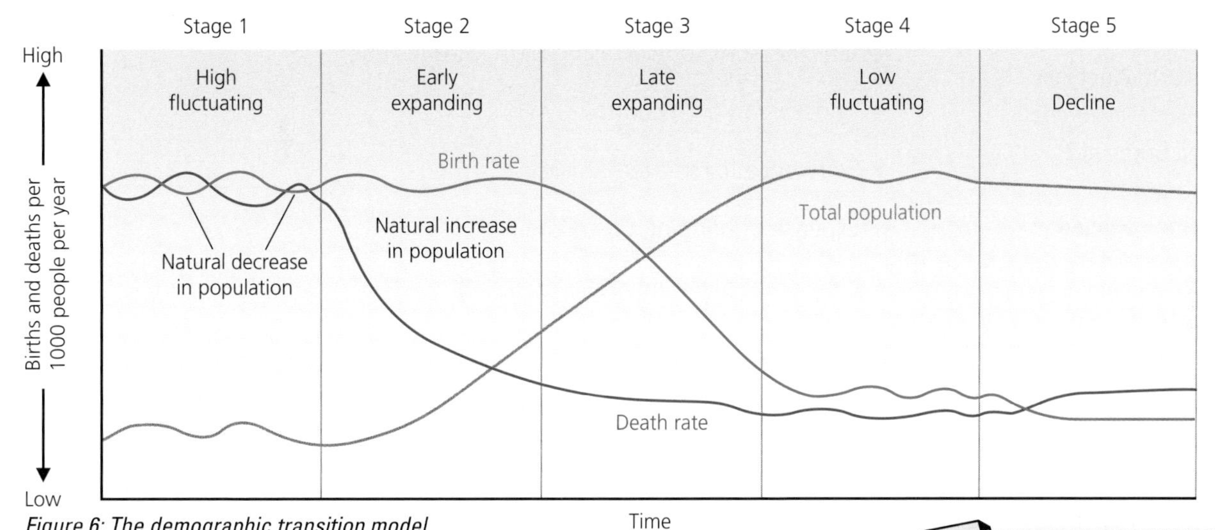

Figure 6: The demographic transition model

The model suggests that countries pass through five different stages:

Stage 1: High fluctuating – A period of high birth and death rates, both of which fluctuate. Natural change hovers between increase and decrease. Reasons for the high birth rate include:

- Little or no birth control
- High infant mortality (death) rate
- Children are seen as an asset and status symbol.

Reasons for the high death rate include:

- High infant mortality
- High incidence of disease
- Poor nutrition and famine
- Poor housing and hygiene
- Little or no health care.

Suggest reasons why the death rate in many LICs has fallen in recent years. (3 marks, June 2007)

How students answered

Some students performed very poorly on this question. They to tended to explain low rates rather than falling rates and some talked about HICs rather than LICs.

20% (0–1 marks)

Many students did reasonably well by recognising one correct reason for falling death rates, with 'better health care' the usual choice.

40% (2 marks)

Many students did very well by identifying two reasons, usually better health care and improved sanitation.

40% (3 marks)

Skills Builder 2

(a) Complete this table by calculating the rates of natural change.

(b) How do you explain the fact that Cambodia and Chile have lower death rates than the UK and Germany?

Birth and death rates in selected countries, 2007

Country	Stage	Birth rate Births/1000	Death rate Deaths/1000	Rate of natural change
Swaziland	1	27.0	30.4	
Cambodia	2	25.5	8.2	
Chile	3	15.0	5.9	
UK	4	10.7	10.1	
Germany	5	8.2	10.7	

Avoid implying that countries in one stage of the demographic transition model will soon enter another stage. That may not happen.

Stage 2: Early expanding – A period of high birth rates, but falling death rates. The population begins to increase rapidly. Reasons for the falling death rate include:

- Lower infant mortality
- Improved health care and hygiene
- Better nutrition
- Safer water and better waste disposal.

Stage 3: Late expanding – A period of falling birth rates and death rates. The rate of population growth slows down as the rate of natural increase lessens. Reasons for the falling birth rates include:

- Widespread birth control
- Preference for smaller families
- Expense of bringing up children
- Low infant mortality rate.

Stage 4: Low fluctuating – A period of low birth and death rates. Natural change hovers between increase and decrease. The population as a whole 'greys' – it becomes older. Death rates are kept low by improving health care. Birth rates are kept low by:

- Effective birth control
- More working women delaying the age at which they start having a family.

Stage 5: Decline – A period during which the death rate slightly exceeds the birth rate. The result is natural decrease and a decline in population. The population becomes even 'greyer'. Modern medicine is keeping elderly people alive longer. Fewer people in the reproductive age range (15–50) means a lower birth rate.

This stage has only recently been reached – by some European countries. It raises some interesting questions. Do populations continue to decline to the point where they disappear altogether? Or will immigration keep up the numbers?

Finally, a word or two of warning about the demographic transition model:

- It is a generalisation
- Not all countries will follow the same pathway
- Countries that do appear to follow the transition will do so at different speeds – some much faster than others. The important factor is the speed of development.

Factors affecting the distribution and density of population

Figure 7 shows a number of factors affecting the distribution and density of population. In the previous section, we dealt with one of those factors: rates of population change.

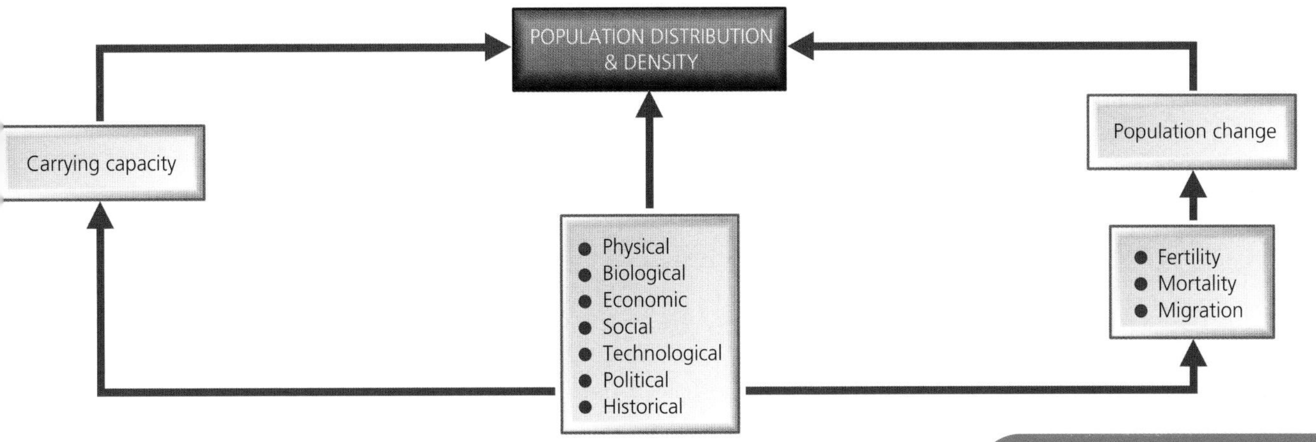

Figure 7: Factors affecting the distribution and density of population

The factors that are listed in the middle of the diagram fall into three groups:

Natural factors	• Physical – relief (slopes), rivers, climate (precipitation, temperatures) • Biological – soils, vegetation
Human factors	• Economic – minerals, energy, industries, services • Social – type of community (urban or rural) • Technological – ability to exploit resources • Political – governments may try to change the distribution
Historical factors	(Discussed below)

We will illustrate some of the natural and human factors in the next two case studies. We will see that population distribution and density can be strongly influenced by natural resources, including such things as fertile soils, plenty of water, minerals and energy.

The historical factors may need a little explanation. Two points need to be understood. First, the present distribution of population is something that has gradually evolved over many centuries. It is mainly an accumulation of the past. Secondly, factors that are no longer significant may have made an impact that still lingers on in the present distribution. For example, spring lines at the base of hills were attractive to early settlement. The settlements are still there and many have grown, but the nearby springs are now irrelevant because, like all settlements today, they rely on water piped in from some distance.

Figure 7 shows one more factor, **carrying capacity.** This is the maximum number of people that can be supported by the resources and technology of a given area. So this particularly affects population density. The greater the carrying capacity, the higher will be the population density.

Activity 2

Brainstorm with a partner the names of places in the world where a population distribution is either favoured or hindered by each of the following: (i) relief (ii) climate and (iii) vegetation.

Activity 3

Read the case studies on pages 210 and 211.

1. Why are upland and mountainous areas often sparsely populated?

2. Suggest at least four different reasons for the concentration of population in the London area.

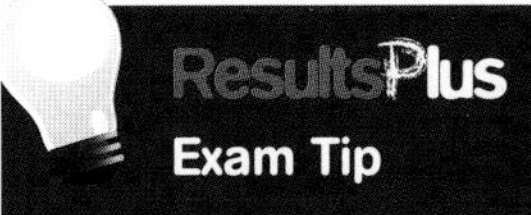

Exam Tip

When explaining the impact of physical factors on the density of the population, always try to go beyond statements such as 'it is colder therefore it is harder to live' and *explain* what is 'harder' about it

Case study: The distribution of population in China

If you look at Figure 8, which shows the distribution of population in China, you will notice the following features:

- Population is concentrated in the eastern half of the country, where densities are everywhere greater than 25 persons per km², and in places exceed 1,000 person per km².
- There is a very sparsely populated belt to the west, where densities are less than 5 persons per km².
- To the north-west of that there is another belt where population densities are between 25 and 250 persons per km².

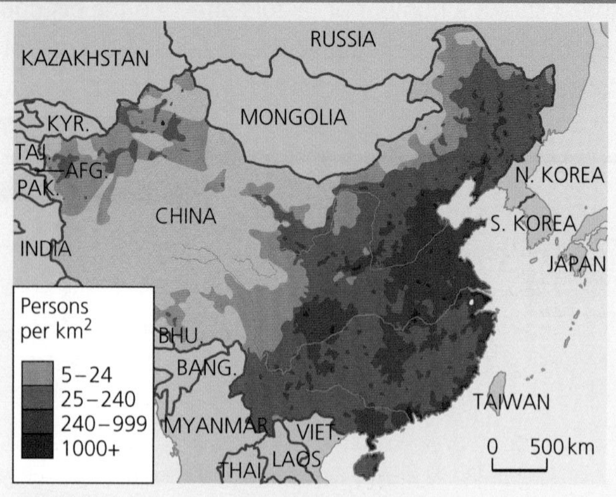

Figure 8: Distribution of population in China

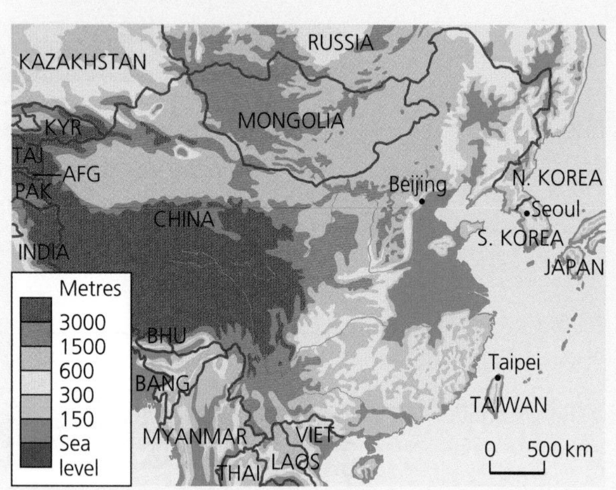

Figure 9: Distribution of relief in China

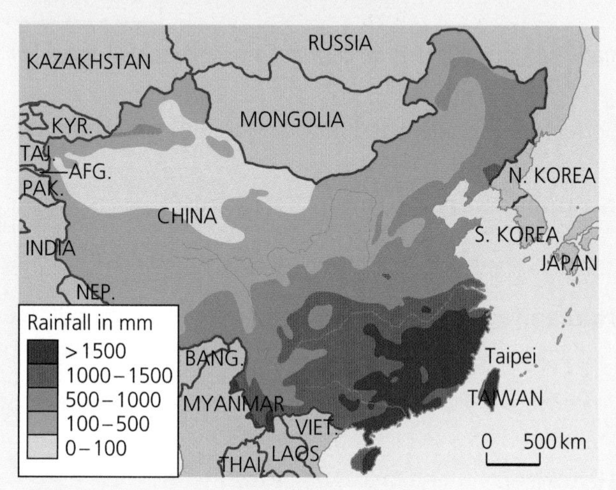

Figure 10: Distribution of precipitation in China

If you now look at the relief map (Figure 9), you will see that the eastern belt of high population densities roughly coincides with the main lowland areas – which probably have the best soils too. High mountains (over 5,000 metres) may well explain the low population densities elsewhere. Similarly, if you look at Figure 10, the eastern belt coincides this time with the highest rainfall (over 50 cm per year). Equally, the rest of the country, with its low densities, appears to be an area of very low precipitation. Given the high mountains, that precipitation is probably in the form of snow. So we can conclude that population distribution in China is strongly influenced in a positive way by lower land and adequate rainfall, and in a negative way by mountains and drought.

But there is one other obvious factor influencing the distribution of population. Note the concentration of population all the way along the coast. What is the attraction of the coast – other than its low relief and abundant rainfall? This is where you find the country's major ports trading with the rest of the world. Ports tend to become centres of industry, and industry attracts people.

Case study quick notes:
Relief and climate have an important influence on the distribution of population.

Case study: The distribution of population in the UK

The influence of relief is also noticeable in the case of the UK. The most sparsely populated areas shown in Figure 11 are the Highlands and Islands of Scotland and the uplands of Northern Ireland, Wales and the north of England. The distribution is dominated by two areas of high population densities, one in the south-east of England and the other to the north-west. What the map is picking out here – as well as in the north-east of England and in central Scotland – are the main concentrations of urban population. But why are these urban concentrations in these locations? The key factor is historical. In all cases apart from the London area, the highest population densities coincide with the coalfields. Although coal is no longer mined, in the past the coal was the energy resource that attracted huge amounts of industry. Despite deindustrialisation, these old coalfield areas still retain large urban populations. Today, however, the people tend to make their living in the tertiary sector.

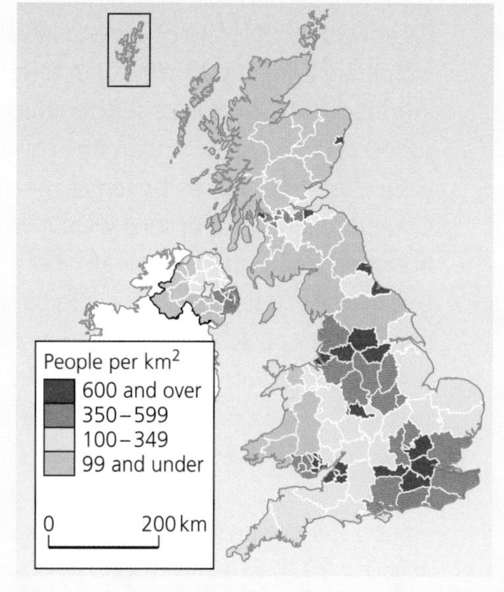

Figure 11: The distribution of population in the UK, 2001

Case study quick notes:
The past can be an important influence on the present distribution of population.

Managing populations – a look at two countries

One of the duties of governments is to keep an eye on what is happening to the population. Is it growing or shrinking? Is it changing at a rate that might create problems? There are many examples where a government has decided that it should intervene – to reduce the rate of population growth or (less often) the rate of decline.

So what can a government do to lower the population? The simplest target is the birth rate. If this can be lowered, then in time the whole population will decrease. The birth rate is usually lowered by encouraging birth control and making it expensive to have too many children. It is important to understand that there are some religions that are strongly against both policies.

When the aim is to increase population, the simplest target again is the birth rate, with couples being offered money or other benefits to have more children. But there is another possible policy – to encourage migrants to come to the country, particularly young adults, who are likely to want to have children eventually.

We will now take a look at two very different case studies – one about dealing with population growth, the other about population decline. They are also different in terms of size. China is the most populous country in the world – and one of the largest – while Singapore is one of the smallest. But they do have one thing in common – most of their people are of the same ethnic origin – Chinese. We have already looked at the distribution of population in China. China covers an area of 9.6 million km² and has a population of 1,320 million. Singapore occupies a small island of just 639 km² and has a population of 4.8 million.

Study Figures 8 and 9 on page 210. These are maps of China's population density and physical features. Explain the relationship between the physical geography of China and its population distribution. (6 marks)

■ Basic answers (Level 1)
Describe the patterns in the figures but do not attempt to use one to explain the other.

● Good answers (Level 2)
Outline the broad pattern and identify that the higher the relief the lower the population density.

▲ Excellent answers (Level 3)
Identify the relationship between the physical geography and population distribution, and also comment on the variations – for example, high density in central China.

Case study: China's one-child policy

Faced with a high rate of growth in its already huge population, the Chinese government introduced 'voluntary' schemes to control birth rate as early as the 1970s. With the birth rate already falling fast, it introduced it's controversial **one-child policy** in 1981. For nearly twenty years after that, no couple was supposed to have more than one child, and those who did were penalised in various ways. All couples were closely monitored by female health workers who were trusted members of the Communist Party. Couples with only one child were given a 'one-child certificate' entitling them to such benefits as cash bonuses, longer maternity leave, better childcare, and preferential access to housing. In return for the certificate, couples would have to pledge that they would not have more children. Unmarried young people were persuaded to postpone marriage. Couples without children were advised to 'wait their turn'. Women with 'unauthorised' pregnancies were pressured to have abortions. Those who already had a child were urged to use contraception or undergo sterilisation. Couples with more than one child were virtually forced to be sterilised.

Since 1996 the policy has been relaxed a little, particularly in rural areas. Birth rate fell from 34 per 1,000 in 1970 to 13 per 1,000 in 2008, and the annual poulation growth rate fell from 2.4% to 0.6%. Even so, the total population has grown from 996 million in 1980 to 1,320 million today. The brake has certainly been put on population growth, and as the cutback in children works its way up the population pyramid, so its effect will become stronger.

One thing seems very clear. The policy has been much more effective in urban areas than in the countryside. In cities, finding enough living space for a family of three is difficult. Raising a child there is much more expensive. In rural areas, however, there is always the need for an extra pair of hands to help on the family farm. In short, there are two very different attitudes towards children.

China's one-child policy remains very controversial. Population growth fell very rapidly before it was introduced in 1981 as a result of changes in Chinese society, land reform and, no doubt, a 'voluntary' policy that may not have been entirely voluntary in practice. The one-child policy had a number of unwanted consequences. We will look at some of them in the next section (see page 217).

Figure 12: Promoting the one-child policy in China

Case study quick notes:
- Population management needs tough government.
- Population management can also be tough on people.

Activity 4

What evidence is there in this case study that the 'one-child policy' has been a success?

Case study: Singapore's 'Have three or more' policy

Since the mid-1960s, the Singapore government has controlled the size of its population. First, it wanted to reduce the rate of population growth, because it was worried that the small island would soon become overpopulated. This policy was so successful that in the mid-1980s the government was forced to completely reverse the policy. The old family planning slogan of 'Stop at two' was replaced by 'Have three or more – if you can afford it'. Instead of penalising couples for having more than two children, they now introduced a whole new set of incentives to encourage them to do just that.
These included:

- Tax rebates for the third child and subsequent children
- Cheap nurseries
- Preferential access to the best schools
- Spacious apartments.

Pregnant women are offered special counselling to discourage 'abortions of convenience' or sterilisation after the birth of one or two children.

Singapore has also used the immigration option as a way of increasing its population. But it has only encouraged graduates to come and settle in the country. Throughout its short history, Singapore has always been keen to raise the education and skills levels of its people. Immigrants who are not graduates take a long time to become Singapore citizens. Once they have, they are classified as 'residents' and can take part in the country's population programme. The 'non-residents' – even if they are living fairly permanently in Singapore – are still subject to the 'Stop at two' policy. The reason for this is that the government wishes to control not just population numbers, but also the ethnic make-up of the population. At present, 75% are of Chinese origin, 14% Malay and 9% Indian. The government wants this mix to stay the same. What this means is that it is particularly difficult for people of Indian origin to become Singapore citizens. Is the message in Figure 13 really being carried out?

Figure 13: Racial equality in Singapore?

Case study quick notes:

- Governments are able to control population numbers in a variety of ways.
- Control is usually achieved by a 'stick and carrot' approach.

Activity 5

Try to find out if the population of the UK has ever been managed? Is it being managed today?

Objectives

- Learn the different characteristics of a population
- Recognise the significance of population pyramids
- Understand the advantages and disadvantages of ageing populations.

Answers to examination questions that ask about the characteristics of a population usually mention age and/ or gender. By adding something else, such as ethnicity or occupation, your answers will have more depth.

Skills Builder 3

Using the table, write a short account highlighting the main population changes. Support your answer with numerical data. This might include calculating percentage changes.

Characteristics of population

Local characteristics

We like to think of ourselves as unique individuals – because 'everyone is different'. But each of us can be 'pigeon-holed' into a number of different slots, based on certain of our characteristics. For example, we can be grouped according to physical qualities, such as gender (male or female) or age (young, middle-aged or old). In terms of **ethnicity**, most of us belong to one of the four main groups based on skin colour (black, brown, yellow or white). We can also be divided on the basis of our religion, the main groupings being Christian, Muslim, Hindu, Buddhist or atheist. Finally, it is possible to group people according to their occupation – what they do for a living. They may be professional, managerial, clerical, skilled manual, unskilled manual or unemployed. (Or they may be 'still in education', like you.)

All this information about people is collected by a census. A census involves literally counting everyone in a country or region and recording their characteristics (gender, age, ethnic origin, etc.). Most countries hold a census once every ten years. When the results are compared with those of the previous census, important information can be obtained about how a population is changing. Is it growing or declining, and at what rate? Is the population as a whole becoming 'greyer'? Is the ethnic or religious mix changing? The first UK census was conducted way back in 1801 and the last was conducted in 2001. In that time, only once (in 1941) was a census not. So, thanks to twenty censuses spread over 200 years, we have a very detailed picture of how the UK's population has changed.

Some aspects of population change in the UK, 1991–2001

Population characteristic	1991	2001
Total population (millions)	55.5	58.8
Sex ratio (males per 1,000 females)	941	946
% of population under 16 years	21	20
% of population over 60	19	21
% of population White	95.4	91.4
% of population Christian	No data	71.6
Clerical workers (%)	10	9

Southampton: highlights from the 2001 census

Southampton is a port city, situated 110 km to the south-west of London on the Hampshire coast. In the census highlights below, the figures for Southampton are compared with those for England and the South East.

- Between the 1991 and 2001 censuses, the population of Southampton increased by 20,581 to 217,445.
- 87.5% of Southampton's residents were born in England, which is very similar to the national proportion (87.4%), but slightly lower than the figure for the South East (88%).
- In 2001 Southampton had a higher proportion of 18–19 and 20–24 year olds than England or the South East, due to the student population in the city (Figure 14).
- 192,970 people identified themselves as White British, 88.7% of the total population, making this by far the largest ethnic group in Southampton (Figure 15). This proportion is lower than that in the South East, which was 91.3%.
- The majority, 65.6%, stated their religion as Christian, which is lower than the national figure of 71.7% (Figure 16). The proportion of Southampton residents having no religion was 21.6%. This was higher than the national (14.6%) and regional (16.5%) figures.
- The largest source of employment for Southampton residents was in the wholesale and retail trade.

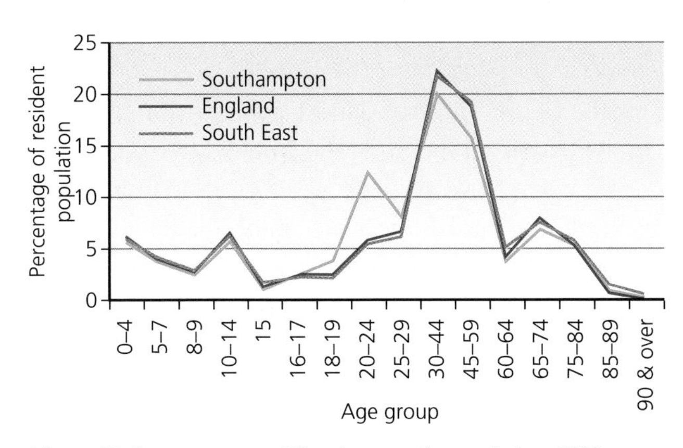

Figure 14: Age structure of Southampton's population, 2001

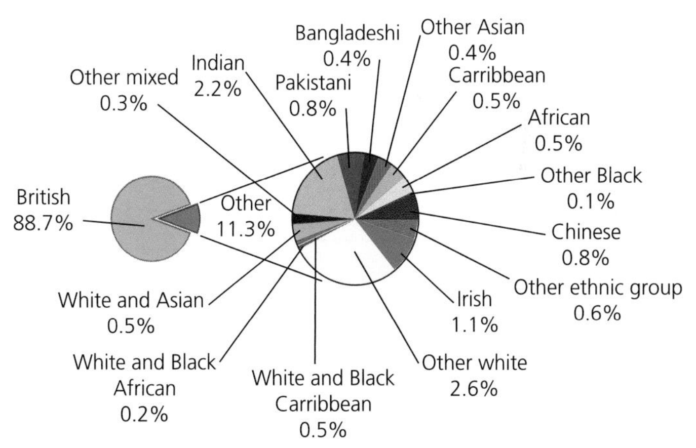

Figure 15: Ethnicity in Southampton, 2001

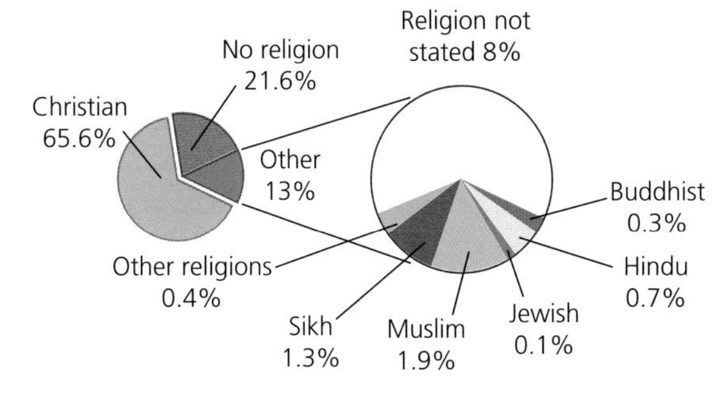

Figure 16: Religious beliefs in Southampton, 2001

Quick notes (Southampton census):
This sort of information can be obtained from the census (Office of National Statistics – www.ons.gov.uk/) about any area, including your home area.

ResultsPlus
Watch out!

- When comparing the results from more than one census be sure that the boundaries of the area you are interested in have not changed.

Population pyramids

Two of the most important characteristics of a population – age and gender – can be shown in a single diagram known as a **population pyramid**. The male population is shown on one side of the pyramid and the female on the other side (Figure 17). The vertical axis of the pyramid is divided up into age groups, usually of five years. The youngest age group, 0 to 4 yrs, is at the bottom and the oldest age group, over 90 yrs, at the top. The number of males or females in a particular age group is shown by a horizontal bar. The bar is drawn proportional in length to either the number (as in Figure 17) or the percentage of all males or females in that age group.

The overall shape of the pyramid tells us about the present balances between the different age groups and between males and females. The diagram can be very helpful when making forecasts about future population totals and population growth rates.

A broad-based and rather squat pyramid shape, like that for Indonesia (an LIC), shows what is called a **youthful population**. There are plenty of young adults in the population and they are responsible for a high birth rate and many young children. But because the death rate is high and **life expectancy** low, there are not many people aged over 55. The 'young' part of the population (under 19) is much greater than the 'elderly' part (over 60).

The population pyramid for Mexico (an MIC) is more even-sided and taller than that for Indonesia. This means that the death rate is lower and life expectancy is greater. The 'young' part of the population is still larger than the 'elderly' but less so than in Indonesia.

The population pyramid for the UK (an HIC) has almost lost its 'pyramid' shape because it bulges in the middle. The base of the pyramid, the young population, has been 'eroded' away. Now the 'elderly' are equal in number to the 'young'. Clearly, the birth rate has declined and so too the death rate. More people are living longer. The pyramid tells us that we have an **ageing** or **greying population**.

From these three population pyramids, we can now understand that their shape is controlled by:

- The birth rate – the higher it is, the broader the base of the pyramid.
- The death rate – the lower it is, the taller the pyramid.
- The balance between the two rates – whether births exceed deaths or vice versa.

We can also relate these three pyramids to the demographic transition model (see Figure 6 on page 207). The population pyramid for Indonesia is typical of a country at Stage 2; that for Mexico typical of Stage 3, and that for the UK typical of Stage 4 (possibly 5). Clearly, as a country develops, its population pyramid changes.

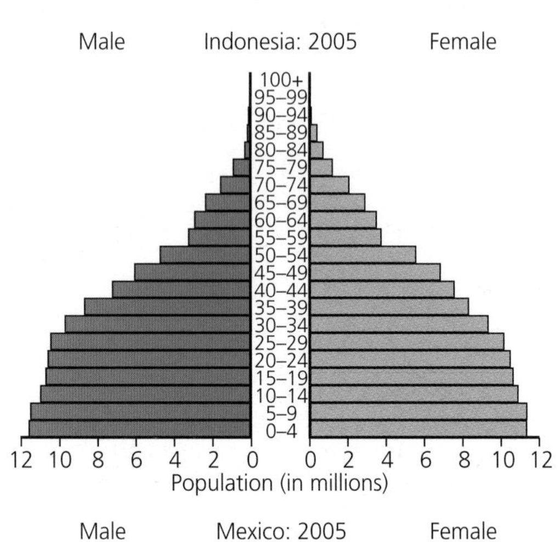

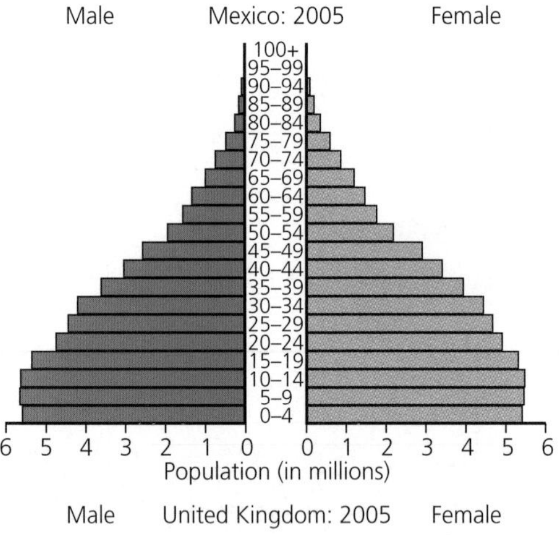

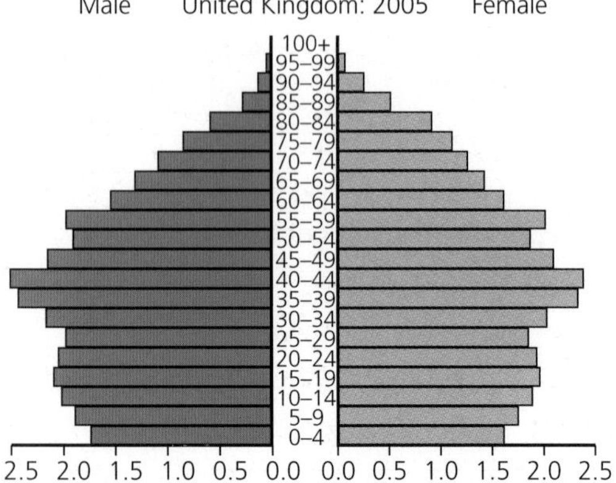

Source: U.S. Census Bureau, International Data Base.

Figure 17: Population pyramids for Indonesia, Mexico and the UK, 2005. Note the different scales used.

China's population pyramid

Figure 18 shows China's population pyramid. The shape is rather odd – a rather 'fatter' version of the UK's. The bulge in the pyramid is greatest for those aged 40 to 44. As you move down from that point, the pyramid becomes narrower. What we are seeing here is the outcome of forty years of the 'one child' policy – a marked lowering of the birth rate.

The Chinese tradition is to prefer sons. So as couples are limited to having only one child, there has been widespread sex-selective abortion. If you look closely at the age bars, you will see there are more males than females below the age of 45. There are now 120 males to every 100 females. This is having two consequences:

- Parents 'spoil' their 'one-boy' child and as a result he tends to be obese, demanding and delinquent. They are referred to as 'little emperors'!
- Because of the increasing shortage of women of marrying age, bartering for brides and 'bride kidnapping' have become common in rural areas. And prostitution has increased in the cities.

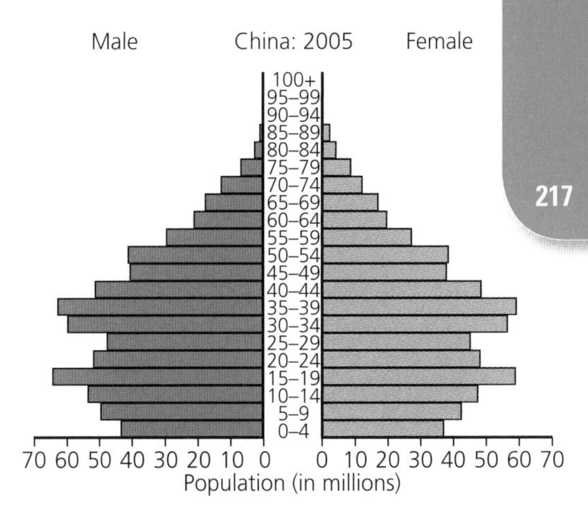

Source: U.S. Census Bureau, International Data Base.

Figure 18: Population pyramid for China, 2005

The consequences of youthful and ageing populations

There are few countries that have what we might call a 'balanced' or 'mature' population structure. Most countries fall into either the youthful or ageing categories. In order to identify the consequences of both types of population, we will explore them under four different headings.

Demographic

In the case of a youthful population, the most obvious consequence is a growth in numbers. More people need to be employed, housed and fed. On the other hand, with an ageing population, the prospect is that numbers will either remain the same or decline. That could be beneficial if there is a shortage of food or other resources.

Economic

Both types of population will need public money to be spent on special services (Figure 20). With a youthful population, this will mean building and running nurseries, schools and children's clinics. An ageing population needs such facilities as care homes, suitable housing and day centres.

Figure 20: Two different services for two different populations

Skills Builder 4

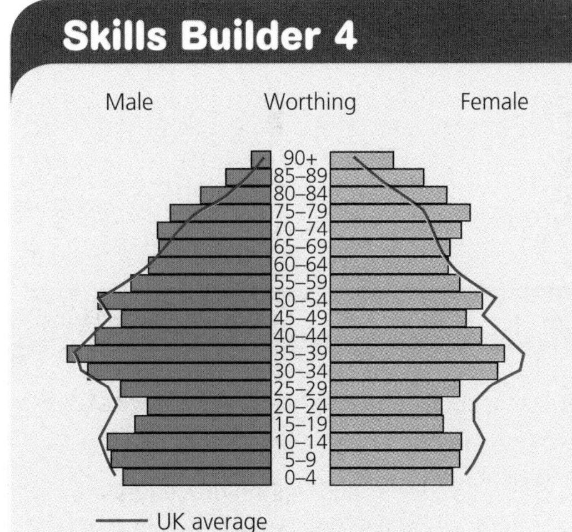

Figure 19: Population pyramid for Worthing, with the UK average shape superimposed

Study Figure 19.

(a) Describe how the Worthing pyramid differs from that of the UK.

(b) As a result of your comparison, what sort of town do you think Worthing is?

Activity 6

Reread the information on China in this chapter and summarise all the consequences of the 'one-child policy' that have been mentioned

With an ageing population comes a reduction in the size of the workforce. A shortage of workers could be a worry to employers. But it is also important that there should be enough jobs for the young people wishing to start work.

Social

Both populations pose a problem in terms of what is known as **dependence**. With a youthful population, dependence will be mainly of children on parents. The challenge will be being able to support growing families. In an ageing population, an increasing number of elderly people become dependent on either their children or care homes.

Political

Politicians are meant to look after the interests of the people. The political priorities of youthful and ageing populations are different. With a youthful population the priorities are likely to be education and job creation. With an ageing population, the priorities may well be pensions, healthcare services and age discrimination.

If we weigh up the consequences, it appears that both types of population have to face particular challenges. In general, we might conclude that the challenges are rather greater in the case of ageing populations. In other words, the negatives are perhaps stronger. But, as the next section will show, ageing populations are not all bad news.

Skills Builder 5

Look again at the UK pyramid in Figure 17 on page 216. Roughly how many people fall in (a) the 'young dependant' category (under 19) and (b) the 'elderly dependant' category (over 64)?

Advantages and disadvantages of an ageing population – a case study of the UK

The UK is one of 61 countries in which not enough babies are being born to replace those people who die. So the population looks set to become 'greyer' and to decline in number. Here are some facts about the UK's ageing population in 2001:

- There were 5.4 million women aged 65 or over, compared with 3.9 million men.
- Of the men aged over 65, and of the women over 60, nearly 10% were still working.
- Seven out of ten pensioners depended on state benefits for over half their income.
- Of all the people aged over 75, nearly two-thirds of women were widowed, compared with around one-third of men.

The ageing population is often portrayed only in negative terms. There is no doubting that it has a downside, but not all is gloom and doom. There is also an upside. Let us illustrate these two sides by identifying pluses and minuses under two of the four headings used previously.

Economic	**Plus:** A combination of good health, paid-up mortgages, plenty of disposable income and 'empty nests' means that many people in their sixties and seventies are helping the boom in the leisure business. A growth in overseas 'SKI' (spending the kids' inheritance) holidays is helping those LICs currently viewed as attractive tourist destinations. That disposable income is also welcomed by UK retailers.
	Minus: With fewer people in the workforce, there is the challenge of raising enough by way of taxes to pay for a growing number of state pensions. Added to that is the fact that the present state pension is inadequate and badly needs to be raised. **Deprivation** is being experienced by a growing number of pensioners – and who is to pay for all those social services?
Social	**Plus:** The creation of retirement resorts in popular locations. It means that pensioners have plenty of opportunities to shake off what is a common problem in old age – loneliness. It also means that these retirement resorts – often referred to as 'grey ghettos' – are going to be geared up and able to provide the social and medical services that are especially needed by the elderly.
	Minus: With people living longer, there is the challenge of who will look after them when they become too frail. It used to be their children's responsibility, but more and more elderly people are being handed over to professional carers. They are being shut away in care homes and sheltered accommodation.

Figure 21: Saga holidaymakers

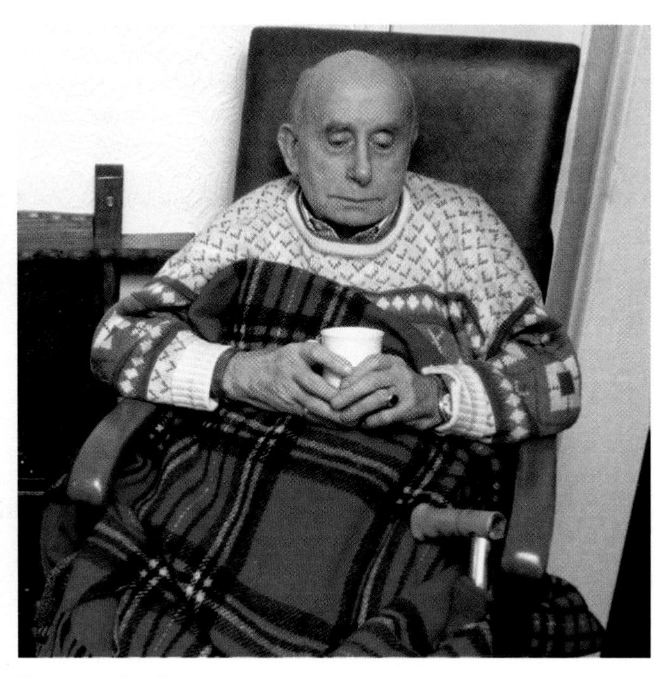

Figure 22: Pensioner poverty

Activity 7

In a small group, make a double-column list of what you think are the positives and negatives of an ageing population. Which column has the longer list?

Know Zone
Chapter 12 Population change

Although population growth is slowing down, this is one of the most challenging issues for the planet. Can we cope with another 3 billion people in the next thirty years and what impact do varying growth rates have on different parts of the world?

You should know...

- ☐ How global population has changed in the last 200 years
- ☐ The distribution of the global population
- ☐ Where the sparsely and densely populated regions are located
- ☐ Why some global regions are more densely populated than others
- ☐ The reasons why birth rates and death rates vary from place to place and from time to time
- ☐ The main features of the demographic transition model and its limitations
- ☐ How physical and human factors affect the population distribution and density in China
- ☐ How physical and human factors affect the population distribution and density in the UK
- ☐ The policies in one country that is trying to increase its population growth
- ☐ The policies in another country that is trying to reduce its population growth
- ☐ How to interpret local census data
- ☐ How to construct and interpret population pyramids
- ☐ How pyramids differ from country to country according to their state of development
- ☐ The positive and negative consequences of an ageing population
- ☐ The positive and negative consequences of a youthful population
- ☐ How an ageing population affects one country

Key terms

Ageing population
Birth rate
Carrying capacity
Death rate
Demographic transition model
Dependence
Deprivation
Emigrants
Ethnicity
Immigrants
Life expectancy
One-child policy
Population density
Population distribution
Population growth rate
Population pyramid
Youthful population
Zero population growth

Which key terms match the following definitions?

A The average number of years a person might be expected to live

B A diagram to show how a population is composed in terms of gender and age

C Where people are located within a given area

D The increase in population over a year, expressed as a percentage

E The number of deaths per 1,000 people in a year

F People who move into an area or country

G When natural change and migration change cancel each other out, and there is no change in the total population

H A population in which there is a high percentage of people aged 65 or over

To check your answers, look at the glossary on page 289.

ResultsPlus
Maximise your marks

Foundation Question: Describe three impacts of an ageing population. (6 marks – 2 marks for each)

Student answer (awarded 3 marks)	Examiner comments	Build a better answer (awarded 6 marks)
Older people cost money so everyone has to work more. *There are more nurses so that costs more.* *There are less people at work so the production of goods might fall.*	• ***Older people cost money...*** This is too vague for a mark. The student needs to say why older people cost money. • ***There are more...*** There probably is a need for more nurses (1 mark), but the answer does not give a reason why. • ***There are less people...*** gets 2 marks because the student makes a point about the workforce and links it to a negative impact.	The more old people there are then the higher the taxes are to pay for pensions. Older people require more health care, especially the over-75s, so more hospitals, doctors and nurses will be needed, which is expensive. There are less people at work so the production of goods might fall. On the other hand, the elderly frequently provide cheap or even unpaid volunteer labour.

Overall comment: Remember that impacts can be positive and negative.

Higher Question: Study Figure 6 on page 207. Explain the changes in **birth rate** in Stages 2 and 3 of the Demographic Transition Model. (4 marks)

Student answer (awarded 2 marks)	Examiner comments	Build a better answer (awarded 4 marks)
Birth rate falls most of the time especially in Stage 3. It is more variable in Stage 2. *There are many reasons for this which might be to do with needing more children to help out.* *Other things got better at this time so with medicine fewer children died so fewer were needed to be born.*	• ***Birth rate falls...***gains 1 mark. It is a *description* of change rather than an *explanation* but it does help the answer. • ***There are many reasons...*** does not score any marks because the need for children does not help explain a fall in birth rate. • ***Other things got better...*** scores a mark because it links infant mortality with fewer births.	Birth rate remains high in Stage 2 but then falls very sharply in Stage 3. Compulsory education and industrial change meant large families were no longer an advantage. Improving diets and better sanitation led to lower infant mortality and so fewer births were necessary.

Overall comment: Be careful to get the focus of the answer correct. In this case, the focus was to *explain* declining birth rate.

Chapter 13 A moving world

Objectives

- Learn the different types of circulation and migration.
- Recognise the significance of distinguishing between different types of migration.
- Understand the reasons for the increasing volume of migration.

ResultsPlus
Exam Tip

Be sure that you understand the difference between 'migration' and 'circulation'.

Population movement

Because of modern methods of transport, people are now able to move much more easily and quickly. Long distances no longer hinder movement as much as they once did.

Types of movement

The top part of Figure 1 shows that there are two main types of population movement – **migration** and **circulation**. With migration, a change of address is involved. But that change must last for at least one year. So someone who works abroad for six months is not a migrant. Circulation is made up of moves that are shorter in terms of time. The shift in location is only temporary, as with the foreign worker on a six-month contract. Shopping, commuting and taking a holiday are three common forms of circulation. These examples also show that we can distinguish between different types of circulation according to their purpose. We can also distinguish, as in Figure 1, according to the time involved:

- Daily – shopping, commuting
- Weekly – longer-distance commuting, holidaymaking
- Monthly – helping with harvesting, fruit picking, etc.
- Seasonal – going away to university, UK pensioners spending the winter in Spain.

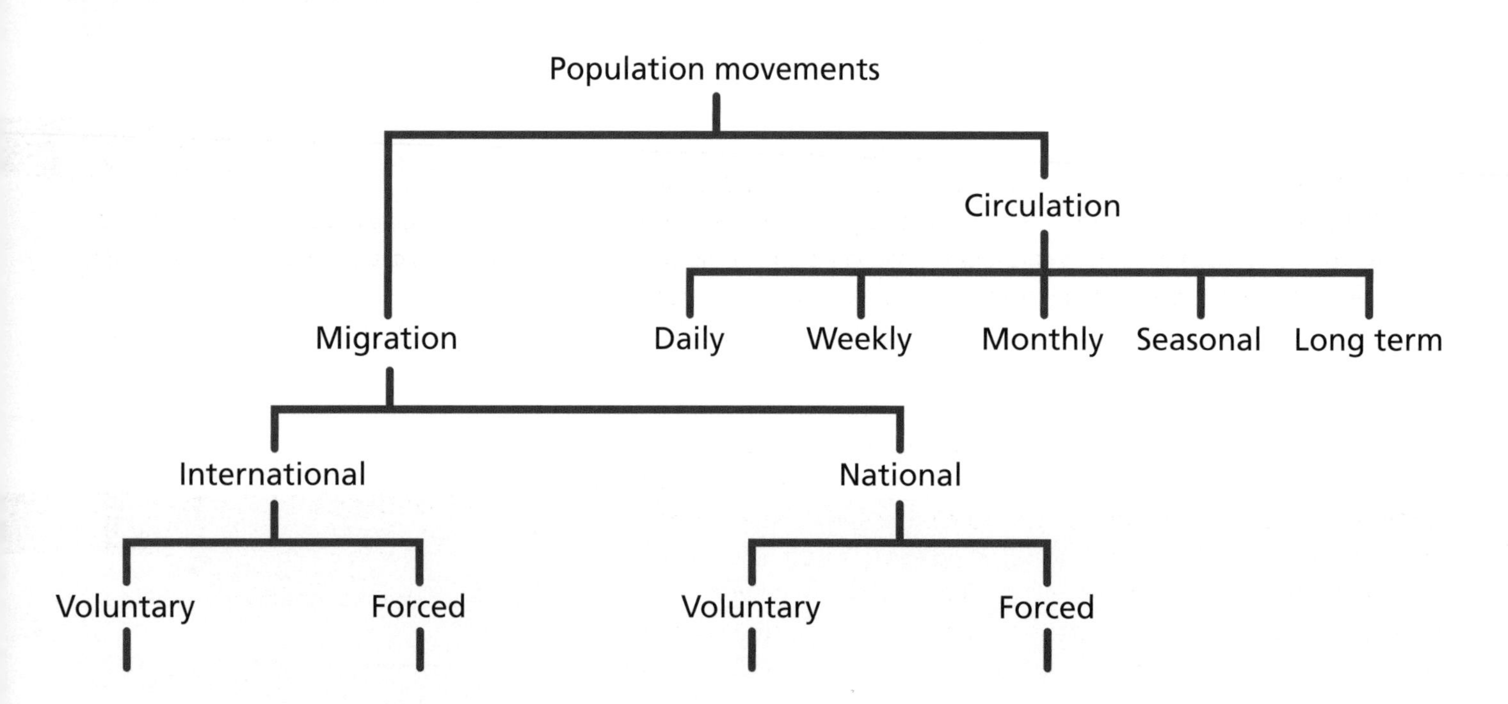

Figure 1: Types of population movement

An important distinction is between national (within a country) and international (between countries) migration. With international migration, another distinction also needs to be made – between those entering a country (**immigrants**) and those leaving it (**emigrants**). The difference between the two during the course of, say, a year is known as the 'migration balance' or the 'net migration'. If the number arriving is greater than the number leaving, then the situation is described as net in-migration. Net out-migration occurs if departures exceed arrivals. So, the migration balance is usually either negative or positive. Figure 5 in Chapter 12 (on page 206) shows that the migration balance is important to a country's population – whether it is growing or declining.

The idea of a migration balance also applies to migration within a country. For example, the rural–urban migration that is so common in many developing countries leads to towns and cities becoming places of net migration gain. As a result, rural areas become areas of net migration loss.

Types of migration

Having made the distinction between **national** and **international** migration, Figure 1 shows that we can recognise two more types under each of those headings – **forced** and **voluntary**. There are other possible ways of classifying migrations. For example, whether they are short-haul or long-haul (distance), short-term or long-term (the number of years spent at the chosen destination), small volume or large volume (number of migrants), legal or illegal. Each migration can show a mix of these features. For example, emigration from the UK to Australia in the middle of the twentieth century was international, voluntary, long-haul, legal, mainly long-term and large in volume.

With forced migration, people have no real choice other than to move. **Refugees** are people who have been forced to flee an area – because, for example, they are persecuted for their religion (Christians in North India), their tribal origins (Tutsi and Hutu in Congo and Rwanda), their politics (opponents of the African National Congress in Zimbabwe) or because their homeland is invaded by a foreign army (Tibetans escaping Chinese troops and settling in Nepal). Refugees mainly fall in the international category. Refugees are also created by natural hazards such as earthquakes, tsunamis and hurricanes. But these refugees rarely cross international frontiers. Eventually, most return home – maybe in a matter of days or weeks – in which case they are not migrants.

With voluntary migration, it is the migrants who choose to move. They do not have to move. Maybe they are looking for a better job and better pay (for example, moving from a remote rural area to a city). Maybe they are wishing to live nearer to family and friends (for example, people from Bangladesh moving to live with relatives in London). Maybe they are wishing to live in a more attractive location (for example, pensioners moving to popular retirement areas). Figure 2 provides a picture of this retirement migration in England and Wales. It is a **flow map**. The flow lines show the direction of movement, and their width is drawn proportional to the number of migrants taking that particular route.

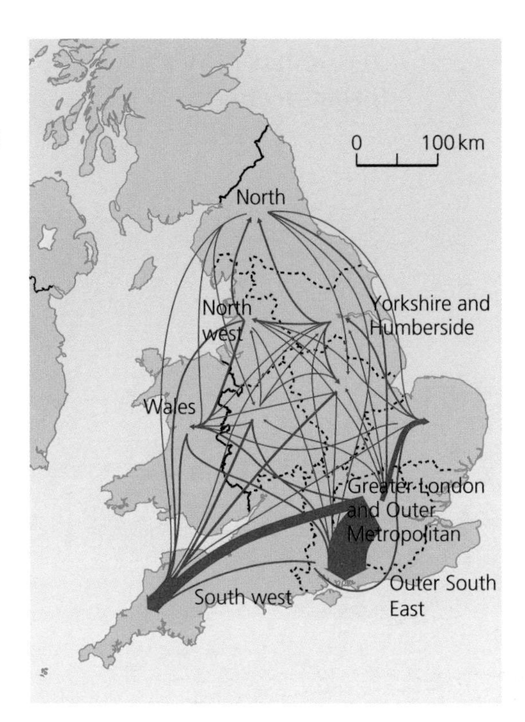

Figure 2: Retirement migration flows in England and Wales

Skills Builder 1

Study Figure 2 and identify the main features of the retirement migration pattern.

Mention both distance and direction in your answer.

■ Be very careful not to confuse internal migration – within a country – and international migration – between countries.

Objectives

- Learn the main flows of people into and within Europe.
- Explain the reasons for those flows.
- Understand why migration has both positive and negative effects.

Figure 3: Immigrants from Jamaica arriving in the UK in 1948

Quick notes (the UK opens its doors):
The UK's need for labour led to it becoming a multi-ethnic society.

Skills Builder 2

Study Figure 4.

(a) By how much did the number of UK residents born abroad increase between 1951 and 2001?

(b) What percentage of the population did they account for in 2001?

(c) What is happening to the rate of increase in this percentage?

Flows of population

International migration into and within Europe

These are two distinct migrations, and they need to be looked at separately.

Migration into Europe

There were large flows of migrants into Europe shortly after the end of the Second World War in 1945. Europe needed a lot of labour to repair the huge amount of bomb damage and to help the economy recover. But Europe was short of labour, because so many people, particularly men, had been killed. The situation was solved by encouraging migrant workers and their families to come to Europe, mainly from Africa, Asia and the Caribbean. France drew migrants from its old colonies in North Africa and Indochina. West Germany (as it was then) attracted many workers from Turkey.

The UK opens its doors

The UK's post-war immigrants came mainly from colonies in the Caribbean, and from what had been the Indian Empire (India, Pakistan and Bangladesh). Immigration was encouraged by an Act of Parliament which gave all Commonwealth (ex-colonial) citizens free entry into the UK. The first ship to bring in immigrants from Jamaica docked at Tilbury (Essex) in June 1948 (Figure 3).

It is estimated that during the 1950s and 1960s over a quarter of a million immigrants came from the Caribbean. Roughly the same number came from what had been the Indian Empire. By 1971 there were over 1 million immigrants from Commonwealth countries. The new settlers took up a variety of jobs. Many found work in textile factories and steelworks. Many drove buses or worked on the railways. Later arrivals, particularly from India, opened corner shops and restaurants or ran post offices. By the 1970s, the UK had more than enough labour, and controls were introduced to reduce the migrant arrivals.

Figure 4 shows the number of UK residents who were born abroad. Despite the controls on immigration, clearly the number has been steadily rising. So too has the percentage of the UK population that they represent.

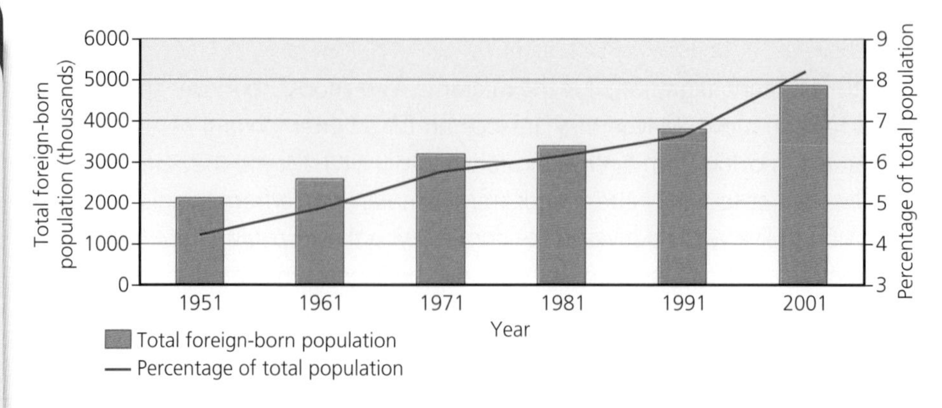

Figure 4: UK residents born abroad, 1951–2001

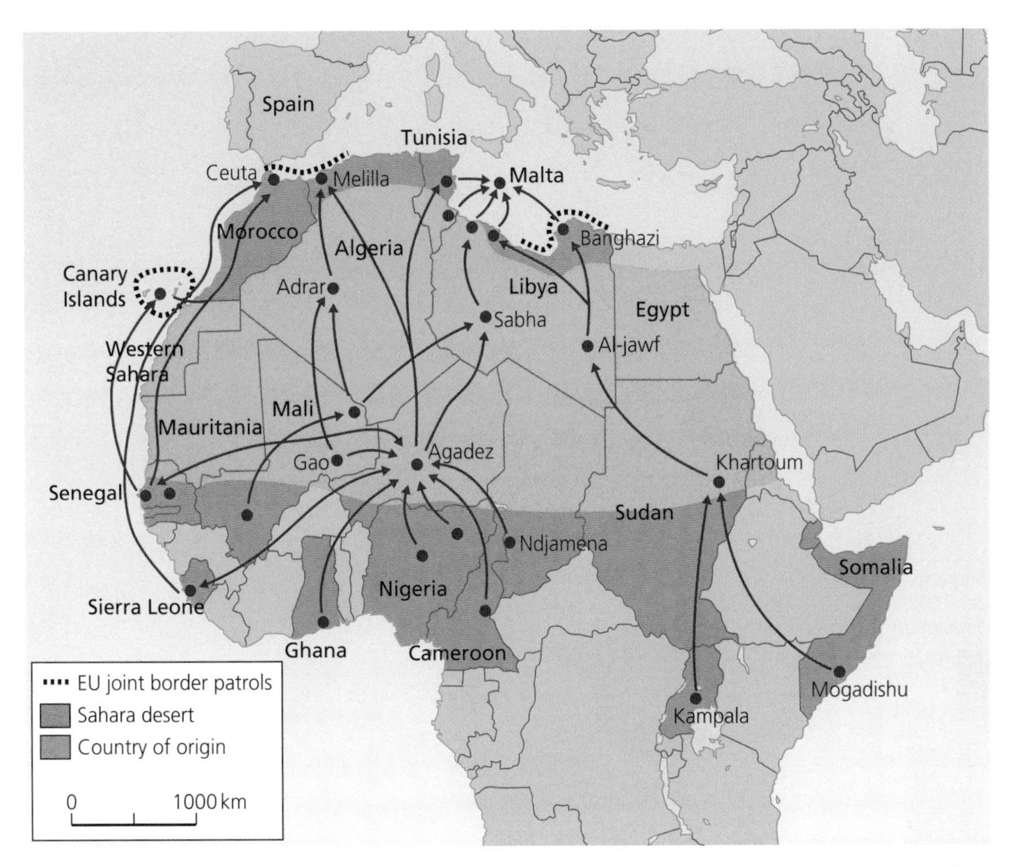

Figure 5: Illegal migrant routes into Europe from Africa

In the last ten years or so, many migrants have been leaving Africa. Poverty, famine, conflict and civil war have been 'pushing' them in the direction of the European Union. However, the EU no longer needs more workers. But the migrants are desperate to leave Africa and are being smuggled into the EU as illegal immigrants. Figure 5 shows the various routes they take. Most enter the EU by boat through the Canary Islands, Malta, Italy or the Spanish territories of Cueta and Melilla.

Migration within Europe

Immediately following the end of the Second World War, there was a large movement of people from Eastern Europe to Western Europe. The East had been occupied by the Soviet Union, and many wished to escape the harsh communist regime. Large numbers of Poles, Ukrainians and Hungarians chose to settle in the UK. But soon the so-called 'Iron Curtain' came down – and escape was no longer possible.

In 1990, after the collapse of the Soviet Union, there was a surge of migration, particularly from the East to Western Europe again. The evolution of the European Economic Community into the European Union in 1993 and its subsequent enlargements have greatly increased migration between member countries.

Suggest reasons why some countries experience much higher rates of in-migration than others. (4 marks)

■ **Basic answers** (0–1 marks)
Make general statements about some countries being 'more attractive', but fail to identify why this is.

● **Good answers** (2 marks)
Identify at least two ideas, such as the availability of jobs and the policies over immigration, but lack the details to develop these ideas.

▲ **Excellent answers** (3–4 marks)
Explicitly identify the pull factors and can describe the type of jobs, the higher wages and the ease of entry with reference to some located data and detail.

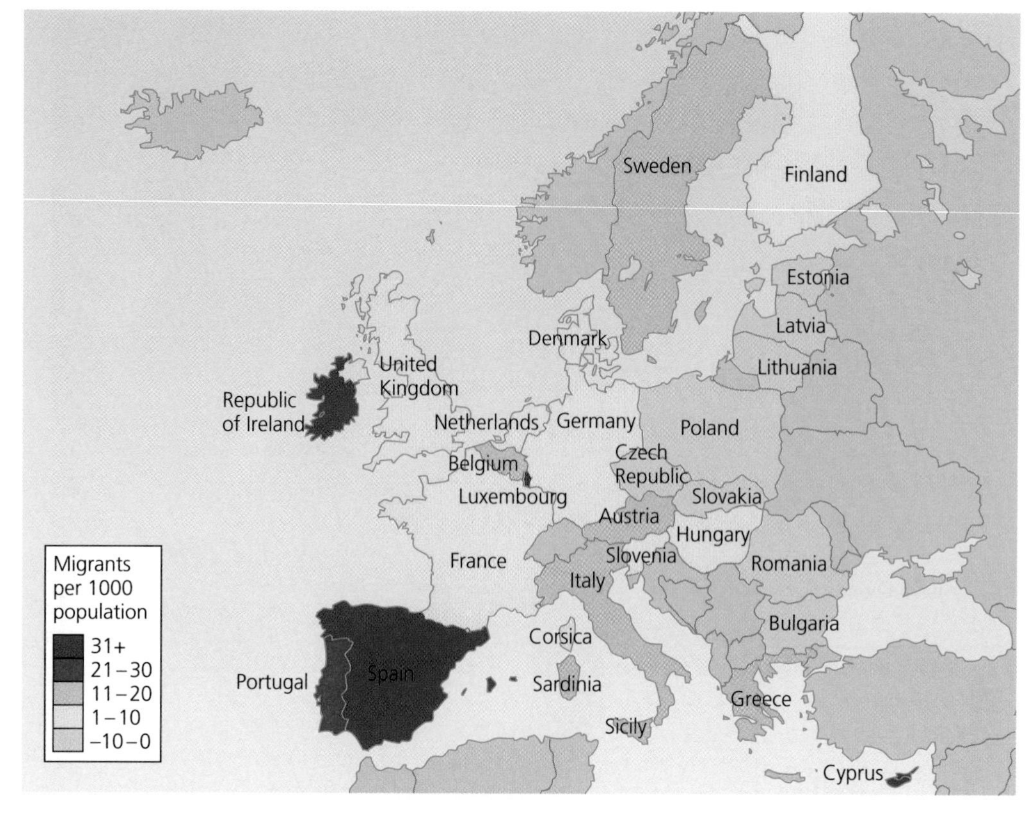

Figure 6: Net migration in the European Union, 2000–2004

Figure 6 summarises the pattern of migration within the European Union between 2000 and 2004. The new member countries in East Europe show negative migration balances. Many adults have left in search of work and better pay. The Mediterranean countries show high net migration gains. This reflects the north to south movement of 'sun-seekers' (especially retired people), plus the illegal migration from Africa (see Figure 5). Contrary to what is often rumoured, the UK is not experiencing a high rate of net in-migration. Thousands of East Europeans may be coming to work in the UK, but many do so for less than one year. So they are not true migrants. Also we need to realise that people are also emigrating from the UK – to France, Spain and Portugal, as well as to Australia and New Zealand.

Activity 1

1. Try and learn the names of all 27 member countries of the EU.
2. Can you suggest reasons for the high rates of in-migration in Ireland shown in Figure 6?

The economic and social impacts of migration

When looking at the impacts of migration, we need to remember that:

- Two locations are involved – the place where the migration starts (the origin) and the place where the migration ends (the destination). In the case of international migration, the terms **country of origin** and **host country** are commonly used.
- In both locations, the impacts can be good (**positive**) and bad (**negative**).
- There are different kinds of impact – on the environment, on the economy and on society.

Impact on the host country

Consider the immigration into the UK that took place between 1950 and 1975. Enough time has passed for us to see and understand its impacts. Its positive economic impacts were:

- It met the shortage of unskilled and semi-skilled labour.
- It played an important part in the post-war reconstruction of the country.

On the negative side:

- Public money had to be spent on meeting the everyday needs of the immigrants and their families – housing, schools, healthcare, etc.
- When the economy went into recession in the 1970s, these immigrants added to the burden of unemployment.

As for the social impacts, it is clear that the immigration created great tensions. The UK was not used to having sizeable ethnic groups in its population. There was hostility towards the immigrants, and they were discriminated against – and abused. As a consequence, they tended to settle in particular areas, for personal security reasons. Better to live with people from the same ethnic group than run the risk of being victimised by white neighbours. Most often, immigrants became segregated, to form 'ghettos' in areas of rundown housing in the inner areas of towns and cities (Figure 7). Clearly, there were some serious negative social impacts.

Figure 7: An inner city area of immigrant housing

Slowly, the social situation has changed. Discrimination has been made illegal. UK law now states that all citizens, regardless of ethnicity, should enjoy equal opportunities. Slowly, most white people have come to realise that they are not threatened by immigrants. They have come to realise that there are positives. Ethnic groups add to the country's skill base and culture. The offspring of the original immigrants have made their way in the UK, and many now occupy well-paid and responsible jobs. They have moved into areas of better housing. They represent the country in a range of sports. They have seats in Parliament. They are now truly UK citizens. The situation is still not one of complete harmony, and many still live in poverty. But the situation is much better than it was forty years ago.

Activity 2

Investigate a town or city you know. Do members of different ethnic groups live in particular areas? Can you think of reasons why this pattern occurs?

ResultsPlus
Build Better Answers

Study Figure 6 which shows net migration in the European Union (2000–2004). Describe the positive and negative impacts of migration on a country (such as Spain) that is receiving net migration gains. (4 marks)

■ **Basic answers** (0–1 marks)
Usually include a simple statement about the negative impact of in-migration, such as 'too many people make the country crowded'. Occasionally, there is also an attempt to illustrate the point.

● **Good answers** (2 marks)
State at least one positive and one negative impact. The positive impact usually focuses on labour supply, while the negative impact usually covers the 'shortages' of housing.

▲ **Excellent answers** (3–4 marks)
Describe several impacts, both negative and positive, and include some facts and figures to illustrate the points made.

ResultsPlus
Exam Tip

▲ Examination questions often use terms such as economic, social and political. Make sure you understand what these mean.

Impact on the country of origin

In order to see the other side of the migration coin, let us look at Jamaica – which has probably supplied more immigrants to the UK than any other Caribbean country (Figure 8). Since it became more difficult for Jamaicans to settle in the UK, large numbers have been emigrating instead to the USA and Canada. Jamaica continues to lose some 20,000 migrants each year.

One of the positive economic spin-offs of this continuing emigration is the increase in 'remittances' – the money sent to family members back home in Jamaica. In 2005 remittances amounted to $1.65 billion (16 per cent of Jamaica's GNI). Between 1990 and 2005, this money helped to cut poverty in Jamaica by half.

Skills Builder 3

Compare Jamaica's population pyramid with the ones shown in Chapter 12. Which one is most like Jamaica's?

An obvious negative economic impact of the emigration is that Jamaica is losing some of its best labour. There is a 'brain drain'. The island's economic development is being held back by this loss of skilled and more enterprising labour. Despite the loss, there is still high unemployment – a sure symptom of the slow rate of economic growth. High unemployment results in two things. It persuades even more people to emigrate. And it encourages people to turn to crime – in particular, there is gang violence related to the drugs trade. Clearly, this is a serious negative social impact.

Exam Tip

▲ Make sure that you know which end of the migration process an examination question is focusing on. Is it the host country or the country of origin or both?

The majority of emigrants are young adults, with women outnumbering men. The loss of the more go-ahead women has a negative social impact. It is depriving Jamaican men of wives and depriving the country of both children and good mothers. The population pyramid (Figure 9) shows a number of features:

- A reducing birth rate – fewer babies
- A relative reduction of the population between the ages of 30 and 60
- Fairly good numbers of elderly people, made up of those who have remained in Jamaica all their lives plus those migrants who have retired and decided to return home.

It is not easy to identify the positive social impacts of the emigration. Can you think of any?

Figure 8: Jamaica's scenic coastal landscape – what the emigrants are leaving behind

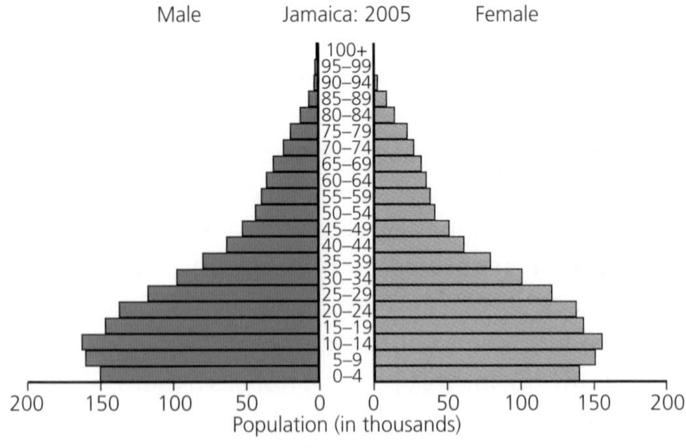

Source: U.S. Census Bureau, International Data Base

Figure 9: The population pyramid for Jamaica, 2005

Factors enabling migration

The decision to migrate is usually the outcome of two forces, known as the **push–pull mechanism**. Figure 10 shows that the push force occurs in the potential migrant's home location. It is something that prods the person to move away. The pull force is something that attracts that same person to a particular destination. Very often the pull factor is the mirror image of the push factor. For example, being out of work sets the person thinking that they must move to find a job. They hear there is a labour shortage in a particular city or country. Thus the combination of the push and pull factors persuades the person to migrate.

Objectives

- Learn some of the factors encouraging people to move.
- Recognise the impact of these factors.
- Understand the particular significance of transport and communications.

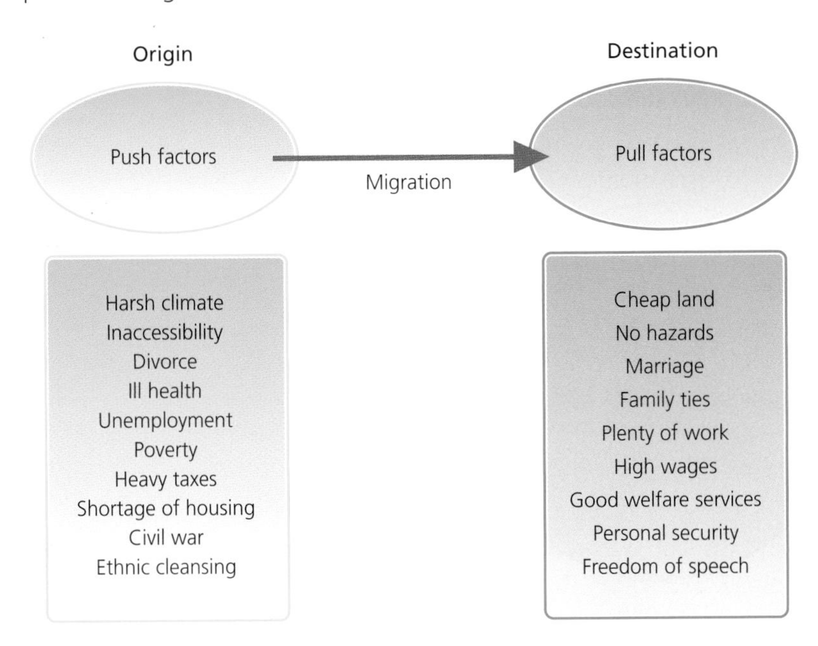

Figure 10: The push–pull mechanism

Modern communications

An important question is this: How does a person find out about the opportunities (the attractions) and the downside of some distant location? How can they be sure that they are moving to somewhere better?

In the past, information was mostly passed by word of mouth or by letter. A migrant would write back home to a relative or friend who was thinking about making the same move. They would give their opinion on such vital things as employment, housing and the cost of living. The decision to migrate or not will have been based on such subjective feedback. The same sort of communication goes on today, except that it is more likely done by telephone or email and is much faster. Thanks to today's mass media, the would-be migrant is now often able to 'see' and 'feel' distant places without taking a step outside the house. Thanks to the internet, the amount of useful information now available is not only much greater, but it is probably more reliable (Figure 11). The abundance of information also allows a person to weigh up and compare a number of possible destinations. Thanks to modern communications technology, therefore, migration today is less likely to be a leap in the dark.

Figure 11: An internet café – one of the many modern means of communication

Modern transport

Once the decision has been taken to move to a particular destination, the migrant is able to take advantage of modern transport. This can move them there quickly and cheaply.

Figure 12 illustrates how the increasing speed of transport has 'shrunk' the world. In recent years, budget airlines, such as easyJet and Ryanair, have made it possible for the not-so-well-off to travel great distances at relatively low cost. High-speed rail services and motorways link major cities. It is now much easier for people to visit and check out possible destinations before deciding if and where to migrate. It has also become easier for migrants to return to their homelands for a holiday or a family reunion. Conversations during such visits can encourage others to make a move.

ResultsPlus
Watch out!

- When answering examination questions about migration do not confuse those things that motivate people to move, such as better jobs, with those things that allow them to move more easily, such as improvements in transport.

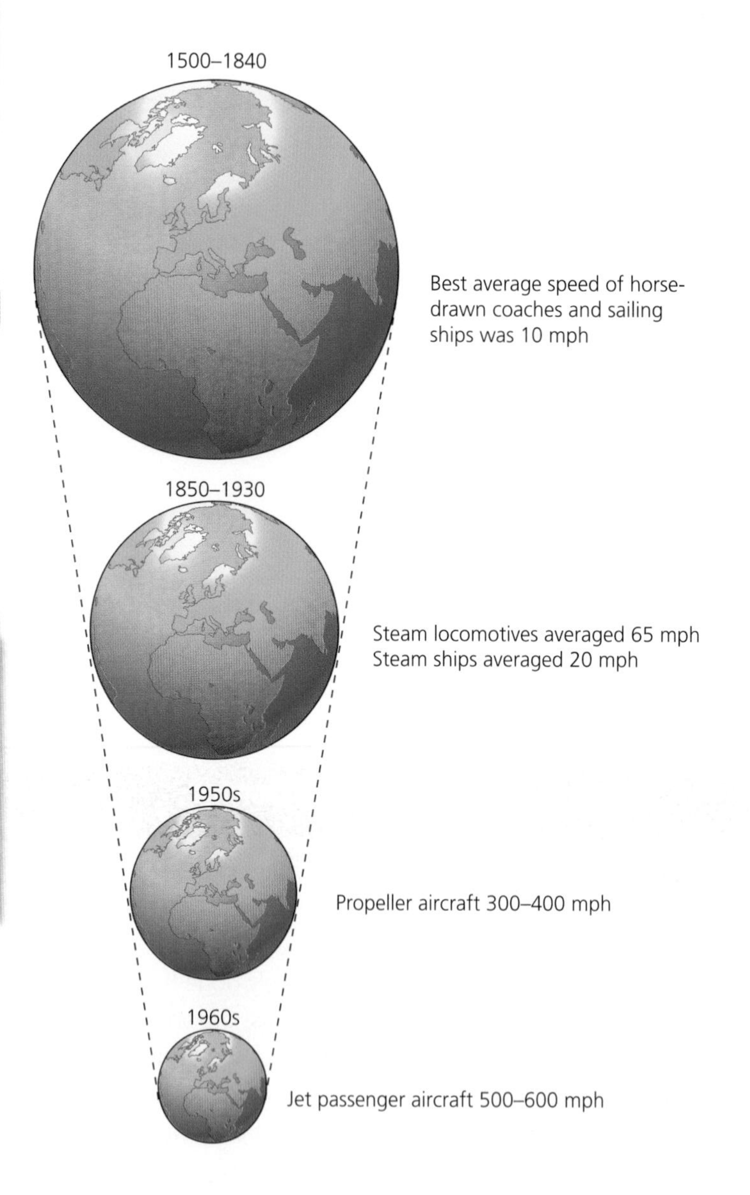

Figure 12: A shrinking world

Activity 3

Explain how both modern transport and modern communications encourage migration. Which do you think is the more powerful influence?

Give your reasons.

Relaxing national boundaries

One of the advantages of belonging to the European Union is that workers are free to move between member countries. All they need is a passport or national identity card (Figure 13). This freedom of movement helps to explain the patterns of migration shown in Figure 6. Whilst it is becoming easier to migrate within the EU, it is becoming more difficult for migrants to enter it from other parts of the world. In order to enter the UK from outside the EU, for example, you need a **visa**. There are various types – visitor, business and working holiday, and they are usually valid for less than a year.

If you wish to come to the UK to work and settle down, then you have to go through the points-based system, which was introduced in 2008. It allows British businesses to recruit the skills they need from abroad. Only migrants with those skills will be able to come and live in the UK. The system recognises five tiers of migrants:

- Tier 1 Highly skilled workers – scientists and entrepreneurs, for example
- Tier 2 Skilled workers with a job offer – teachers and nurses, for example
- Tier 3 Low skilled workers filling specific temporary labour shortages – construction workers for a particular project, for example.
- Tier 4 Students
- Tier 5 Youth mobility and temporary workers – musicians coming to play in a concert or people recruited to help with a particular project, for example.

As national boundaries are tightened, so the volume of illegal immigration increases. There are two 'porous' frontiers through which most illegal immigrants enter the EU – in the Mediterranean countries (see Figure 5) and along the eastern border with Belarus, Russia and Ukraine.

Figure 13: UK border control

Objectives

- Learn some examples of short-term population flows.
- Recognise the different reasons for such flows.
- Understand the reasons why economic migrants are not always made welcome.

Reasons for short-term population flows

Short-term population flows are often described as '**temporary migration**'. There is no permanent change of address, but push and pull forces are at work. The case studies below illustrate three different types of temporary migration. All involve movements into and out of the UK. A fourth example is tourism (see Chapter 14).

East European workers come to the UK

In 2004 the East European states of the Czech Republic, Estonia, Hungary, Latvia, Lithuania, Poland, Slovakia and Slovenia joined the EU. Since then, many of their citizens have come to work in the UK. Figure 14 shows the push and pull factors. In most cases, these **economic migrants** intend to stay only until they feel they have made enough money to take home. Figure 15 shows that over half of the East European migrants come from Poland, the largest of the new member states. The vast majority of migrants are young and single, with over 80% of them aged between 18 and 34.

Some UK newspapers (and citizens) take a very negative attitude towards these economic migrants. They are accused of depriving UK workers of jobs and taking advantage of our state benefits system. Figure 14 shows these as two of a number of issues relating to these economic migrants. But the critics choose to ignore four important facts:

- The migrants contribute to the UK's economy by the taxes they pay.
- The jobs that many of them take up are mainly low-paid (see table). Such jobs are often avoided by UK workers.
- The migrants have a strong work ethic, which can directly benefit employers. They are efficient and polite. Sadly, there are employers who unfairly exploit these qualities.
- Less than 5% of them receive any sort of state benefit.

Many of these workers are now returning home – persuaded by the economic recession and better employment prospects back home.

Employment of East European migrants

Employment type	%
Hotels and catering	28
Factory worker	20
Farming	14
Food processing	5
Cleaning	5
Care assistant	3
Sales assistant	3

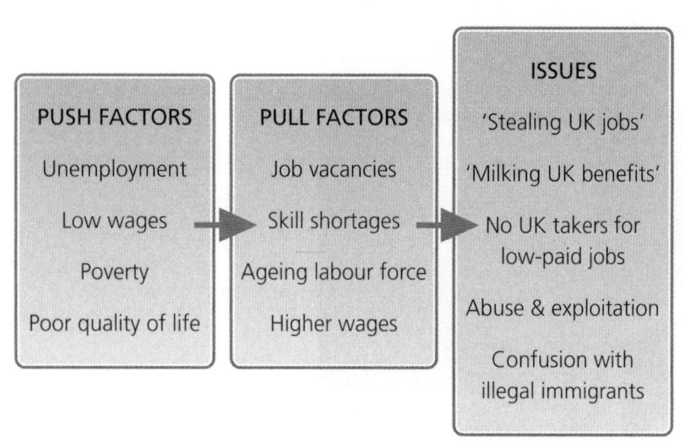

Figure 14: The push factors, pull factors and the issues of East European workers in the UK

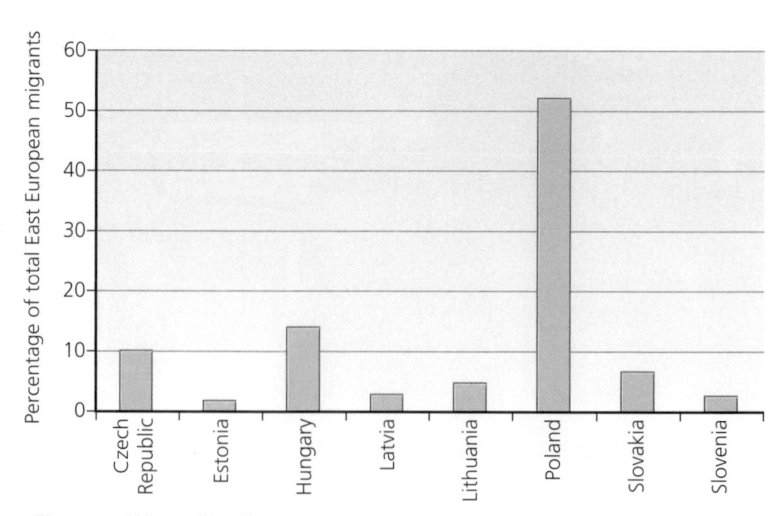

Figure 15: Where East European economic migrants come from

Going abroad for treatment

Each year, long hospital waiting lists and the high cost of private medicine are persuading around 100,000 UK citizens to become **medical tourists**. They go abroad for a variety of treatments – infertility, dental, cosmetic and orthopaedic. Orthopaedic treatment includes such things as hip and knee replacements. The costs of a hip replacement in ten countries is shown below. Given the £8000 cost in the UK, much money can be saved by having the operation abroad. Even allowing for the costs of travel and a week's accommodation, there are still great savings to be made on the total package. Notice that not all medical tourism destinations are in HICs – India, Malaysia and Tunisia are all popular, for example.

A comparison of the costs of having a hip replacement

Country	Treatment price	Treatment saving	Package saving
Bulgaria	£2,000	87%	69%
Cyprus	£4,100	49%	43%
France	£5,689	29%	23%
Germany	£5,296	34%	26%
Hungary	£4,450	44%	40%
India	£3,547	56%	49%
Malaysia	£2,205	72%	60%
Tunisia	£3,000	63%	56%
Turkey	£4,725	41%	36%
UK	£8,000	–	–

Playing the game

The English Premiership started in 1992 with 22 foreign footballers. Today there are well over 300. They are 'pulled' by the global reputation of the Premiership as one of the best football leagues in the world. They are also attracted by the extremely high wages and the glamorous lifestyle (Figure 16). The Premiership now has players from every continent (except Antarctica).

There are many other examples of this globalisation of sport. Professional golfers, tennis players and racing drivers are amongst the greatest movers. They are playing the game and living away from home for much of the year.

Activity 4

1. Identify the push and pull factors illustrated in the three case studies (East European workers come to the UK, Going abroad for treatment and Playing the game).

2. With a partner, see if you can name a current Premiership football player from six different countries.

Figure 16: Christiano Ronaldo from Madeira, Portugal

Quick notes (going abroad for treatment and playing the game):

- Temporary migration is a response to push and pull factors.
- In all cases, the pull seems stronger than the push.

Objectives

- Learn that retirement migration takes place at three different scales.
- Recognise the reasons for these three different scales of retirement migration.
- Understand the impacts that retirement migration has on its destinations.

Retirement migration

UK residents are living longer. The average life expectancy for women is now 81 years, and for men it is 76 years. Most people can expect to enjoy ten or more years of retirement. With this prospect, more and more people are moving after they have retired. They are doing this for a number of reasons:

- It is no longer necessary to live close to what was their place of work
- To downsize into a smaller home
- To sell their home for something cheaper and use the difference in price as a pension
- To move into a quieter, calmer and more attractive environment.

Three main types of **retirement migration** may be recognised:

- Local – where people stay in the same locality, but move house.
- Regional – where people stay within the UK but move to what they think is a more attractive location.
- International – where people make the bold decision to move to another country.

Local retirement migration

Local retirement moves are usually driven by three wishes: (i) to stay in a familiar location, (ii) to continue to live close to family and friends and (iii) to downsize into a smaller, cheaper and easier-to-run dwelling. Many of the new apartment blocks to be seen in UK towns and cities today have been built to meet the third of these wishes (Figure 17).

Figure 17: A hoarding advertising retirement flats

There is at least one more form of local retirement move. This occurs when people are no longer able to be independent or on the death of their partner. Such people tend to end up living either with children and grandchildren or in some form of sheltered or wardened accommodation. The migration trail does not necessarily end there. For some, there may be a final move into a care home or hospice.

Regional retirement migration

Figure 2 (on page 223) shows three very strong retirement flows out of Greater London – to the Outer South East, the South West and East Anglia. Figure 18 shows that many make for coastal areas where the attractions are milder winters, pleasant scenery and a slower pace of life. Coastal resorts such as Yarmouth (on the East coast), Worthing (on the South coast) and Lytham St Anne's (on the West coast) are also popular for another reason. They are already well geared up to meet the needs of a 'greying' population. They are sometimes referred to as the 'grey' resorts.

When you are old, you may not be able to afford to drive a car – or you may not be fit enough. The most important thing is that you live close to shops and healthcare services. That is why more retired people, if they do decide to move, are choosing to move to country towns and coastal resorts – or even abroad.

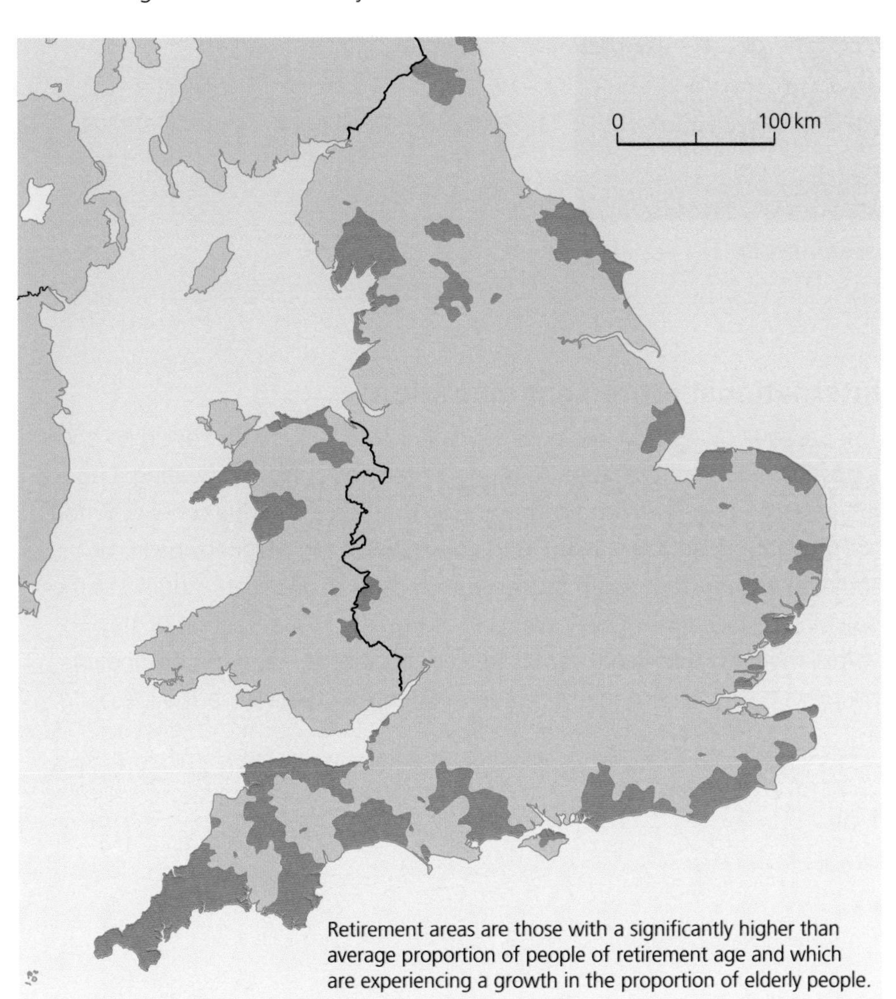

Figure 18: Popular retirement areas in England and Wales

Skills Builder 4

Study Figure 18.

(a) 'Most of the retirement areas are on the coast.' Can you improve on this rather general description?

(b) Identify an inland area that is popular with retired people.

Case study: The 'best place' in the UK to retire to

The magazine, *Yours*, has ranked retirement destinations according to a range of factors – house prices, council tax levels, shopping facilities, crime rates, hospital waiting times, the availability of NHS dentists and the weather. A seaside resort famous for having the world's longest pier has been named as the best retirement place in the UK. Southend-on-Sea in Essex is described as a 'bargain' retirement location (Figure 19). It was ranked top because it is relatively flat, with a pedestrianised centre, a low violent crime rate and a council tax that is almost £100 lower than the UK average. It has 10 km of award-wining beaches, more than eighty parks and lots of activities for older people. It is also close to London. Southend lies within the 'Outer South East' and contributes to the big migration flow shown in Figure 2 (on page 223).

Poole in Dorset was placed second, because of its natural harbour and because waiting times for hip operations there are lower than the national average. Whitehaven in Cumbria was third, with below-average waiting times for a hip replacement and house prices of nearly half the national average. Other places to make it into the top ten include Swansea in Wales, Clacton-on-Sea in Essex, Stirling in Scotland, Leamington Spa in Warwickshire, Skegness in Lincolnshire, Weymouth in Dorset, and Southport on Merseyside.

Figure 19: Southend-on-Sea – the top place to retire to

Case study quick notes:
This case study illustrates the factors that create the 'pull' of a retirement location.

Activity 5

Explain how each of the factors used to rank the retirement destinations in the case study would be a 'pull' for an elderly person.

International retirement migration

The previous case study mentioned weather as a factor when deciding about a possible retirement location. Old people feel the cold, and heating a house in the UK is expensive. It is for this reason that increasing numbers of Britons are retiring abroad. Most head in the direction of the Mediterranean, but some go even further south to places such as the Caribbean. Adding to the pull of such locations is the lower cost of housing. An attraction of staying within the EU is that healthcare is free to UK citizens – and the destination country's health service might be even better than the NHS in the UK.

Case study: Is Spain really so sunny?

With retirement looming, you sell up in Britain, buy your dream villa in Spain and set off to live out your golden years in the sunshine. The idea is so tempting that three-quarters of a million Britons have already done just that. In some areas of the Spanish coast, one in ten residents is now of British origin.

Many of these retirement migrants probably gave a lot of thought to the view from the villa (Figure 20) and how far they are from the nearest golf course or beach. Many probably did not think about falling ill or the increasing problems that age can bring. The consequences can be catastrophic. Some migrants are forced to sell up and come back to the UK, often with no savings or property to fall back on. That is partly because the huge influx of elderly settlers is putting a massive strain on Spain's health service. Some Spanish doctors are refusing to treat anyone who does not speak Spanish unless they have an interpreter present. They fear they might make a wrong diagnosis and be sued.

Another problem is that the Spanish system of caring for the elderly is different from that in the UK. In Spain, it is traditionally a family responsibility. As a result, there are very few services for the elderly. Most British couples retiring to Spain have no family living close by to help them should they fall ill. In the UK there is after-care treatment when the elderly come out of hospital. There are old people's homes, meals-on-wheels and care in the community. A British Consul in Spain has said:

> *Sometimes a partner has died and the other is too old or too ill even to go out and buy food. Some have not budgeted their pensions properly and are living in extreme poverty. British retirees need to realise that some European countries do not have the welfare provision available in the UK.*

Retiring to another country is very different from holidaying there. The key to a successful retirement move is knowing what to expect. For some, the reality will not live up to the dream.

Figure 20: Spanish retirement villas

Case study quick notes:

- There are risks involved in moving to a foreign country, particularly for the elderly.
- Migrants often tend to be blinkered – they see only the attractions of their destination and ignore the negatives.

Consequences for the destination

So to sum up, for the destination area, the consequences of retirement migration – whether it is regional or international – include:

- Top-heavy age structures
- The need to provide a range of health, welfare and social services for the elderly
- Plenty of work for a sizeable labour force of carers
- The need to provide appropriate accommodation
- A relatively high percentage of people living on or close to the poverty line.

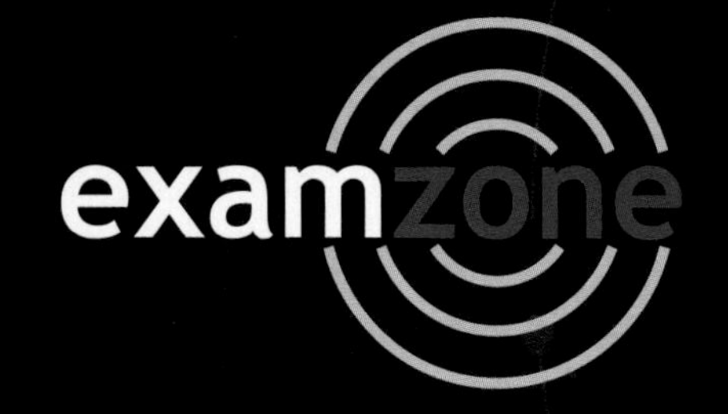

Know Zone
Chapter 13 A moving world

As the world has globalised, the movement of people has increased. In the UK people move home once every seven years on average. Crossing international borders is not always easy, but people are likely to move towards jobs and security if they can. The impact of these movements can be both positive and negative.

You should know...

- ☐ How to define different types of population movement
- ☐ The difference between short-term movements and longer-term migration
- ☐ A good example of a migration within a nation-state
- ☐ A good example of an international migration
- ☐ The difference between forced and voluntary migrations
- ☐ The main patterns of population movement within Europe since 1945
- ☐ The impacts of these movements on both the host countries and the countries of origin
- ☐ The role of new technology in both stimulating people to move and allowing them to do so
- ☐ The changing rules controlling the movement of people across international borders within Europe
- ☐ Why short-term population movement has increased
- ☐ A number of examples of such short-term movements
- ☐ A case-study of a short-term movement that covers the motives of the migrants as well as the problems caused

Key terms

Circulation
Country of origin
Economic migrant
Emigrants
Forced migration
Host country
Immigrants
International migration
Medical tourism
Migration
National migration
Refugees
Retirement migration
Temporary migration
Visa
Voluntary migration

Which key terms match the following definitions?

A Migration between different countries

B People who have been forced to flee an area – because of persecution or war, for example – and have had to take refuge elsewhere

C Endorsement on a passport giving permission to enter a country for a specified time

D Migration between different areas of one country

E A shift in location that is only temporary, such as a foreign worker on a six-month contract

F A type of voluntary migration where people choose to move when they retire

G People who move out of an area or country

H Where a migrant comes from – their source

To check your answers, look at the glossary on page 289.

Foundation Question: Describe three reasons for retirement migration. Use a case study that you studied of either retirement migration within a country or overseas. (6 marks)

Student answer ■ (awarded Level 1)	**Examiner comments**	**Build a better answer** △ (awarded Level 3)
I have chosen retirement migration from the UK to Spain. Many people go to Spain because the weather is loads nicer.	• *Many people go...* gains 1 mark. The quality of language could be improved.	I have chosen retirement migration from the UK to Spain. Warmer summers and much milder winters make Spain attractive and cheaper with reduced heating bills.
There are lots of houses there too and people can buy them really easily.	• *There are...* gets no credit. The quality of language makes it hard to understand what is meant.	Property developers have built purpose-built houses for the elderly.
These days you can get there really easily on EasyJet and this makes it safer.	• *These days...* scores 1 mark because the idea is right although *making it safer* is unclear.	Easy and cheaper travel to and from the UK makes people less isolated from their friends and families.

Overall comment: The student would have done much better if they had used the correct geographical language and avoided descriptions that are not meaningful without some further development.

Higher Question: Describe the reasons for short-term population flows. Use examples in your answer. (6 marks)

Student answer ● (awarded Level 2)	**Examiner comments**	**Build a better answer** △ (awarded Level 3)
There are many reasons for short-term population flows and they involve many different forces and have many different impacts.	• *There are many reasons...* This gives no detail to the answer – they are all obvious points. No marks are awarded.	There are many temporary workers in countries like Saudi Arabia from places like Pakistan. They have moved country in order to work in construction during the 'new building' boom.
Some people move for work when they come to a country that has jobs and better money than their own country.	• *Some people move...* gets 1 mark but would be better if it was illustrated with an example.	Polish migrants come to the UK for work. Some of them travel to the UK on a weekly basis to fill skills gaps, such as in medicine and dentistry.
There are lots of footballers in the Premiership, like Ronaldo, who move to the UK to earn lots of money. However, most of these footballers move to another club or retire, probably not in Manchester.	• *There are lots of footballers...* The idea is fine and the example and appropriate location scores 2 marks	There are lots of footballers in the Premiership, like Ronaldo, who move to the UK to earn lots of money. However, most of these footballers move to another club or retire, probably not in Manchester.

Overall comment: Try not to rewrite questions or state the obvious. Statements such as 'this is complicated' are not going to be rewarded.

Chapter 14 A tourist's world

Objectives

- Learn the different types of holiday and holiday attraction.
- Recognise that holidays are only a part of tourism.
- Understand the reasons for the growth of tourism.

■ Be sure that you understand the difference between 'leisure' and 'tourism'.

Growth of the tourist industry

Tourism is probably the biggest cause of population movement in the world today. Tens of millions of us become tourists each year. Maybe we are just visiting friends for a weekend or we may be going to France for a couple of weeks during the summer holidays. Tourism is also one of the most global industries today. There are few places in the world that are not touched by tourism in some way or another – from the vast empty spaces of the Antarctic to the crowded streets of central London. And millions of people make a living by being involved in tourism in some way or another.

Before looking at the tourist industry in detail, let us first be clear about the meaning of 'tourism'. A tourist is anyone who stays away from home for at least one night. It might be to stay with relatives, or on a campsite or at a hotel. But tourism is not just about holidays and **leisure**, and enjoying oneself. As the table below shows, tourism also includes some 'business' activities such as going on a school field trip, attending a conference or working away from home for a short while. Going on a pilgrimage and going abroad for medical treatment are also included.

Types of tourism

Main categories	Types
Leisure	Holiday
	Sporting event
	Festival
	Pilgrimage
Visiting friends and relatives	Stay with friends
	Stay with relatives
Business visits and trips	Business meeting
	Conference or exhibition
	Educational trip
	Medical treatment

Causes of the growth of tourism

All tourism falls into one of two categories. With international tourism (sometimes called overseas tourism), the tourist travels to another country, whilst with domestic tourism, the tourist remains within their country of residence.

The growth of international tourism has been spectacular. Figure 1 shows that the number of international tourist arrivals – foreign tourists coming into a country – increased nearly thirty times between 1950 and 2002. It also tells us about the relative importance of the continents for international tourism. Europe clearly has the most foreign tourists, but a high proportion of these are between different European countries. Figure 2 breaks down the international tourist arrivals in 2005 in terms of the trip purpose. Four categories of visit are recognised.

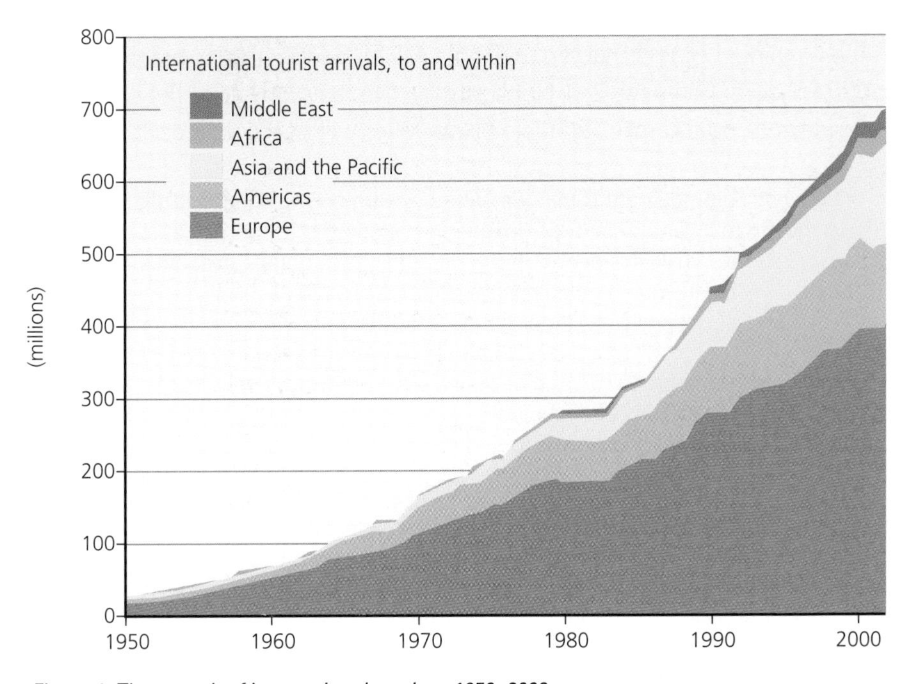

Figure 1: The growth of international tourism, 1950–2002

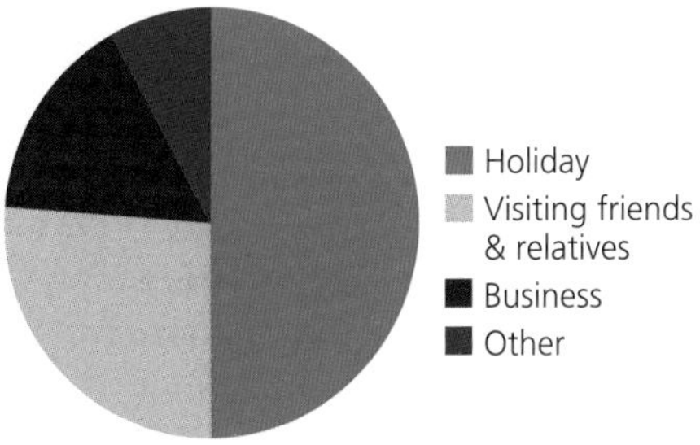

Holiday
Visiting friends & relatives
Business
Other

Figure 2: Purpose of visit, international tourism, 2005

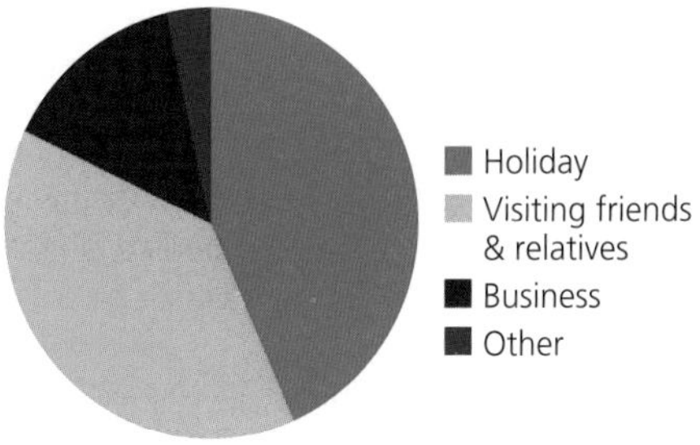

Holiday
Visiting friends & relatives
Business
Other

Figure 3: The purpose of UK domestic tourist trips, 2007

UK tourism

A UK Tourism Survey shows that just over 70 per cent of the holiday trips taken by Britons are to destinations within the UK. This might seem to contradict the popular belief that most of us just take one holiday – usually an overseas holiday – each year. But remember that the domestic figure includes all the weekend breaks and the stays with friends and relatives (see Figure 3).

Holidaymaking tourism

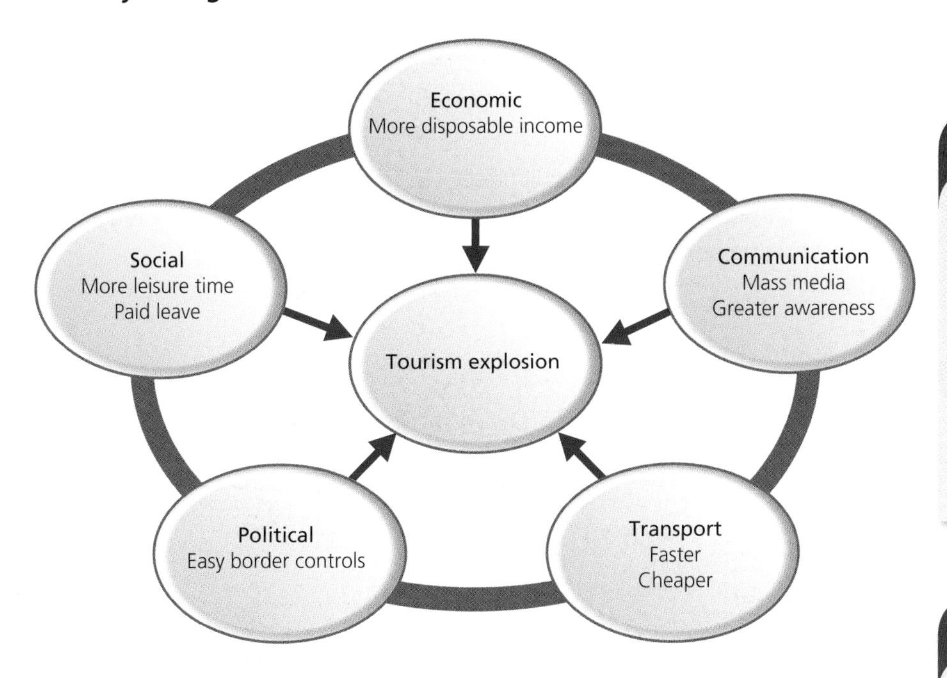

Figure 4: The causes of the tourism explosion

Skills Builder 1

Study Figures 1 and 2 and make notes about what they show under the headings: (a) the growth of international tourism; (b) the relative importance of the continents for international tourism, and (c) trip purpose. Ask your teacher to check your notes, and then keep them for when you revise for the examination.

Skills Builder 2

Compare Figures 2 and 3. How and why do the patterns of trip purpose differ between international and domestic trips?

In the remainder of this chapter, the focus will be on the holiday branch of tourism. Figure 4 on page 241 highlights the five main causes of the explosion in holidaymaking tourism. The table below takes a closer look at each of them.

Economic	One of the benefits of economic development is that it increases the amount of **disposable income** that people have. This is money that can be spent on 'luxuries' – including tourist activities such as a cruise or an adventure weekend. Whilst it is people in the HICs that have the greatest amount of disposable income, much of it is being spent in the LICs on exotic holidays. As a result, the growth in tourism is helping the economic development of those poorer countries.
Social	Most workers in HICs now work less than 40 hours a week and enjoy up to six weeks of paid annual leave. This combination of more leisure time and paid holidays has given a powerful boost to tourism. The **consumer culture** is also encouraging people to find new tourist destinations and to try new leisure experiences.
Transport	Developments in transport have revolutionised travel, both between and within countries. Journey times have been dramatically reduced, long journeys have become more comfortable, and the relative costs of travel have become lower. The introduction of modern forms of transport, such as wide-bodied aircraft, hotel cruise ships, cruiser coaches and high-speed rail networks, have all helped to make global travel a reality for increasing numbers of people.
Communication	The mass media, especially TV and the internet, have increased people's awareness of faraway places and potential tourist destinations. They have also opened people's eyes to the range of leisure activities that might call for a weekend away from home, from birdwatching to visiting stately homes, from fell walking to collecting antiques.
Political	More and more countries are realising the benefits of being a tourist destination, and relaxing their border controls. Governments stand to make money from tourist visas and departures taxes. Even the European Union, which makes it difficult for workers to enter, warmly welcomes tourists. Within the EU tourists can move from country to country with a minimum of fuss. Since the collapse of the Soviet Union around 1990, Eastern Europe and Russia have become major sources of tourists, now that their citizens are free to travel the world.

Examination questions will often ask about the impact of tourism. Make sure that you cover both the positive and negative effects.

Holiday attractions

Like other forms of economic activity, tourism exploits resources and it is created by a demand and a supply. Like other forms of population movement, tourism is created by push and pull forces. All four elements are shown in Figure 5. The pull is much stronger than the push, because it is the resources of a particular destination that attract the tourists.

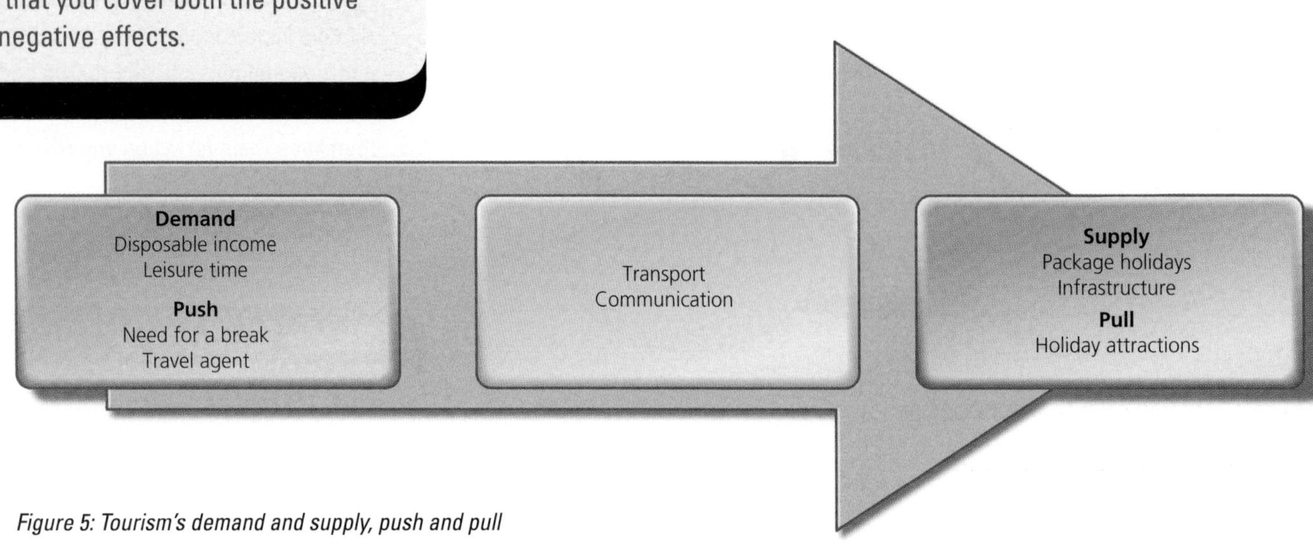

Figure 5: Tourism's demand and supply, push and pull

The attractions or resources of tourist areas are of two broad types – environmental and human. The environmental, mainly physical, attractions include:

- Climate – plenty of sunshine, good snow conditions
- Wildlife – game animals, birds, whales
- Scenery – attractive, spectacular
- Protected areas – national parks, nature reserves
- Beaches – clean sand, safe swimming, good surf.

The human attractions include:

- Cultural heritage – museums, galleries, temples, festivals, archaeological sites
- Accommodation – quality hotels, comfortable lodges, well-equipped chalets
- Local cuisine – special food, good restaurants
- Recreational facilities – golf courses
- Access – quality of transport connections, ease of local travel
- Personal security – low crime rate.

Activity 1

1. Can you add two more physical and two more human attractions to the list opposite?

2. Which of the attractions listed opposite would be important to you when choosing where to go for a week's holiday? Give your reasons.

3. Which two attractions are not important in choosing your own ideal holiday? Give your reasons.

Different types of holiday

There are many different types of holiday. They may be distinguished in a variety of ways. For example, we have already identified overseas and domestic holidays. What we do on holiday is also an important difference. Some just go on holiday to 'chill out' – to relax on the beach and soak in the sunshine. Others prefer more active holidays focused on particular specialist activities, such as scuba diving, skiing or rock climbing. Holidays may also be classified according to the holiday environment – whether it is coastal, rural or urban. The main types of holiday are described below.

A **package holiday** consists of transport and accommodation advertised and sold together by a tour operator. Other services may be provided during the holiday, such as a rental car, special activities or excursions. Package holidays are put together for both domestic and overseas destinations. The overseas destinations are usually the population holiday resorts (Figure 6), such as the Spanish 'Costas' (Brava, del Sol, Blanca, etc.) and islands (Majorca, Minorca, Ibiza and the Canaries), the Algarve and Madeira (Portugal), the French Riviera, Cyprus and the Greek islands. The appeal of package holidays is that:

Figure 6: A typical Spanish resort

- They are relatively cheap.
- Everything is organised.
- There is plenty of entertainment and socialising.
- They go to popular destinations.

Figure 7: One of the latest generation of cruise liners

The package holiday is the key part of what is known as **mass tourism** – a form of tourism in which large multinational companies shape developments according to global demand. It is large-scale, highly commercial, focused on popular destinations and pays little regard to local communities. The cruise is fast becoming another part of mass tourism. At one time, the cruise was the expensive holiday for the rich. Today, there is a large and growing fleet of huge cruise ships that are literally floating hotels (Figure 7). For Britons, popular cruise venues include the Caribbean, the Mediterranean, the Norwegian fjords and, for the more wealthy, Antarctica.

Another development in mass tourism has been the development of theme parks. Perhaps the best known are the Disney World parks in Florida and on the outskirts of Paris. A visit to the smaller theme parks is a day out, but the larger ones require a stay of a few days if all that they offer is to be enjoyed.

Figure 8: A beach wedding

At the other end of the holiday market, there is what is known as 'alternative tourism', which, unlike mass tourism, is about individuals and small groups. The holidays have varying degrees of organisation – some are led by guides, while others are 'do-it-yourself'. The different types of holiday under the 'alternative tourism' heading include:

- **Adventure holidays** – bungee-jumping, snorkelling, backpacking
- Wildlife holidays – birdwatching, whale-watching, safaris
- **Educational holidays** – learning to cook, wine tasting, pottery
- Shopping holidays – in duty-free zones
- **Self-catering holidays** – often in remote rural areas.
- Community or conservation holidays – in which tourists give their time, labour and expertise to help particular projects.

One form of tourism that has increased in popularity is couples going abroad to get married and spend their honeymoon in some exotic location (Figure 8). Beach weddings in Mauritius or in the Seychelles in the Indian Ocean are particularly popular.

Another category of tourism – **eco-tourism** – overlaps with many of these different types of alternative tourism. We will look at an eco-tourist destination later but first we should identify the main features of eco-tourism:

- It is based on natural resources, such as wildlife and wilderness.
- It focuses on experiencing and learning about nature.
- It does not consume non-renewable resources or damage the environment.
- It is locally oriented – controlled by local people, employing local people and using local produce.
- Its profits stay in the local community.
- It is sustainable and it contributes to the conservation of areas.

Resort development

As we have seen, there are many different types of tourism and, as a result, there are different types of tourist destination. They can range from a coastal or winter sports resort to a clinic or a conference centre, from a rainforest campsite to a stately home. In this section, we will deal with just one kind of tourism (holidaymaking) and one type of destination (the resort).

Objectives

- Learn that holiday resorts go through a sequence of changes.
- Recognise that resorts develop at different speeds.
- Understand the reasons for the changing fortunes of resorts.

The Butler model

One of the most obvious consequences of holidaymaking is the growth of resorts in the more popular destination areas. **Resorts** are most often towns that provide for the needs of visiting holidaymakers (Figure 9). Hotels and guest houses, restaurants and bars are the most common elements in all resorts, followed by facilities such as souvenir shops, parks, theatres and cinemas. In the case of coastal resorts, piers, seafront promenades and funfairs are added ingredients. With winter sport resorts, there will be chairlifts, ski runs and ice rinks.

It is important to remember that even the most famous resorts in the world today – such as Monte-Carlo, Las Vegas or Zermatt – had humble beginnings. They all grew from very small settlements, often mere clusters of dwellings. This then raises some interesting questions:

- What was it that triggered their early growth as resorts?
- What has kept them growing and booming as resorts?
- Once a resort becomes well known, popular and prosperous, does it stay that way for ever?

The Butler model, which suggests that all resorts or tourist areas follow the same broad sequence of changes, might help us answer some of these questions. According to Butler, resorts all follow a sort of pathway, which he called a **life cycle**. Figure 10 shows this life cycle to be made up of six stages.

Figure 9: Monte-Carlo – one of the famous resorts

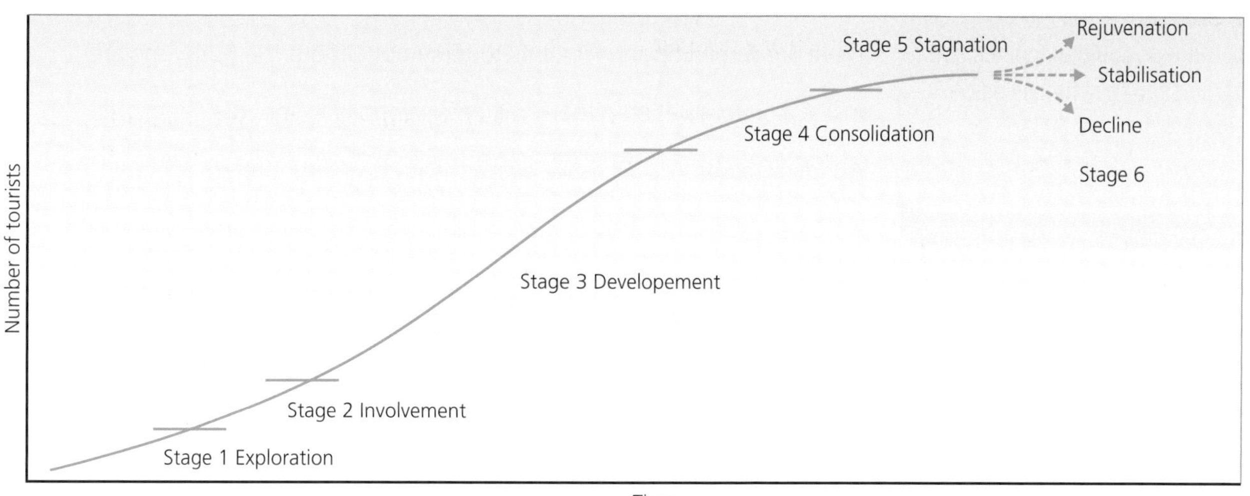

Figure 10: Butler's life cycle of a resort

Activity 2

Practise drawing the shape of the Butler model curve, making sure that you can correctly locate and label each of the six stages on it.

Read the case studies on pages 247 and 248. Describe the options faced by holiday resorts when they reach stagnation. (3 marks)

■ **Basic answers** (0–1 marks)
Offer a statement about how decline is likely, but lack detail about what might follow.

● **Good answers** (2 marks)
Identify at least one strategy that could be adopted, but lack detail of the possible management of that strategy.

▲ **Excellent answers** (3 marks)
Describe two or more possible outcomes and relate these outcomes to the choices made by the key players.

Stage 1: Exploration – There are very few tourists and few facilities for them. Those who come are attracted by the quiet and the undeveloped nature of the location. There is much interaction with local people. Tourism type: discovery or adventure.

State 2: Involvement – Tourist numbers increase steadily, and facilities begin to be provided. A tourism 'season' may be defined and low-key advertising may start. There is pressure on local government to improve access and to be involved in the promotion of tourism. Tourism type: independent travellers.

Stage 3: Development – Numbers of tourists rise rapidly. Up-to-date facilities are provided but are not run by local people. Tensions begin to build between the locals and the outsiders who have now taken over control of the resort's development. Much change. Tourism type: early package holiday.

Stage 4: Consolidation – Maximum tourist numbers. Peak number of facilities. The pace of change slows. Tourism type: package holidays.

Stage 5: Stagnation – The resort begins to lose its popularity due to changing fashions and ageing tourist facilities. The resort is working below its capacity. Tourism type: cheap and old-fashioned package holiday.

Stage 6: Decline, Stabilisation or Rejuvenation – At this point, the resort is faced with three options. It may go into complete decline, because it cannot provide new attractions, although it may still be used by weekenders or daytrippers. Much property is up for sale and prices fall. The resort drifts away from tourism. Hotels may be converted into old people's homes. It has a rundown appearance and both residents and tourists decline in number. The second option is stabilisation – the resort continues but in a rather downgraded form. The third option is a deliberate rebranding. Investment in some new attractions and facilities can bring rejuvenation. The resort might move into the 'business' sector of tourism as a conference centre or a resort for medical tourism. Perhaps it might even break away from tourism altogether.

There are some important points to remember about the Butler model:

- It is a generalisation – there will be exceptions to the rule.
- Individual resorts will move along the pathway at different speeds.
- Some resorts may stay at a particular stage for a long time.
- When a resort reaches the stagnation stage, what happens to it will depend on a number of **players** – planners, local government, investors and business people – and how ambitious and enterprising they are.

A tale of the development of two resorts

The case studies on pages 247 and 248 apply the Butler model to the growth of two different coastal resorts. Blackpool in Lancashire has been a resort for over 150 years. By comparison, Benidorm, located on Spain's Costa Blanca, is a relative newcomer, having only been a resort for about 50 years.

Case study: Blackpool

During the eighteenth century, it became fashionable for well-to-do people in Britain to go to the seaside in the summer. In those days, it was believed that bathing in sea water could cure diseases. Some of those wealthy people came to Blackpool on the Lancashire coast, but the settlement remained a hamlet until the early nineteenth century. So it was at Stage 1 in the Butler model.

The coming of the railway in 1846 had a huge impact on Blackpool. The train cut both the costs and time taken to reach seaside resorts. Many people were now able to travel to the coast. Huge numbers of working-class visitors began coming to Blackpool every weekend. So Blackpool had reached Stage 2 of the model.

Another boost to business came in the 1870s, when workers were granted annual holidays. Each of the cotton textile towns of Lancashire began declaring its own 'wakes week', when all the mills would be shut. Thousands of people from these towns were soon pouring into Blackpool during the holiday period. The amenities and attractions of the resort were gradually extended. The tower, promenades, piers, amusement arcades, theatres and music halls were built. By the First World War, the number of visitors during the high season had risen to 4 million. Blackpool was now at Stage 3.

During the interwar years (1918 to 1939), Blackpool's prosperity continued, helped by a new law that gave workers holidays with pay. The town's permanent population reached nearly 150,000 (Figure 11) and Blackpool was recognised as one of Europe's leading coastal resorts. This was the time that it passed through Stage 4 of the model.

After the Second World War, Blackpool's fortunes began to stagnate (it was now at Stage 5). The main reasons were the advent of the package holiday and cheap air transport. The keenest competition came from the coastal resorts of the Mediterranean. They were able to deliver, not just sea and sand, but most importantly sun. Blackpool has tried to offset its decline by developing other attractions, such as conference facilities and casinos. However, the town is still struggling to survive. So it is still poised between the decline and rejuvenation paths of Stage 6.

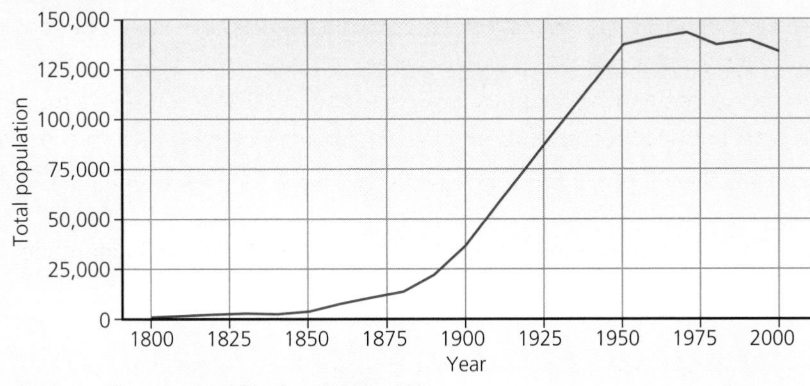

Figure 11: The growth of Blackpool, 1801–2001

Figure 12: South Promenade, Blackpool, in the interwar years

Figure 13: Blackpool's Pleasure Beach today

Case study: Benidorm

Benidorm is one of the most famous of the modern Mediterranean seaside resorts. It has a permanent population of about 70,000, but during the summer peak season the population is more than half a million.

The development of Benidorm as a coastal resort started in 1954, when its then young mayor drew up an ambitious plan of urban development. The development really took off in the 1960s when it became popular with British tourists on summer package holidays. Today, Benidorm's tourist season is all year round, and its attractions are now much more than sand, sea and sun. The nightlife – based on a central concentration of bars and clubs – is a strong pull, especially among the younger set. Within a few decades, Benidorm has been transformed from a small sleepy village into a modern pulsating urban area of skyscraper hotels and apartment blocks, theme parks, pubs, clubs and restaurants. In other words, it has moved at great pace through the first three stages of the Butler model.

It says much for the original plan that the built-up area today retains much greenery. And Benidorm is immensely proud of its three main beaches, all of which have 'blue flag' status. At the height of the season, they must be some of the most densely populated parts of the world. At the moment, there are no signs that Benidorm is stagnating. It is clearly at Stage 4. But it may be running out of building space – and perhaps this will cause it to move on to Stage 5.

Figure 14: The beach in Benidorm

Activity 3

1. Apart from a lack of space, can you think of anything that might cause Benidorm to enter Stage 5 of the Butler model?

2. Why are the lengths of Blackpool's and Benidorm's tourist season different?

Case study quick notes:
These two case studies, of Blackpool and Benidorm, illustrate the following points:

- Some resorts do seem to follow the Butler model quite closely.
- Resorts differ, however, in their speed through the life cycle.
- The same factor that caused the stagnation of Blackpool (overseas package holidays) also caused the spectacular rise of Benidorm.

The effects of the growth of tourism

In this section we will examine the impacts that tourism has on the tourist destination. The impacts fall under three headings – economic, social and cultural, and environmental. Under each of those headings, we will distinguish between positive (good) and negative (bad) impacts.

Objectives

- Learn some of the effects of the growth of tourism.
- Recognise that those effects can be both positive and negative.
- Understand the reasons for those different effects.

Economic impacts

There is no doubt that tourism has positive impacts. Most important are its **multiplier effects**, the direct and indirect consequences it has, as shown in Figure 15. Tourism is labour-intensive, and it creates many jobs, not just in hotels and restaurants, but in other tourist services, such as transport. Whilst tourism is a tertiary sector activity, it has indirect impacts on the other two sectors. Tourists need food, so that is potentially good for agriculture. Tourists buy souvenirs and that can be good for manufacturing. The hotel staff, the ice-cream sellers and the souvenir shop owners then spend their money in the local shops. Tourism puts money into many people's pockets and, through the multiplier effect, the whole local economy can be lifted. Few would disagree that tourism can do much to help economic development in LICs.

However, there are some negative aspects. Much of today's international tourism is in the hands of big companies, such as TUI, Thomas Cook and First Choice. This means that the profits made in a particular country 'leak' out to the country where the tour operator has its headquarters. This is money that could be used to help the development of the country where it was earned. For example, in Vanuatu – a 'hot' destination in the Pacific Ocean – 90 per cent of the profits go to foreign companies. Tourist destinations, particularly in LICs, can become very dependent on foreign companies.

Other serious negatives are to do with the basic nature of tourism. For example, its fortunes fluctuate from year to year, depending on the state of national economies. When an economy is booming, there is plenty of disposable income to be spent on holidays. But when times are hard, the tourist market can shrink considerably. Even in good times much tourism is seasonal – especially seaside and winter sports holidays – resulting in high rates of unemployment during the slack season.

Activity 4

List three direct and three indirect jobs associated with tourism.

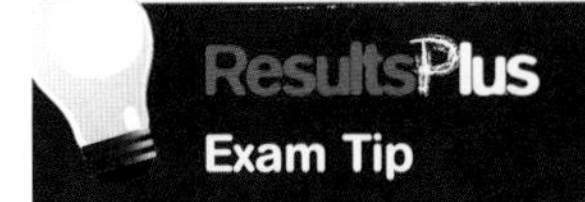

Examination questions will often identify economic, social, cultural and environmental causes and consequences. Make sure that you really understand these terms.

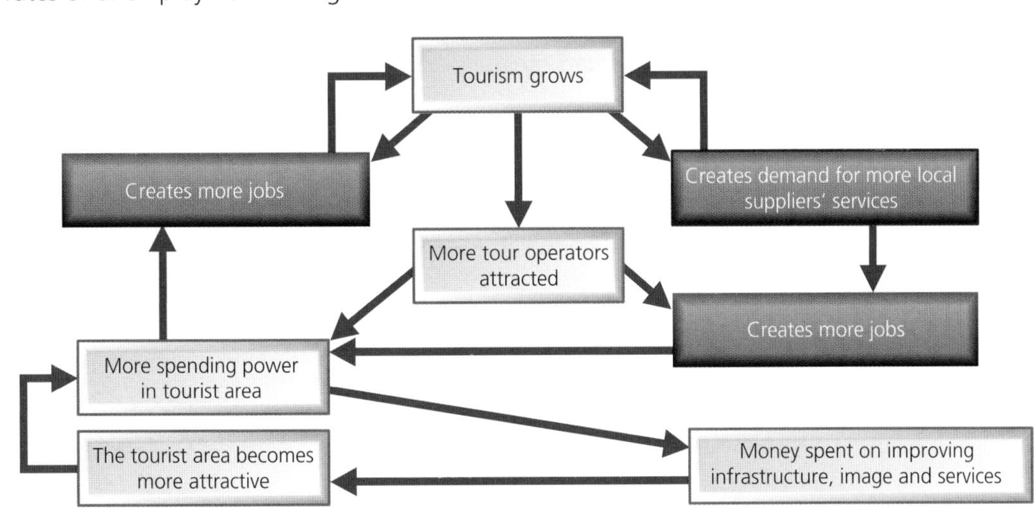

Figure 15: The economic multiplier effect of tourism

Social and cultural impacts

The degree to which tourism impacts on people and their traditional ways of life – their culture – depends on the type and volume of tourism. The independent tourist is likely to have little impact. But the same cannot be said for mass tourism. It might be claimed in some places that tourism has helped to revive local handicrafts (Figure 16) as well as the performing arts and rituals – if only as a commercial entertainment for visitors. But, generally speaking, the socio-cultural impacts of tourism are mostly negative. The greater the number of tourists converging on a location, the more likely there is to be tension with local people. Tourists can easily offend the traditional values of local people and their codes of behaviour, in a number of ways:

Figure 16: The commercialisation of local culture

- Drinking too much alcohol and becoming loud and offensive
- Ignoring local dress codes and revealing too much flesh
- Encouraging prostitution and, unintentionally, crime
- Eroding the local language by relying too much on English
- Failing to behave in churches, temples and mosques in the proper way.

It would be good to think that international tourism provides the opportunity for people of different cultures to mix and learn about each other. Sadly, that positive effect is rarely realised. Indeed, in some parts of the world, tourists are deliberately kept away from local people (or is it the other way round?). This happens, for example, in Cuba (for political reasons) and in the Maldives (for religious reasons). In the UK, a high proportion of holiday lets and second homes in villages and rural areas can often lead to communities becoming fractured into two – natives and incomers.

Environmental impacts

It is also difficult to identify any positive environmental impacts. Again, it depends on the nature and the volume of tourism. Alternative tourism does provide some opportunities for people to learn about the environment and to become supporters of environmental conservation. But it is easy to compile a long list of negative impacts (Figure 17):

Figure 17: More hotels and apartments – a negative environmental impact

- The clearance of important habitats, such as mangrove and rainforest, to provide building sites for hotels
- The overuse of water resources
- The pollution of the sea, lakes and rivers by rubbish and sewage
- The destruction of coral reefs by snorkellers and scuba divers
- The disturbance of wildlife by safari tourism, hunting and fishing
- Traffic congestion, air and noise pollution.

Two examples will help to illustrate some of these different types of impact.

The impacts of tourism in the Khumbu region of Nepal

For many mountaineers and trekkers, the dream location would be the Himalayas and the world's highest mountain, Mount Everest (8,848 metres). The eastern border of the Khumbu region reaches up to the top of Everest (Figure 18). Every year, well over 100,000 tourists visit this part of Nepal, which is one of the world's poorest countries. Traditionally, the people of Khumbu are subsistence farmers, growing crops and rearing livestock, but they now provide most of the Sherpas who carry the packs and guide the trekking and mountaineering expeditions (Figure 19).

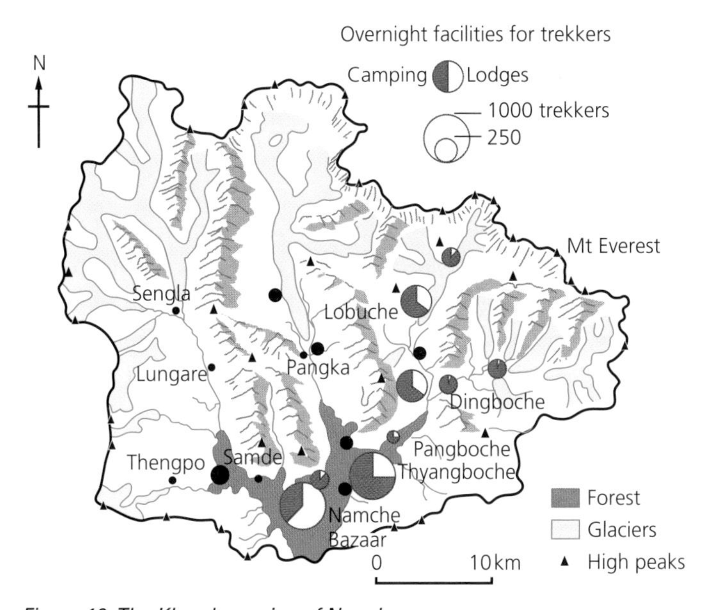

Figure 18: The Khumbu region of Nepal

Figure 19: Trekking in the foothills of Everest

Involvement in the trekking and mountaineering business has brought many changes to Khumbu:

- Young men have left their villages to become porters or guides.
- Some of the more enterprising men have moved to the Nepalese capital, Kathmandu, where they have set up businesses that organise and advertise expeditions. Others have set up small camps and lodges used by the trekkers.
- Sherpa wages are good, and young men earn enough money to rebuild and improve their houses.
- There is not enough male labour to work the small farms, and women are now having to take over.
- Basic food prices are being pushed up by the tourists.
- Tourists are bringing in Western foods, so diets are more varied and health has improved.
- The large number of visitors is causing fuelwood (used for cooking and heating) to become scarce and expensive – forested areas are being cleared.

Activity 5

1. Make a table like this one, and place each of the bullet points listed in the Khumbu case study in the appropriate space.

Impact	Positive	Negative
Economic		
Social and cultural		
Environmental		

2. What conclusion do you reach – do the positives outweigh the negatives?

Quick notes (the Khumbu region):
- Even alternative tourism has its negative impacts.
- Clearly, the impacts are a mix of positives and negatives.

- Electricity is now supplied to a growing number of homes by a mini-hydro scheme paid for by money made from trekking.
- Schools have been built and education improved.
- Many teenagers drop out of education early to take up jobs in tourism.
- Traditional garments are no longer made, and Western-style clothes are now common.
- Many porters and guides suffer serious injuries from carrying very heavy loads. There is too little attention paid to safety and there are too many deaths.
- Families are breaking up, with men living away from home for months on end.

Source: Sue Warn, Recreation and Tourism, Stanley Thornes, 1999

Activity 6

1. Suggest ways in which Ibiza might rebrand its tourism.

2. Can you think of any ways in which mass tourism on the island might be made 'greener'?

Ibiza

Ibiza is a small Spanish island in the Mediterranean. It has been involved in mass tourism since the 1950s, when it was one of the early package holiday destinations. Ibiza today is well known and popular – but perhaps for the wrong reasons. It is mostly renowned for its nightlife and clubs, as well as for the anti-social behaviour of its young adult tourists, drawn from the UK and northern European countries (Figure 20). Because of Ibiza's apparent tolerance towards bad behaviour from tourists, it has acquired the reputation as the 'Gomorrah of the Med' (a reference to the town which, in the Bible story, was destroyed by fire from heaven because of the wickedness of its inhabitants).

Thankfully, Ibiza is not all bad news. The holiday season is a short one – from June to October. During it, many of the residents move off the island, to be replaced by workers drafted in to work in the hotels, bars, cafés and clubs. Outside the season, the island is empty of tourists, the migrant workers have gone home, and most of the clubs are closed. Though largely known for its party scene, parts of the island are designated World Heritage Sites. These sites, which are mainly cultural, are therefore protected from development. And, despite the tourist developments, there are still areas of great scenic beauty on the island which attract artists, photographers and film companies (Figure 21).

Quick notes (Ibiza):
Remember that the seasonal nature of many forms of tourism means that destinations undergo twice-yearly transformations.

Figure 20: Clubbing in Ibiza

Figure 21: Ibiza's beautiful scenery – a view to Dalt Vila

Eco-tourism

Objectives

- Learn some of the features of eco-tourism.
- Recognise what is involved in promoting eco-tourism.
- Understand the reasons why eco-tourism projects are small scale.

The essential character of eco-tourism was described on page 244. We might re-emphasise four characteristics before looking at an example of eco-tourism in action. Eco-tourism:

- Involves areas that are in some way 'special' or 'precious', because of their scenery, wildlife, remoteness or culture.
- Aims to educate people and increase their understanding and appreciation.
- Minimises the impact on, and damage to, the environment and local community.
- Maximises local involvement, local control and local benefits.

As people become more aware of the economic, social and environmental costs of mass tourism (discussed earlier in this chapter), the tourism marketplace is becoming 'greener'. It is estimated that 85% of people in HICs believe that the environment is the number one issue facing the world. Tourists are beginning to translate their concerns about the environment into appropriate action by choosing holidays that are environmentally and ecologically 'benign' (kind rather than damaging). More tourists are asking the sorts of question shown in 'the eco-tourist's questionnaire' below.

The eco-tourist's questionnaire

Before booking your holiday, ask yourself the following questions:

1. *What is the environmental impact of tourism on the country I want to visit?*

2. *Have people been forcibly resettled to make way for tourist developments?*

3. *By travelling to this country, am I supporting a repressive regime?*

4. *Are my needs as a tourist increasing the demand for goods and services supplied by HICs?*

5. *Is my presence as a tourist likely to have an adverse effect on local society and culture?*

Case study: Eco-tourism in Trinidad

Trinidad, in the West Indies, is classified as a middle-income country. The first eco-tourism venture in the West Indies – the Asa Wright Nature Centre – was built in 1967 in the upper Arima valley, which cuts into the rainforest-clad Northern Ranges of the island (Figure 22). The Nature Centre was started by a non-profit-making trust which bought a former mixed plantation estate of 78 hectares, located at an altitude of 370 metres. The main aims of the venture were to:

- allow the cleared areas of the plantation to revert to mountain rainforest
- conserve this part of the Arima valley for the protection of wildlife and for the enjoyment of local people
- promote public awareness of the value of the rainforest and its wildlife
- provide accommodation for visitors.

The idea was that profits derived from providing guest accommodation should be used to finance the conservation and education work of the Trust, as well to purchase adjacent areas of primary (relatively untouched) rainforest. Twenty-four chalets mean that around 2000 visitors stay at the Centre during the course of a year. There are also larger numbers of day visitors.

In many ways, the Asa Wright Nature Centre has fulfilled virtually all of the criteria of eco-tourism and sustainable tourism identified earlier in this chapter. For example:

- Most of the fifty or so staff live in the Arima valley.
- Staff members are helped in a variety of ways, from training to interest-free loans for building or renovating homes.
- There is on-site recycling of refuse and waste water.
- Much of the food served to guests is either grown on the estate or purchased from local producers.
- Only 10% of the original estate is accessible to tourists via five designated trails. The remainder is insulated – so wildlife is left undisturbed. The total area of the estate has now been extended to 600 hectares, with the gradual acquisition of more tracts of primary forest.

The Asa Wright Nature Centre is a small-scale venture. It is feared, however, that if it were to be made larger, it would become more commercial and less 'green' and environmentally friendly.

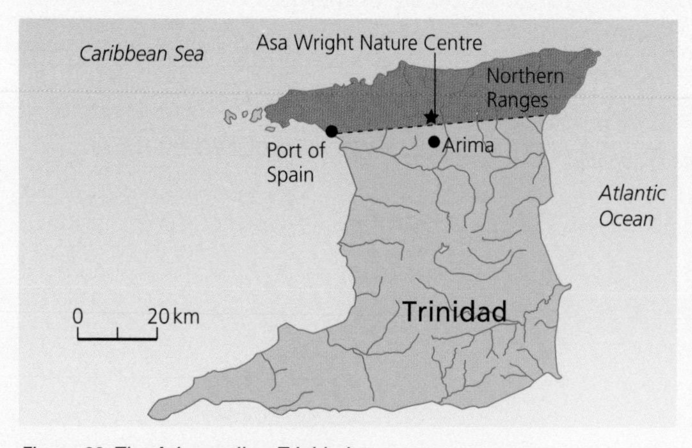

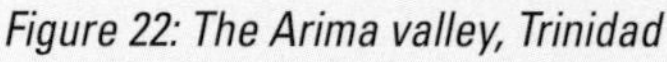

Figure 22: The Arima valley, Trinidad

Figure 23: The Asa Wright Nature Centre in Trinidad

Case study quick notes:

- Eco-tourism can be made to work but it will take tens of thousands of Asa Wright Nature Centres before eco-tourism begins to make a mark on global tourism.

Finally, it is worth making a brief comment about those who are supporting eco-tourism – the eco-tourists. A survey of tourists in the UK who have recently undertaken a long-haul eco-holiday came up with an interesting demographic profile. The survey showed that nearly half were either retired or working part-time. Nearly 70% were either married or living with a partner. There were slightly more male eco-tourists than female. Just over 20% had university degrees.

But what was it that drew these people to take an eco-holiday? Two-thirds said that it was the chance to walk or watch wildlife in untouched natural environments (Figure 24). Half wanted to visit renowned ecological or heritage sites. However, the big surprise of the survey was that the top eco-holiday destinations were in Australia (39%), Canada (25%) and New Zealand (22%). In other words, HIC destinations appear to be outshining those in LICs. This result is rather disappointing to those who believe that many LICs have much to gain from being involved in eco-tourism. So the vital question is – what do they need to do to attract more eco-tourists?

Figure 24: An eco-tourism destination – watching polar bears in Canada

Eco-tourism has many positives. It is much less exploitive of resources than mass tourism. It shows greater respect for both the natural environment and local culture. But so long as the key to successful eco-tourism remains 'small is beautiful', it is hardly ever likely to challenge mass tourism.

Activity 7

1. Do you think that eco-tourism has a downside? Give your reasons.
2. What might LICs do to attract more of the UK's eco-tourists?

Tourism in fragile environments requires sustainable management. Choose a case study that you have studied. Describe the problems that tourists could cause or have caused in this area and explain how these problems have been managed. (8 marks, June 2006)

How students answered

Many students answered this question poorly. They tended to describe a few features of 'fragile' tourist environments, but did not add detail or explain how the problems could be managed.

32% (0–3 marks)

Most students answered this question reasonably well. They identified some of the problems caused by tourists, such as waste creation and their social impact, but the management techniques were not addressed.

61% (4–6 marks)

Very few students answered this question well. Those that did described management methods and were able to explain the links between the problems caused by tourists and the management response to these problems.

7% (7–8 marks)

Know Zone
Chapter 14 A tourist's world

Tourism is the fastest-growing industry in the world. It has a range of positive and negative impacts both on people and the environment. For LICs it can offer one of the very few sources of foreign currency.

You should know…

- ☐ The social, economic and political causes of the growth of tourism
- ☐ How higher incomes and greater leisure time have allowed the growth of tourism
- ☐ The variety of holiday destinations and their attractions
- ☐ A range of different holiday types, including package trips, adventure, weddings and backpacking
- ☐ How tourism has developed over time (as in the Butler model)
- ☐ A case study of an EU resort that has changed and developed over time
- ☐ The social, economic and environmental effects of tourist growth
- ☐ How these effects can be both positive and negative
- ☐ How these effects can be economic and environmental
- ☐ A range of these impacts drawn from both LICs and HICs
- ☐ A case study for eco-tourism, showing how tourism can protect the environment and benefit the local community

Key terms

Adventure holiday
The Butler model
Consumer culture
Disposable income
Eco-tourism
Educational holiday
Leisure
Life cycle
Mass tourism
Package holiday
Resort
Self-catering holiday
Tourism

Which key terms match the following definitions?

A A type of holiday that involves activities such as bungee-jumping and snorkelling

B A form of tourism that tries to minimise the environmental impacts of the tourists, by using local providers and resources and by keeping profits within the local area

C The amount of money which a person has available to spend on non-essential items, after they have paid for their food, clothing and household running costs

D A holiday in which travel and accommodation are put together by a tour operator and sold as a relatively cheap package

E Use of free time for enjoyment

F A settlement where tourism is the main function

G A model showing the sequence of changes experienced by holiday resorts

H The large-scale movement of tourists to popular destinations, most often promoted through package holidays

To check your answers, look at the glossary on page 289.

ResultsPlus
Maximise your marks

Foundation Question: Tourism can have positive and negative impacts on an area. Describe one negative impact and one positive impact. (4 marks)

Student answer (awarded 2 marks)	**Examiner comments**	**Build a better answer** (awarded 4 marks)
The tourists can spoil an area by being a nuisance to the locals not being respectful. *Tourists often visit places and help the economy of these places because they come there with money.*	• *The tourists can...* scores 1 mark. By being a nuisance is unclear. The student needs to detail in what way they are a nuisance and to whom. • *Tourists often visit...* This part of the answer does not make it clear what the tourists do with their money. It makes a point, so is awarded 1 mark, but needs to be clearer in order to get any extra marks.	The tourists can spoil an area by being a nuisance to the locals with late night noise and drunkenness or by wearing the wrong clothes when visiting churches and other sites. This is a negative impact. Tourists often help the economy of the places they visit because they spend money in local restaurants and gift shops. This also creates jobs in these areas. This is a positive impact.

Overall comment: Try to balance your answers when asked to describe the impact of something. Decide whether there are positive and negative things to be said. If so, then make this clear.

Higher Question: Explain how eco-tourism can protect the environment and benefit the local community. Refer to a case study you have studied. (6 marks)

Student answer (awarded Level 2)	**Examiner comments**	**Build a better answer** (awarded Level 3)
Eco-tourism tries to be sustainable and to make an area last longer into the future. There are many eco-tourist resorts in the Amazon. *In some places there is a lot of recycling, such as using refuse and waste water.* *In some areas like in Trinidad most of the estate is restricted. Only 10% is open to visitors using designated trails.*	• *Eco-tourism tries...* is not awarded any marks. This is a long and rather weak statement about sustainable tourism which is not asked for in the question. • *In some places...* This response is worth 1 mark. The idea is good, but this is not developed or located – either would improve the answer. • *In some areas...* is awarded 2 marks. The idea is clear and is illustrated with an appropriate case study.	Eco-tourism involves choices about areas. Tourists now take more care about where they visit so this can protect sensitive environments. In Trinidad the Asa Wright Nature Centre is a non-profit venture and the money made from tourists is ploughed back into the environment. In some areas like in Trinidad, most of the estate is restricted. Only 10% is open to visitors using designated trails.

Overall comment: When you are asked for a case study example, you need to offer some information that is well located. Your point will always be stronger if you can add detail.

Unit 4 Investigating geography

Your course

This is the controlled assessment part of your course. The controlled assessment is an investigation in which you will be asked to undertake a fieldwork task. Having completed your fieldwork, you will then be required to write a report based on your results, completed under examination conditions and supervised by your teacher.

Your assessment

This list shows you how the marks (50 in total) are allocated between the different elements, the suggested time to be spent on each, and *what is actually being assessed.*

The purpose of the investigation (6 marks)
Suggested time: 3 hours

Your definition of the question you will be investigating, including the location of your fieldwork.

The methods of collecting data (9 marks)
Suggested time: 1 day for collecting your data, plus 3 hours for your write-up

Your description and explanation of the methods of data collection you will use.

The methods of presenting data (11 marks)
Suggested time: 6 hours

The data-presentation techniques you will use and their quality.

Analysis and conclusions (9 marks)
Suggested time: 5 hours

How the findings of your investigation are brought together and conclusions are drawn.

Evaluation (9 marks)
Suggested time: 2 hours

Your evaluation of your methods of data collection and presentation, and the analysis and conclusions drawn.

Planning and organisation (6 marks)
Suggested time: throughout your investigation, with 1 hour to organise your final report

The planning and organisation of your report, including your use of geographical terminology and the quality of your writing.

What are the controls in the controlled assessment?

You need to be aware of the level of control which occurs at each stage in the investigation. There are two levels which affect the way you must work:

High level of control

Certain stages of your investigation must be completed individually – by yourself without help from others, in the classroom, under the close supervision of your teachers. This work cannot be taken home – it has to be handed in at the end of each lesson, so that you cannot continue working on it in your own time. These stages are:

- Deciding on the focus of your investigation. The list of possible investigations is created by Edexcel, and your teacher will help you decide the focus of your investigation.
- The completion of your final report, including the writing up of your analysis, conclusions and evaluation.

Limited level of control

For some stages of your investigation, you will be able to work in your own time – at home, in the library or elsewhere. You will also be able to work in groups. These stages are:

- Any secondary research you need to do, such as reading about the processes involved in your work or the location of your sites.
- The writing up of your purpose and methods of data collection.
- The collection of your data. You can work in groups to collect data, but any extra research should be completed individually.
- The development of your data presentation.

Chapter 15 Your fieldwork investigation

Objectives

- Identify geographical questions suitable for small-scale fieldwork.
- Develop clear methods to gain data in response to the question asked.
- Be able to present geographical data.
- Understand the geographical processes responsible for patterns seen in the data.
- Be able to analyse and explain geographical processes.
- Be able to evaluate the strengths and weaknesses of the investigation when it has been completed.

How you will be marked

The purpose of your investigation (6 marks)

Mark	Reasons for marks given
0	No location or issue is identified.
1–2	Issue or question is weakly identified. Location is mentioned but unclear.
3–4	A clear statement identifies the issue or question. Location is mentioned.
5–6	A focused statement identifies and evaluates the issue or question. Location is focused on the place of the investigation.

An important element of geography is its use in the real world. It is a subject which allows us to investigate issues and problems with the intention of making a situation better, or developing a better understanding of a chosen issue. Most geographical investigations cannot be achieved by the use of a laboratory or computer simulation alone – they require us to collect information through the use of fieldwork. This 'controlled assessment' is an opportunity to develop the skills which are important in becoming a geographer, skills which are also transferable to other situations.

Purpose of your investigation

All investigations require a purpose. Geographical enquiry is not simply a case of walking around a place and seeing what we can see. In this assessment, you will be asked to react to a question which is deliberately written to allow you to develop a more focused purpose of your own. For example, a typical task question might read: *How effective is the coastal management at your chosen location?*

This is obviously a very wide-ranging question – which would need a great deal of research and a huge amount of writing to answer properly. You therefore need to choose a smaller focus from such a question – a specific element which you can investigate. This is what we call a 'focus concept' – a specific issue, given in the form of a question. In the case of the coastal management question above, you might ask a specific question about the environmental damage caused by the management of the coast at a location, or about whether its benefits outweigh its costs.

One of the first things that you must do in developing your work, is to identify a focus which is clearly linked to the general question given to you, and then give a clear explanation as to why your focus is important and helps to answer the general question.

Locating your investigation

To make your investigation informative for the person reading it, you must first locate it by describing where you carried out your fieldwork. This description will normally include various pieces of information (Figure 1) and maps to show the location graphically.

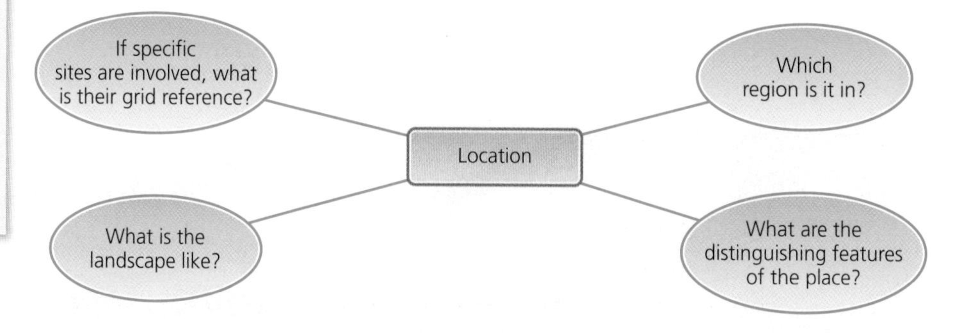

Figure 1: Questions to consider when writing a location description

You need to think carefully about the level of detail given in locating your investigation. If your work is focused on the question 'How has service provision changed in your chosen rural area?', for example, you might simply be locating and describing one or two villages, and the areas within them which were investigated. However, if you were focused on the question 'How do channel characteristics vary along your chosen river?', you would need to identify and describe specific points along a river, using a map and perhaps some grid references to pinpoint the location of all the individual sites used. This shows that it is important to consider the scale at which you locate your investigation. The most important factor is that your description needs to be clear to others so that they have a clear idea of where your investigation took place.

The location of your investigation is an ideal opportunity to use websites such as Google Earth or Multimap. Figure 2 shows a possible pair of images for showing the location of two villages used to investigate service provision. Used with a short written description, this would lead to a clear impression for the reader. Figure 3 does the same thing for an investigation of a river. Note the difference in the scale of the images used, and how in both cases, the images have been annotated so that they have some clear information regarding the locations of the investigation.

Control

The topics available to you are controlled by Edexcel who will publish a list each year from which your teacher will select relevant questions.

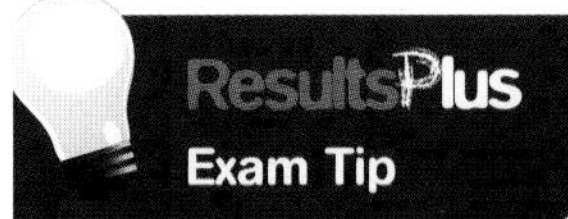

Exam Tip

Make sure you annotate any maps you include, giving details about the locations which are important, and perhaps the data collection methods you have used at particular locations.

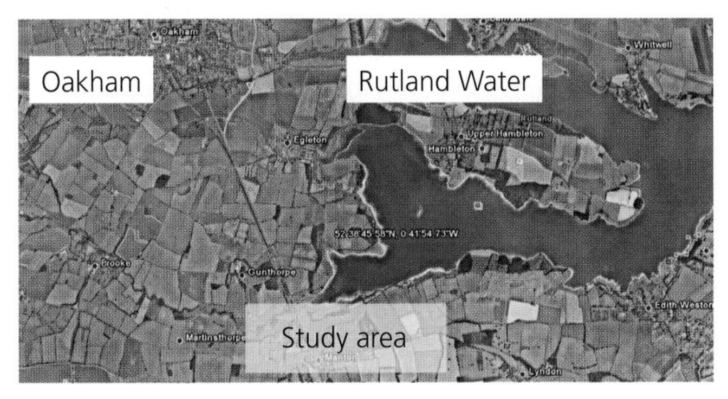

Figure 2: Images taken from Google Earth used to locate study villages

Figure 3: An image used to locate sites for a river investigation (note the difference in scale from those images used in Figure 2)

Control

You will be able to research some information about your location in groups, but when writing up this section as part of your report you will need to do this individually.

How you will be marked

The methods of collecting data (9 marks)

Mark	Reasons for marks given
0	There is no evidence of data collected or method(s) of collection.
1–3	There is limited evidence of data collected. The methods used to collect the data are briefly described. There is no explanation of the methods used to collect the data.
4–6	The evidence collected is appropriate for the investigation. The method(s) used to collect the data are clearly described. There is some explanation of the method(s) used.
7–9	The evidence is accurately collected and is appropriate for the investigation. The method(s) used to collect the data are described in detail. There is detailed explanation of the method(s) used.

Control

When you collect your data, there is only limited control. This means you can work as part of a group if appropriate.

Remember to explain why you have used a particular sampling method.

Methods of collecting data

When you have explained the purpose, and described the locations to be used, you will then need to decide which methods you are going to use for collecting data. This requires careful planning, and includes a consideration of the types of data that you will use – primary or secondary data, or both.

Primary data	This is data which is collected first hand. In the case of a school-based investigation this is the data which is collected by a group of students whilst undertaking an exercise of fieldwork.
Secondary data	This is data which has already been collected by others for a particular purpose, and then 'published' on websites, in books, official reports, etc. It can be included to support primary data in an investigation. A good example is the UK census, which is taken every ten years and provides a large amount of data about the population for academic studies.

Sampling

As well as deciding on the types of data which are to be collected, the sampling method must be considered. Sampling is nearly always necessary because it is just not possible to measure every item or interview every person. A carefully chosen sample will be much easier and quicker to investigate and will still give fairly accurate results. Choosing a representative sample – the *when*, *what*, *which* and *where* of the data collection – is an important part of the process. If you were using questionnaires, for example, you would need to make sure that the right people were targeted. There is little point conducting a survey to find out which services young people in a town would like to see developed if you only go to a skate park for views, as only a single interest group is likely to be asked. This would give you a very untypical data set and would call into question any eventual analysis of that data. You need to decide on the timing of your data collection as well as being sure of the groups you wish to measure to make sure that the results gained are not biased. There are a number of ways in which sampling can be carried out, and you should decide which is most appropriate for your investigation.

Sampling methods
There are a number of ways in which samples can be taken, so you should be careful to plan – and identify – how you have carried out your sampling.

Random sampling – the locations for data collection are chosen by chance. An example might be the use of a quadrat to count the number of plant species at a number of sites. As you finish at one site, you would throw the quadrat and count the next site as being where it landed.

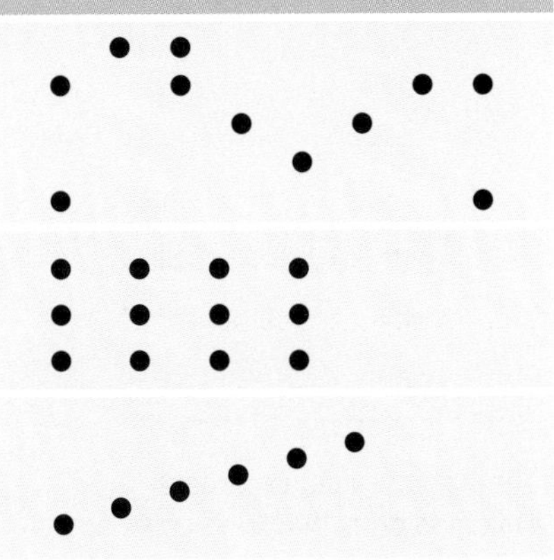

Systematic sampling – the locations of the sites used are found at equal intervals from each other. This might be each fourth shop within a shopping area, or at points on a grid if sampling the size of pebbles on a beach.

Line-intercept sampling (also known as a transect) – sites are sampled along a line. This might be used to measure the environmental quality across a city centre, or a river at points along its course.

Having decided on the sampling strategy of the investigation, and having perhaps gained some secondary data which can be used as background information to support your work, you need to decide on the primary data collection methods you wish to use. In geography there are a whole range of techniques which can be used to collect data, and you should look beyond the few introduced below for further ideas.

Sketching/photos

It is often useful to include a clear picture of your main sites, or close-up details of the ideas which you discuss, e.g. photos of graffiti. Sketches should be annotated to show features, processes, etc., giving a clear description and explanation of important features (see Figure 5).

Figure 4: A typical beach

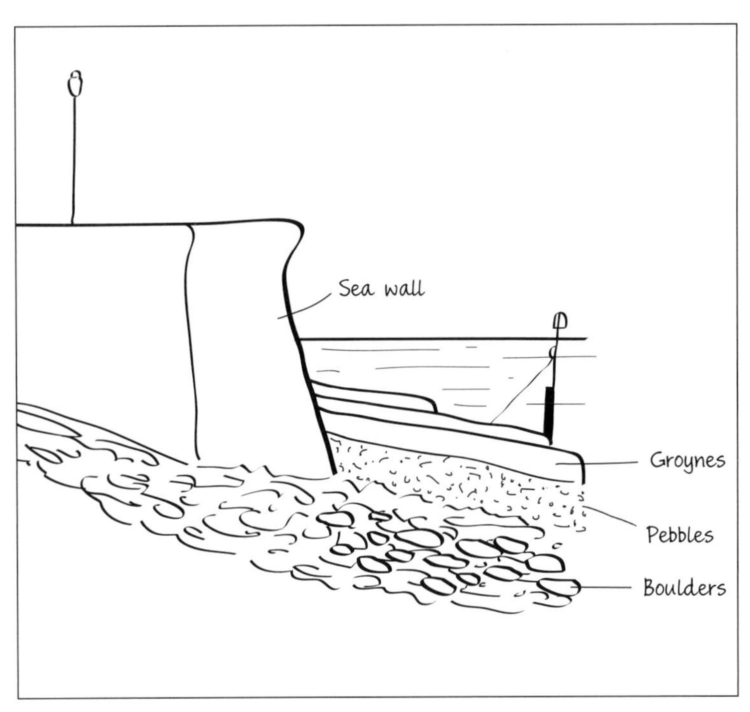

Figure 5: An annotated sketch of Figure 4, emphasising its important features

Mapping

Geography uses mapping a great deal. There are many different types of mapping. You might be looking at parking provision, and need to shade in areas of car parking on a street map. You might be looking at the pattern of plant growth on some sand dunes, and use a map of the dunes to shade in where different plant species are found. A frequently used method is that of land use mapping, where different types of shop or other building use are identified by differently shaded colours. You could look up different types of map in books to see what might be possible, and then consider the type of map you need. At this stage, you may well need to ask for help in finding the correct map. You might use:

- an Ordnance Survey map
- a town street plan
- a sketch map that you have created yourself.

ResultsPlus
Exam Tip

As with sampling, you will need to be able to justify why the methods you have chosen are appropriate.

Questionnaires

Questionnaires are a very useful way in which to collect data – but you need to consider carefully what you are trying to find out. There are a number of choices you need to make:

- Which questions will allow you to collect the information you need, without making the questionnaire too long?
- Will the questions be *open* (allowing respondents to offer opinions or supply information) or *closed* (requiring respondents to choose from a set of fixed alternatives, such as Yes/No, 1/2/3/more than 3)?
- Can some questions be answered without being asked, such as the sex of the person you are interviewing, and possibly estimating their age rather than asking for it?
- Will you fill in the questionnaire yourself, from someone's answers, or will they fill it in by themselves?
- How will you introduce yourself when you first ask someone to respond to your questionnaire?

Once you have thought about these issues, you should draft a questionnaire and then ask a friend to fill it in, to make sure that it works as you want it to.

Measuring physical features

Measurement is central to the study of physical geography. It includes activities such as:

- measuring the width of a river
- measuring the density of species found in a woodland
- measuring wind speed.

In each case it is important to make sure that the measurements are taken accurately. By using appropriate sampling strategies, and measuring particular features, a good understanding can be gained as to the processes occurring in a physical environment (see Example 1).

Example 1: Basic measurements used to investigate river characteristics

When investigating the changing characteristics of a river as it flows downstream, a number of different measurements can be taken:

- Channel width, by using a tape measure
- Channel depth, taken in several equally spaced places across the stream so that a cross-section can be drawn. This can allow you to calculate the cross-sectional area.
- Water velocity, using a flow meter or float and stop watch
- Sediment size, measuring the size of sediment on the river bed. This might be done using a ruler (where the sediment is large) or sieves (where the sediment is small).

Figure 6: Location of Site 3

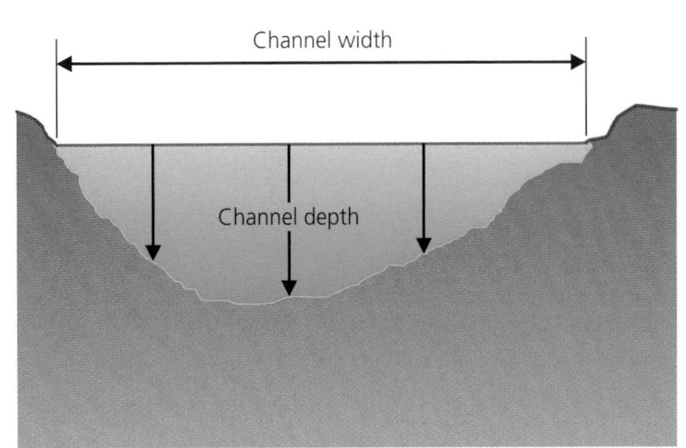

Figure 7: Cross-section of the river channel at Site 3

Data sheets

You should be confident about developing a clear and efficient data sheet from your work on your local area. Remember that you should leave plenty of space to ensure that results can be clearly accommodated. If this means you need to use more than one sheet of paper, then do so.

How you will be marked

The methods of presenting data (11 marks)

Mark	Reasons for marks given
0	There is no evidence of data presentation.
1–4	A limited range of basic presentation techniques is used. The methods used are usually not appropriate.
5–8	A range of mainly appropriate data-presentation techniques are used. Techniques are well presented, with scales and titles present on most techniques. At the top of this level, some of the techniques should be more sophisticated.
9–11	A wide range of presentation techniques are used, which are well presented and appropriate. There are scales and titles present on most techniques. A number of the presentation methods will be more sophisticated.

When you work with the data you have collected, make sure you use different, but appropriate, ways of presenting your data.

Methods of presenting data

Once you have collected your data, you then need to decide how to present it, so that those reading your investigation can fully understand it. As with data collection, there are a number of ways in which you can present your data.

Tables and graphs

The simplest way of representing data is by using tables with the numeric results in them. Graphs and charts are more visually interesting and they can make a description and analysis of data very clear, but you need to be careful that you use the correct type for the data.

Some people simply work their way along a spreadsheet toolbar to provide some variety in their presentation, but this shows a lack of understanding of the use of graphs. You should use the type of graph that does the job most accurately and efficiently. Some basic types are explained below.

Line graphs

Line graphs can be used to plot continuous data, such as a population size increase over a number of years, or the temperature of a particular location over a number of hours (Figure 8). The variable should be plotted on the y-axis, whilst the time should be plotted on the x-axis. It is also possible to plot more than one set of a particular data type on one graph – for example, the temperature change at three sites – allowing for comparison.

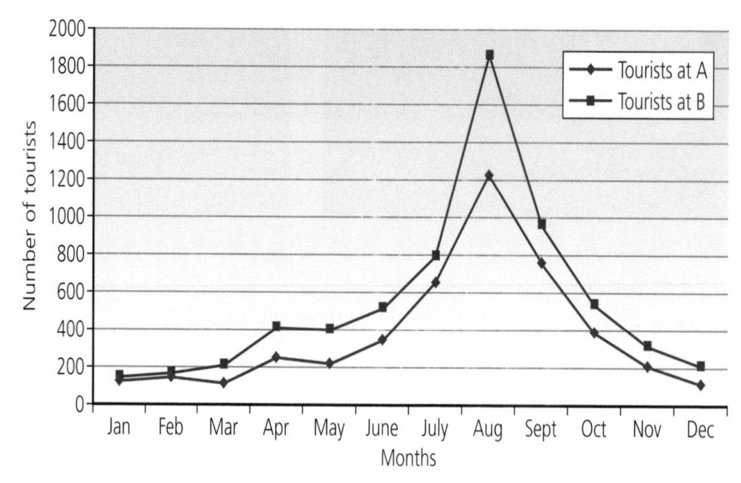

Figure 8: A line graph showing the distribution of monthly tourist numbers at two sites

Pie-charts

This graph type is used to present group values such as the number of different transport types observed at a road junction over the course of a day (Figure 9). The values are first converted into percentages, and then into degrees to allow the plotting of the data. If you use a graphics package on a computer, this will normally be done for you automatically. Pie-charts tend to be overused, and are best used when there are several categories of data in the group.

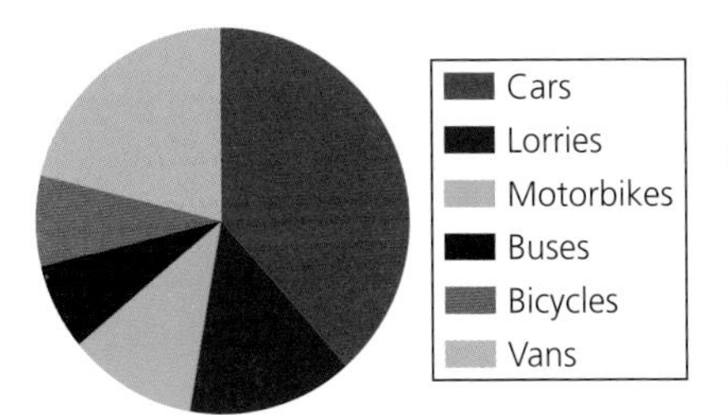

Figure 9: A pie-chart showing traffic composition at an urban location

Bar charts

As with pie-charts, bar charts are a common form of graph used in investigations. The y-axis is used for a numeric scale, such as the frequency of an event (Figure 10) while the x-axis is used to identify the categories of the data.

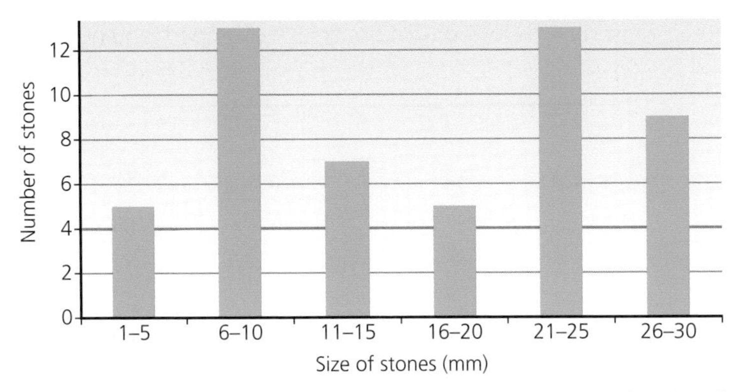

Figure 10: A bar chart showing the number of stones of different sizes found on a stream bed

Scattergraphs

Scattergraphs are more complex than the other types of graphs described here, as they do more than just present data visually. Scattergraphs are used to plot two sets of data to find out if there is a link between them. For example, you might count the number of services or amenities in twelve settlements, and find out what the population of each is. You would then plot the results as a number of points (Figure 11). These show a pattern from bottom left to top right. This is what is called a *positive correlation*, because as one variable (population size) gets larger, so does the other (number of services).

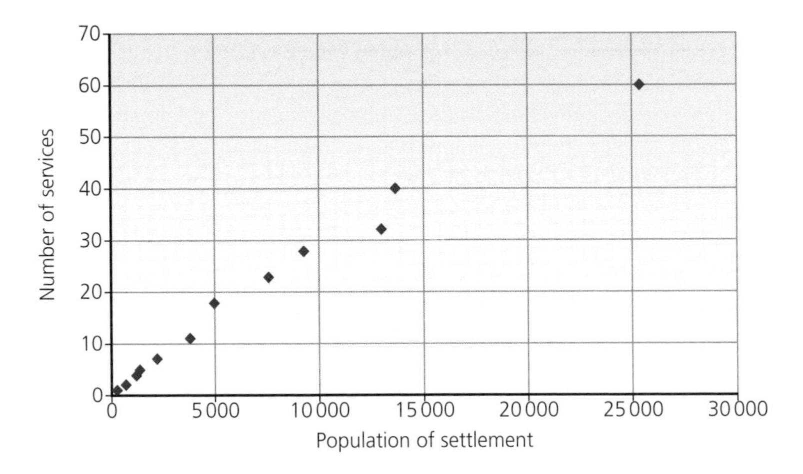

Figure 11: A scattergraph showing the relationship between settlement population and the number of services. (In this case, a positive correlation is shown.)

Control

You will be able to discuss your fieldwork and write about your purpose and methods of data collection in class under limited control to take into the final analysis.

If a scattergraph shows a clear pattern of points from top left to bottom right (Figure 12), this shows a *negative correlation* – as one variable gets larger the other gets smaller. Both these patterns would show that there is probably some link between the two variables being plotted. But if the points are random and show no pattern (Figure 13), there is no apparent correlation, so the two variables are probably not linked.

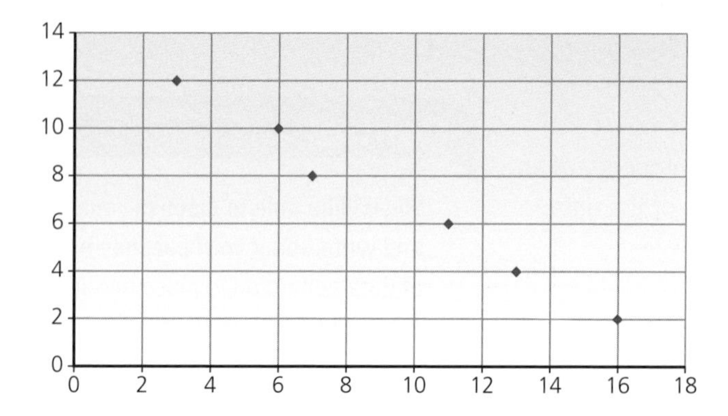

Figure 12: A scattergraph showing a negative correlation

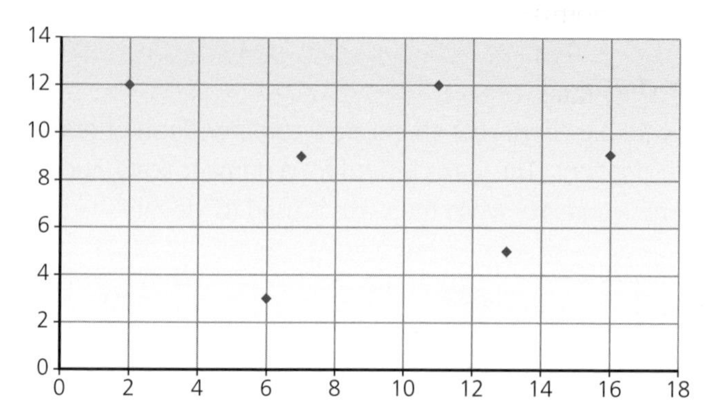

Figure 13: A scattergraph showing no correlation (i.e. random results)

Simple statistics

You may also want to calculate some simple statistics using your data. The most frequently used statistics are the *mean*, *mode*, and *median*. These can help in describing either the most frequently occurring or largest group of data in a data set, and can therefore help quantify and make clearer the patterns in the data you have collected.

Mean	The **mean** is the statistical average of a set of numbers. This is found by adding the values together and dividing the result by the number of values present. For example, the average of the four values 3, 4, 6 and 8 is 3+4+6+8 divided by 4 = 21 divided by 4 = 5.25.
Median	The **median** is the middle value in a set of numbers. It is found by arranging the values in order, and identifying the middle value. If the number of values is even, the two middle values are added together and divided by two. For example, with values 3, 6, 2, 7, 9 and 4, the median is found by first ordering the values: 2, 3, 4, 6, 7, 9, and then – because there are an even number of values – adding the middle two together (4+6) and dividing the result by two. Hence the median value is 10 divided by 2 = 5.
Mode	The **mode** is the most frequently appearing value in a group of numbers. For example, if a set of values is 5, 3, 4, 8, 5, 7, 5, 2, 3, 5 and 1, then the mode value is 5, because it appears more frequently than any other value.

Maps

Maps are another valuable way of presenting your data. You should try to make them clear and colourful. As with graphs, there are a number of different types of map which can be used.

Choropleth maps (Figure 14) use shading to show patterns in data, with shading normally becoming darker with larger numbers in the data. This type of map is used to compare areas in terms of grouped values, such as infant mortality rates, or house prices. Data must be sorted into groups, with clear boundaries, such as county boundaries, or regional boundaries.

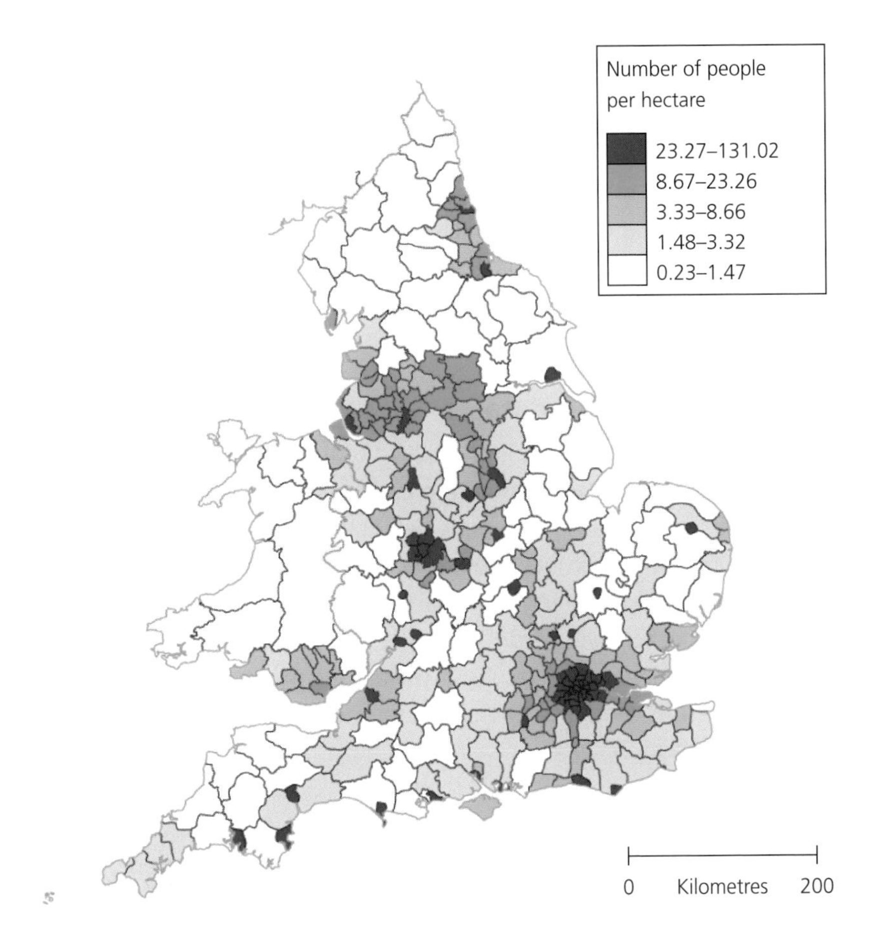

Figure 14: A choropleth map of population density in the UK

Flow lines (Figure 15) are used to show data relating to movement. Examples might include the number of pedestrians who walk down a certain street or the amount of traffic on a road. The direction in which the flow is moving is indicated by an arrow head at one end, and the width of the flow line is dependent on the volume of the flow. The scale used to determine the width of the lines must be chosen carefully, to allow the lowest and highest value to be shown clearly on the map.

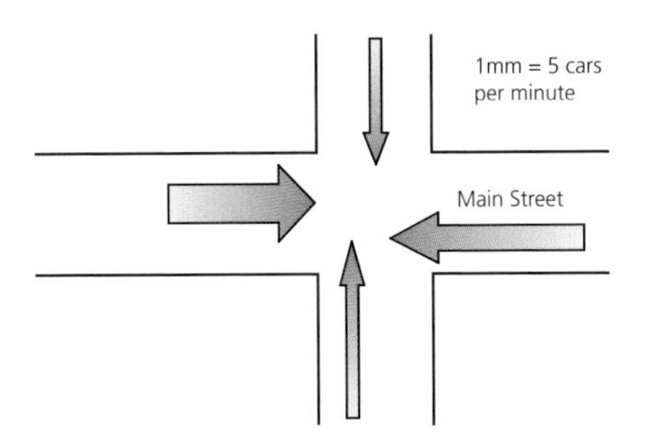

Figure 15: A flow line diagram showing traffic numbers at a crossroads

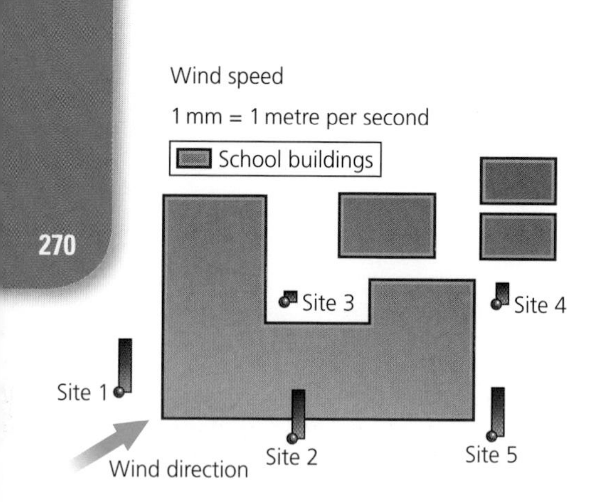

Figure 16: A map of a school's buildings, with wind speeds plotted at five sites

How you will be marked

Analysis and conclusions (9 marks)

Mark	Reasons for marks given
0	There is no analysis or conclusion.
1–3	Data has been extracted and described. Some basic conclusions have been drawn which vaguely relate to the question or issue investigated.
4–6	Data is described in some detail with analytical comments. Plausible conclusions are reached using the evidence which is presented in the investigation report.
7–9	There is analysis, which draw together the student's findings. The conclusions are accurate and proven and refer to the correct theory where appropriate.

Control

You must complete the write-up of your analysis, conclusions and evaluation individually and in class time under exam-style conditions.

Where data cannot be easily grouped and shown on a choropleth map, you may want to plot separate graphs or charts on to a map (Figure 16). This is a way of showing simple data distributions relative to each other in space, such as environmental quality values at given sites in an area. Again, you need to consider scale carefully so that the highest and lowest values can be clearly shown on the same map.

If you are using maps to present data, this might be a good opportunity to use GIS applications. These packages are designed to allow for the presentation of spatial data, and will also make decisions on issues such as shading and scale far easier to handle.

Developing your analysis and reaching your conclusions

Once you have presented your results, you will need to consider how to make the best use of them in explaining the patterns you see. For students, the analysis of data can be the most challenging part in the process of completing a fieldwork-based project – often because they find it hard to distinguish between *describing* the data and *explaining* the data. Analysis always includes explaining what you have found.

Making sense of your data

It is useful to start an analysis by laying out all of the results which you gained during the data collection and data presentation phases. Make a list of the data collected. Next to each set of data, describe what you think it shows (using any graphs, tables or maps you have produced to help). Finally, try to explain the data, highlighting *why* you think the results appear as they do. Remember that you will need to *describe* your results – *what* do they show – before *explaining* your results – *why* are the results as they are? The description of results should be completed fairly quickly, merely outlining what a group of data shows.

If you look back at Figure 14 on page 269, for example, you might describe the population density data given by highlighting the general pattern, i.e. that in general terms the population density is greater in and immediately surrounding large conurbations such as London, Birmingham and Liverpool, while rural areas have a much lower population density. This should then be supported by some specific examples such as Greater London having the highest population density of between 23.27 and 131.02 people per hectare, while North Devon has a low density of 0.23 to 1.47 people per hectare. Therefore, any description of the data should give a general impression of the patterns and include specific examples such as numbers, or quotes from interviews.

Explaining the patterns in your data

Your description of the data should be brief, while a greater focus should be given to explaining their patterns. This section of your project – where you give reasons for the patterns in your data – is crucial. You have presented and described your data, but can you now suggest why the results and associated patterns look like they do?

Look again at a question given at the start of this section – *'How do channel characteristics vary along your chosen river?'*. The results for the size of the sediment in the channel from several sites might appear like this.

Site	Average size in millimetres	Sediment shape
1 (upstream)	12	Angular
2	8	Sub-angular
3	7	Sub-angular
4	5	Sub-rounded
5 (downstream)	2	Rounded

Size of sediment found at five different sites along a stretch of river

We can *describe* the pattern as one showing decreasing sediment size as we move downstream, from an average of 12 mm at Site 1 to 2 mm at Site 5. We then need to explain this. There might be a number of explanations, but we could argue that the decrease in size is due to erosive processes in the channel, causing pebbles and gravel to constantly hit each other, leading to attrition of the sediment. It is at this point that we might then draw together some of our other results to make our argument stronger. If we have collected data on the shape of sediment at the five sites and can show that the sediment is becoming increasingly rounded, this might give extra evidence that attrition is occurring and wearing down the pebbles/gravel. Your explanations therefore need to give reasons for the data presented and, at the same time, they should also attempt to make links between different elements of your data, rather than simply reading it like a detailed list.

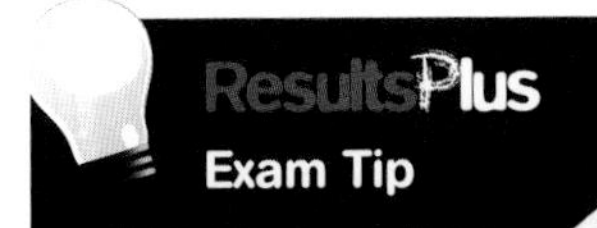

Remember that it is important to explain your results, not just describe them. Make sure you understand the difference between the two.

Concluding your study

Having explained your results, you should write a conclusion. This should summarise your main findings, and include the following two elements:

1. You should summarise your main findings in relation to how far they answer the question you posed at the start. Having posed that question, have you been able to answer it to some extent, and what is your answer?
2. You should refer back to any theory which is related to your study. For example, if studying service provision in rural areas, theory would suggest that smaller settlements will have fewer services. Do your results agree with theory, or are they different? If they are different can you explain why they might be different?

Therefore, having presented your findings in graphs, maps, etc., you then need to use these results to describe and explain what you have found, before concluding your study by relating what you have found to your original question, and to relevant geographical theories.

How you will be marked

Evaluation (9 marks)

Mark	Reasons for marks given
0	There is no evaluation.
1–3	There is limited evaluation of the investigation. Either all aspects of the investigation have been evaluated in limited detail or some aspects of the investigation have been evaluated in more detail.
4–6	There is evaluation of the investigation which varies in completeness between the aspects. Some of the limitations of the evidence collected have been recognised.
7–9	There is detailed evaluation of the investigation which reflects on the limitations of the evidence collected.

Evaluating your study

The evaluation of a piece of work is another area which students often find very difficult. This is the part of a study which aims to reflect on the process of collecting data, and how that process might have impacted on the quality of the results gained. An evaluation should consider the collection and analysis of data and how these might impact on the conclusions made at the end of the study. It is very important that you accept that no study is perfect – even those carried out by university academics – and it is therefore perfectly reasonable to highlight where you think the shortcomings of your work are.

What is an evaluation?

An evaluation should be based on three basic questions:

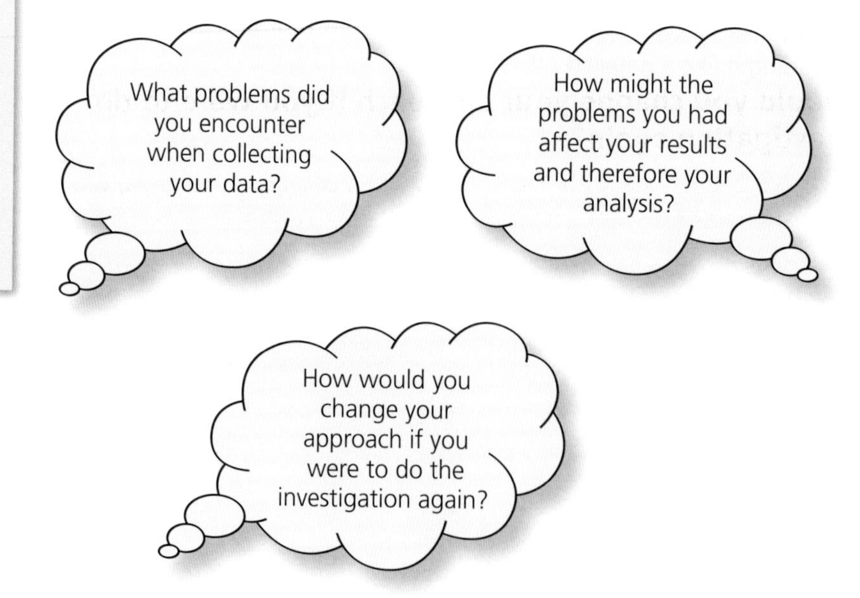

What problems did you encounter when collecting your data?

When collecting data, you need to be aware of the possible problems involved. If you were measuring the impact of tourism on a local environment, collecting your data on a bank holiday Monday might have given unusual pedestrian flow counts. While this is useful in showing the extremes of use that the environment sees, it might not be representative. If you were focusing on traffic volumes at locations around a CBD, you would not have been able to be in all the locations at once. This means that some of the differences in traffic might be due to time differences. For example, you might have visited two locations during the rush hour but, by the time you reached the third, rush hour might have been over.

Whatever your focus, you should consider the shortcomings of your data collection. Remember that this evaluation does not make your results incorrect – it demonstrates that you have a clear understanding of the difficulties involved in any data collection exercise.

How might the problems you had affect your results and therefore your analysis?

If you have identified any problems experienced in collecting your data, you next need to suggest how they might have affected your results. If we take the example of traffic flows around a CBD, having described the problems in collecting the data, you might then go on to say that, given that the third location was visited after the end of rush hour, its results might have been lower because of the time difference. Hence, your analysis that the third location is much quieter than the other two may be correct, but you have to accept that it may in part be inaccurate and it is possible that the site might be much busier during rush hour, much like the other two sites. Having explained this, you should finally explain how this might impact on your conclusions. Again, you should remember that this should not be seen as showing that you have done a poor piece of work, but that you are aware of the impact of any problems on your results.

How would you change your approach if you were to do the investigation again?

Finally, having identified the problems you had when collecting your data, you should now suggest how you would try to alter your collection methods if you did the study again. Therefore, if you identified the problems with collecting traffic flow data within a CBD, you might suggest that you would alter the locations used to ensure that all of them could be covered in the rush hour, or that you would ask a friend or parent to collect data in one or more locations, so that the timings were as close to each other as possible. This section, therefore, is focused on developing solutions to the problems identified in the first part of the evaluation.

Planning and organising your investigation

As you complete your fieldwork investigation, you will begin to gain a lot of paper, data, photos, and other evidence. During the process it is important that you organise your work so that you do not lose information, and when it comes to writing up your study, it is very important that you organise your work and thoughts into a well-planned and coherent end product.

Putting a study together

When an investigation is planned, it needs to be clear what the parts of the finished product will be and how they will fit together. It is possible to complete the fieldwork investigation using more than one medium but, if that is your plan, you should be clear before you start about how the different elements will come together to make sense.

The elements of a completed fieldwork investigation are shown in the bullet points below and should act as your basic framework when planning and organising your work. Remember that the word limit for the controlled assessment is 2,000 words. This means you should carefully consider the amount you write for each element of the report, and in each case make sure your writing is focused.

- Report outline
- Purpose of investigation
- Methods used to collect data
- Presentation of data
- Analysis and conclusions
- Evaluation.

You must remember that it will only be possible for each element to be worked on for a limited period of time. So you should plan your time carefully, to ensure you finish each element – you do not want to run out of time, leaving some elements unfinished. You should always allocate time to check each element – making sure that you have presented your information well, and that the spelling and grammar is accurate (see the section on Spelling, punctuation and geographical terminology) as well as ensuring that everything flows properly from one section to the next.

ResultsPlus
Exam Tip

When planning your investigation, remember to refer back to pages 258 and 259 to give you an idea of how long your teacher will ask you to work on particular parts of your work, and the level of control which you will be asked to work under.

How you will be marked

Planning and organisation (6 marks)

Mark	Reasons for marks given
0	The investigation report lacks any planning or organisation.
1–2	The work may be incomplete and not organised into a clear sequence. Geographical terminology may not be used accurately or appropriately.
3–4	There is a sequence of enquiry in the investigation report. Content is clear, for example page numbers are all present. Spelling, punctuation and the rules of grammar are accurate. Geographical terminology is used appropriately in the investigation report.
5–6	The sequence of the enquiry in the investigation report is clear. Diagrams are integrated into the text with appropriate sub-headings. Spelling, punctuation and the rules of grammar are accurate. There is accurate and appropriate use of geographical terminology.

Including diagrams

A well-presented project will have a series of illustrations – graphs, photos, and perhaps maps. You should ensure, however, that you only include illustrations which show useful and relevant information. And it is important that they are properly integrated into the written element of the work – you should not simply paste them in and assume that the reader will understand how they relate to the text. You should give each one an appropriate title, starting with 'Photo', 'Figure' or 'Table' and numbered – Photo 1, Photo 2, Figure 1, Figure 2, etc. The title should then describe what the illustration is showing, in a way that helps the reader understand it. In your text, when you describe or explain results, you should link the writing to the relevant diagram by referring to it in brackets, as shown in Example 2 below.

Example 2: Labelling a photo and referring to it in the text

The village had a small number of shops, such as a newsagent (Photo 1), and a chip shop, but few other services.

There are a number of reasons for the lack of services in the village . . .

Photo 1: Local services in village A

Spelling, punctuation and geographical terminology

Finally, you need to be careful with your spelling and punctuation. If you are using a word processor, the software may pick up on spelling, grammatical and punctuation 'errors', but do not assume that the computer is always right. You must always read through what you have written and check it carefully. Remember that marks are obtained for good use of language, and that poor spelling and punctuation can lead to a loss of marks.

You should also use geographical terminology where possible. For example, rather than writing 'amount of water flowing through the river', use 'discharge'. You will gain credit for using geographical terminology – if it is used correctly – because it shows a higher level of understanding on your part, and can often lead to briefer, but better explanations.

Variety of report formats

It is important that you carefully consider how you intend to present your fieldwork investigation. You have the opportunity to present your work in a number of different formats – just as professionals in a commercial setting might use different ways to present their information and ideas.

The most obvious format to use is a written report of approximately 2,000 words. Where this is chosen, it is simply a case of using a word processor to write up the various sections of your report. However, alternative formats can be used:

- DVDs – perhaps used where interviews are part of the data collection, or filming yourself explaining some of your results, perhaps carried out whilst collecting data on a stretch of river, or explaining changes in a transect across a city centre.
- PowerPoint presentations – which might be written in a similar fashion to a word processed piece, but completed as a presentation (Figure 17).
- Personalised GIS maps – these could be used in data presentation to show results, and possibly in annotating satellite images to develop issues or ideas in a more graphic form.
- Web pages – these can be developed to create a website format rather than a simple word processed project. One advantage of this format might be the ability to link different parts of the report together so that the reader can move backwards and forwards between elements. You might also decide to include hyperlinks to other websites which might help give background information or provide sources of secondary data (Figure 18).

When deciding on a report format, remember that it is the geography which is being assessed, not the standard of graphic ability.

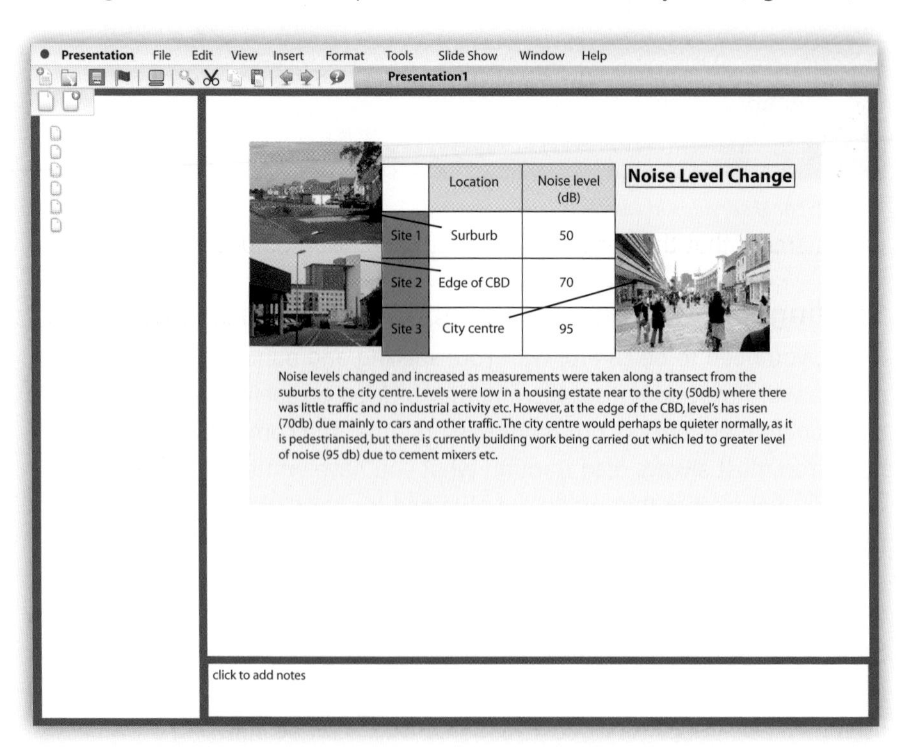

Figure 17: Example of a PowerPoint slide

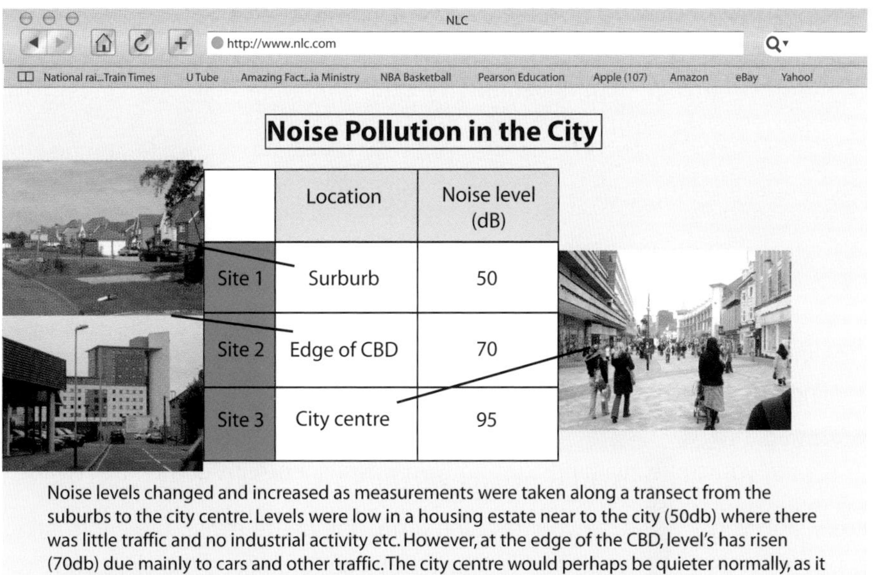

Noise Pollution in the City

	Location	Noise level (dB)
Site 1	Surburb	50
Site 2	Edge of CBD	70
Site 3	City centre	95

Noise levels changed and increased as measurements were taken along a transect from the suburbs to the city centre. Levels were low in a housing estate near to the city (50db) where there was little traffic and no industrial activity etc. However, at the edge of the CBD, level's has risen (70db) due mainly to cars and other traffic. The city centre would perhaps be quieter normally, as it is pedestrianised, but there is currently building work being carried out which led to greater level of noise (95 db) due to cement mixers etc.

Figure 18: Example of a website used to discuss noise level change

You can use more than one format, but your teacher may decide to use particular formats having considered both the available technology within your school and/or any other organisational restrictions which might apply.

If you are given a choice of formats, you should consider the following points in planning and developing your ideas:

- You need to understand how the elements and formats of your report will fit together. Will the reader understand how to 'read' your work? Are you trying to use too many formats?
- Given that you have limited time, you should be confident that you can use the format you have chosen efficiently. A website might sound like an exciting idea, but if you do not know how to write one, you will not have enough time to teach yourself.
- Remember that the geography is what is important. Students can often get carried away in spending time on the design aspect of a PowerPoint presentation or website, and forget that it is the geographical content that they are gaining credit for.

You should decide on a format, if given a choice, at the initial planning stage and discuss it with your teacher so that you are both happy with your decision.

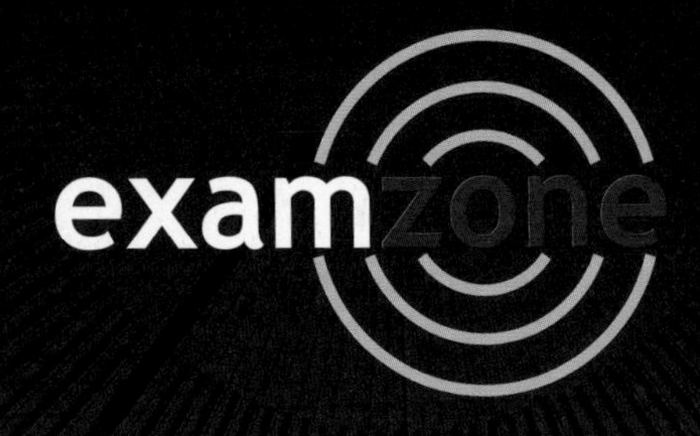

Welcome to ExamZone! Revising for your exams can be a daunting prospect. In this section of the book we'll take you through the best way of revising for your exams, step-by-step, to ensure you get the best results that you can achieve.

Zone In!

Have you ever become so absorbed in a task that it suddenly feels entirely natural? This is a feeling familiar to many athletes and performers: it's a feeling of being 'in the zone' that helps you focus and achieve your best. Here are our top tips for getting in the zone with your revision.

UNDERSTAND IT

Understand the exam process and what revision you need to do. This will give you confidence but also help you to put things into proportion. These pages are a good place to find some starting pointers for performing well at exams.

BUILD CONFIDENCE

Use your revision time, not just to revise the information you need to know, but also to practise the skills you need for the examination. Try answering questions in timed conditions so that you're more prepared for writing answers in the exam. The more prepared you are, the more confident you will feel on exam day.

DEAL WITH DISTRACTIONS

Think about the issues in your life that may interfere with revision. Write them all down. Think about how you can deal with each so they don't affect your revision. For example, revise in a room without a television, but plan breaks in your revision so that you can watch your favourite programmes. Be really honest with yourself about this – lots of students confuse time spent in their room with time revising. It's not at all the same thing if you've taken a look at Facebook every few minutes or taken mini-breaks to send that vital text message.

FRIENDS AND FAMILY

Make sure that they know when you want to revise and even share your revision plan with them. Help them to understand that you must not get distracted. Set aside quality time with them, when you aren't revising or worrying about what you should be doing.

GET ORGANISED

If your notes, papers and books are in a mess you will find it difficult to start your revision. It is well worth spending a day organising your file notes with section dividers and ensuring that everything is in the right place. When you have a neat set of papers, turn your attention to organising your revision location. If this is your bedroom, make sure that you have a clean and organised area to revise in.

KEEP HEALTHY

During revision and exam time, make sure you eat well and exercise, and get enough sleep. If your body is not in the right state, your mind won't be either – and staying up late to cram the night before the exam is likely to leave you too tired to do your best.

Planning Zone

The key to success in exams and revision often lies in the right planning. Knowing what you need to do and when you need to do it is your best path to a stress-free experience. Here are some top tips in creating a great personal revision plan:

JUNE

1. Know when your exam is
Find out your exam dates. Go to www.edexcel.com/iwantto/Pages/dates.aspx to find all final exam dates, and check with your teacher. This will enable you to start planning your revision with the end date in mind.

2. Know your strengths and weaknesses
At the end of the chapter that you are studying, complete the 'You should know' checklist. Highlight the areas that you feel less confident on and allocate extra time to spend revising them.

3. Personalise your revision
This will help you to plan your personal revision effectively by putting a little more time into your weaker areas. Use your mock examination results and/or any further tests that are available to you as a check on your self-assessment.

4. Set your goals
Once you know your areas of strength and weakness you will be ready to set your daily and weekly goals.

5. Divide up your time and plan ahead
Draw up a calendar, or list all the dates, from when you can start your revision through to your exams.

6. Know what you're doing
Break your revision down into smaller sections. This will make it more manageable and less daunting. You might do this by referring to the Edexcel GCSE Geography A specification, or by the chapter objectives, or by headings within the chapter.

7. Link it together
Also make time for considering how topics interrelate. For example, when you are revising your case studies for Unit 1 it would be sensible to cross-reference these to other parts of your work. If you have studied the Wasteful World section then the case study on Germany's waste disposal would obviously support your work on recycling for the Unit 1 topic of sustainability. You could draw up a mind-map of the case studies showing how and where they link together.

8. Break it up
Revise one small selection at a time, but ensure you give more time to topics that you have identified weaknesses in.

9. Be realistic
Be realistic in how much time you can devote to your revision, but also make sure you put in enough time. Give yourself regular breaks or different activities to give your life some variety. Revision need not be a prison sentence!

10. Check your progress
Make sure you allow time for assessing progress against your initial self-assessment. Measuring progress will allow you to see and celebrate your improvement, and these little victories will build your confidence for the final exam

Finally – stick to your plan!

Know Zone

Remember that different people learn in different ways – some remember visually and therefore might want to think about using diagrams and other drawings for their revision, whereas others remember better through sound or through writing things out. Think about what works best for you by trying out some of the techniques below.

REVISION TECHNIQUES

Highlighting: work through your notes and highlight the important terms, ideas and explanations so that you start to filter out what you need to revise.

Key terms: look at the key terms highlighted in bold in each chapter. Try to write down a concise definition for this term. Now check your definition against the glossary definition on p289.

Summaries: writing a summary of the information in a chapter can be a useful way of making sure you've understood it. But don't just copy it all out. Try to reduce each paragraph to a couple of sentences. Then try to reduce the couple of sentences to a few words!

Concept maps: if you're a visual learner, you may find it easier to take in information by representing it visually. Draw concept maps or other diagrams. These are particularly good at showing links. For example, you could create a concept map which shows how to learn about sustainability.

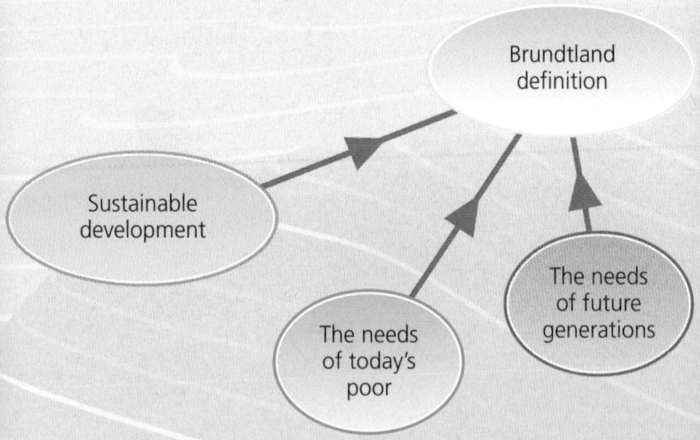

Mnemonics: this is when you take the first letter of a series of words you want to remember and then make a new word or sentence. An example of this is SAGA. This stands for Slope, Aspect, Ground conditions and Altitude.

Index cards: Write important events, definitions and processes on index cards and then test yourself.

Quizzes: Learning facts can be dull. Why not make a quiz out of it? Set a friend 20 questions to answer. Make up multiple-choice questions. You might even make up your own exam questions and see if your friend can answer them!

And then when you are ready:

Practice questions: go back through all the ResultsPlus features with questions to see if you can answer them (without cheating!). Try writing out some of your answers in timed conditions so that you're used to the amount of time you'll have to answer each type of question in the exam. Then, check the guidance for each one and try to mark your answer.

Use the list below to find all the ResultsPlus questions.

Chapter 1: p 13, 18, 23, 25 and 28

Chapter 2: p 30, 32, 39, 43, 45 and 52

Chapter 3: p 60, 62, 69 and 70

Chapter 4: p 75, 77, 85 and 86

Chapter 5: p 89, 90, 96 and 98

Chapter 6: p 104, 111, 113 and 114

Chapter 7: p 119, 125, 130 and 134

Chapter 8: p 139, 150 and 152

Chapter 9: p 160, 164 and 170

Chapter 10: p 173, 180 and 186

Chapter 11: p 195, 200 and 202

Chapter 12: p 207, 211 and 220

Chapter 13: p 225, 227 and 238

Chapter 14: p 246, 255 and 256

Don't Panic Zone

Once you have completed your revision in your plan, you'll be coming closer and closer to the big day. Many students find this the most stressful time and tend to go into panic-mode, either working long hours without really giving their brain a chance to absorb information, or giving up and staring blankly at the wall. Follow these tips to ensure that you don't panic at the last minute.

TOP TIPS

1. Test yourself by relating your knowledge to geography issues that arise in the news – can you explain what is happening in these issues and why?

2. Look over past exam papers and their mark schemes. Look carefully at what the mark schemes are expecting of candidates in relation to the question.

3. Do as many practice questions as you can to improve your technique, help manage your time and build confidence in dealing with different questions.

4. Write down a handful of the most difficult bits of information for each chapter that you have studied. At the last minute focus on learning these.

5. Relax the night before your exam – last-minute revision for several hours rarely has much additional benefit. Your brain needs to be rested and relaxed to perform at its best.

6. Remember the purpose of the exam – it's for you to show the examiner what you have learnt.

LAST MINUTE LEARNING TIPS FOR GEOGRAPHY

- Remember that an intelligent guess is better than nothing – if you can't think of an example of an LIC city then take a guess – you cannot lose marks.
- Know your categories – don't go into the examination unclear about basic definitions: LICs and HICs, urban and rural, erosion and weathering. Check out the glossary.
- Many examination questions ask you to interpret resources. Make sure that you revise the skills that help you do this effectively.

Exam Zone

Here is some guidance on what to expect in the exam itself: what the questions will be like and what the paper will look like.

UNIT	% OF OVERALL GCSE	MARKS	DESCRIPTION	KNOWLEDGE AND SKILLS
Unit 1: Geographical skills and challenges You will study both these sections: **Section A – Geographical skills (Ch 1)** **Section B – challenges for the planet (Ch 2)**	25	50	• You will be entered for either the Foundation level paper or the Higher level paper • Unit 1 is a compulsory unit • It is an assessed by an exam that is marked by an Edexcel examiner • The exam is 60 minutes • All questions must be answered	**Section A – Geographical skills** • Basic skills • Cartographic (map) skills • Graphical skills • Geographical enquiry skills • ICT skills • GIS skills **Section B – Challenges for the planet** • The causes, effects and responses to climate change • Sustainable development for the planet
Unit 2: The natural environment **Section A – The physical world** You will study one topic from: 1. Coastal landscapes (Ch 3) 2. River landscapes (Ch 4) 3. Glaciated landscapes (Ch 5) 4. Tectonic landscapes (Ch6) **Section B – environmental issues** You will study one topic from: 5. A wasteful world (Ch 7) 6. A watery world (Ch 8)	25	50	• You will be entered for either the Foundation level paper or the Higher level paper • The exam is 60 minutes • Your exam paper will have a resource booklet • You must answer **one** question from Section A and **one** question from Section B • There are four questions in Section A, one on each topic (chapter) • There are two questions in Section B, one on each topic (chapter) • Each question is worth 25 marks and is broken up into seven or eight sub-sections – 2 (a) (i) and so on • The marks available for each sub-section vary • Extended writing questions examine the quality of your written communication	The following are tested: • knowledge of the course content • understanding of the processes • ability to interpret the questions correctly • map interpretation • analysis of photographs • ability to read graphs and tables correctly • understanding the limitations of the data collection methods

UNIT	% OF OVERALL GCSE	MARKS	DESCRIPTION	KNOWLEDGE AND SKILLS
Unit 3: The human environment **Section A – The human world** You will study one topic from: 1. Economic change (Ch 9) 2. Farming and the countryside (Ch 10) 3. Settlement change (Ch 11) 4. Population change (Ch 12) **Section B – People issues** **You will study one topic from:** 5. A moving world (Ch 13) 6. A tourist's world (Ch 14)		50	• You will be entered for either the Foundation level paper or the Higher level paper • The exam is 60 minutes • Your exam paper will have a resource booklet • You must answer **one** question from Section A and **one** question from Section B • There are four questions in Section A, one on each topic (chapter) • There are two questions in Section B, one on each topic (chapter) • Each question is worth 25 marks and is broken up into seven or eight sub-sections – 2 (a) (i) and so on • The marks available for each sub-section vary • Extended writing questions examine the quality of your written communication	The following are tested: • knowledge of the course content • understanding of the processes • ability to interpret the questions correctly • map interpretation • analysis of photographs • ability to read graphs and tables correctly • understanding the limitations of the data collection methods
Unit 4: Investigating Geography Your fieldwork investigation will be on one of these themes: 1. Coasts 2. Contemporary 3. Countryside 4. Environmental 5. Rivers 6. Tourism 7. Transport 8. Urban	25	50	• This is assessed in school under controlled conditions and moderated by an Edexcel examiner • You will carry out your Fieldwork and produce a write-up of approximately 2000 words • Your work will be evaluated according to: a) the purpose of the investigation b) methods of collecting data c) methods of presenting data d) analysis and conclusions e) evaluation f) planning and organisation.	This unit examines your ability to select and use a variety of skills and techniques to investigate, analyse and evaluate questions and issues. It is important that you are aware of the limitations of any fieldwork and research assignment, and can properly assess these weaknesses in your evaluation.

ASSESSMENT OBJECTIVES

The questions that you will be asked are designed to examine the following aspects of your geography. These are known as Assessment objectives (AO). There are three AOs.

AO1	Recall, select and communicate knowledge and understanding of places, environments and concepts.
AO2	Apply knowledge and understanding in familiar and unfamiliar contexts.
AO3	Select and use a variety of skills, techniques and technologies to investigate, analyse and evaluate questions and issues.

THE TYPES OF QUESTION THAT YOU CAN EXPECT IN YOUR EXAM

The examination papers are designed so that the opening part of each question is the easiest part. The difficulty becomes progressively harder as you move through the question. The level of difficulty is controlled by the command word and content required in your answer. The Foundation tier papers have questions which have more 'scaffolding' (helping you to structure and develop your answer) to make these papers more accessible.

There are four different types of question:

TYPE
Short – single word answers or responses involving a simple phrase or statement.
MCQ – Multiple Choice Question.
Open – free-response questions that involve a limited amount of continuous prose.
Long – free-response questions where candidates have the opportunity for extended writing and allow opportunities for assessing the quality of your written communication.

UNDERSTANDING THE LANGUAGE OF THE EXAM PAPER

It is vital that you know what 'command' words ask you to do. Common errors are:

1. Confusing *describe* with *explain*.
2. Adding *explanation* when you are only asked to *describe*.

Identify…	Name a process or a location
Complete	Finish of a task that has already been partly done
Name	Like 'identify'
Describe…	Give the main characteristics of a topic or issue
Explain..	Give reasons why something is as it is
Examine..	Describe something with some detail
Outline..	Give the main features of something
Define..	Say what something means
Suggest reasons…	Say why something might have happened or occurred
Give the reasons…	Say why something happened or occurred
Comment on…	Give some reasons why or how something is as it is
State…	Like 'name' or 'identify'

Meet the exam paper

This section shows you what the exam paper looks like. Check that you understand each part. Now is a good opportunity to ask your teacher about anything that you are not sure of here.

Print your surname here, and your other names in the next box. This is an additional safeguard to ensure that the exam board awards the marks to the right candidate.

Here you fill in your personal exam number. Take care when writing it down because the number is important to the exam board when writing your score.

Write your name here

Surname | Other names

Centre Number | Candidate Number

Edexcel GCSE

Geography A

Unit 3: The Human Environment

Foundation Tier

Sample Assessment Material
Time: 1 hour

Paper Reference
5GA3F/01

You must have:
Resource Booklet (enclosed)

Total Marks

Ensure that you understand exactly how long the examination will last, and plan your time accordingly.

Here you fill in your school's centre number. You will be given this by your teacher on the day of your exam.

Instructions

- Use **black** ink or ball-point pen.
- **Fill in the boxes** at the top of this page with your name, centre number and candidate number.
- Answer **one** question from Section A and **one** question from Section B.
- Answer the questions in the spaces provided – *there may be more space than you need.*

Information

- The total mark for this paper is 50.
- The marks for **each** question are shown in brackets – *use this as a guide as to how much time to spend on each question.*
- Questions labelled with an **asterisk** (*) are ones where the quality of your written communication will be assessed – *you should take particular care with your spelling, punctuation and grammar, as well as the clarity of expression, on these questions.*
- The following abbreviations are used: LIC – low-income country, HIC – high-income country

Advice

- Read each question carefully before you start to answer it.
- Keep an eye on the time.
- Try to answer every question.
- Check your answers if you have time at the end.

Turn over ▶

Ensure that you read the instructions carefully and that you understand exactly which questions from which sections you should attempt.

Note that the quality of your written communication will also be marked. Take particular care to present your thoughts and work at the highest standard you can for maximum marks.

N35662A
©2008 Edexcel Limited.
3/1

edexcel
advancing learning, changing lives

Edexcel GCSE in Geography A | Sample Assessment Materials | © Edexcel Limited 2008 | 155

If Economic change is the topic that you have studied in class and you wish to answer the question on it, remember to indicate this where you are asked on the paper.

It is not always one examiner who will mark your entire paper. Sometimes, one examiner will mark one question and another will mark a different question. So, you must indicate which question you have answered so that your paper is sent to the correct examiner!

SECTION A – THE HUMAN WORLD

Answer only ONE question from Section A.
Indicate which question you are answering by marking a cross in the box ☒. If you change your mind, put a line through the box ☒ and then indicate your new question with a cross ☒.

Topic 1: Economic change

If you answer Question 1 put a cross in the box ☐.

1 Employment structure varies from place to place and has varied over time.
Complete the following sentences that describe some of these trends.
Use some of the words below.

(5)

increased	**decline**	**secondary**	
demand	**primary**	**profit**	**manufacturing**
services	**raw materials**	**extraction**	

(a) In most HICs (high-income-countries) there has been a .... in the numbers of people employed in the primary and .... sectors. At the same time the numbers employed in the tertiary sector have .... The primary sector is concerned with the extraction of .... ; an example is mining. The secondary sector is the manufacturing of goods for sale. The tertiary sector is very varied and includes both low paid jobs in sectors such as retailing but also highly paid professionals. It involves the offering of ....

(b) The following questions are multiple choice. Read the questions carefully and then put a cross in the box of the answer that you select.
There is only one correct answer to each question.

(i) A good example of a primary activity that is declining in HICs is

(1)

A ☐ tourism
B ☐ banking
C ☐ agriculture
D ☐ shipbuilding

Read the instructions each time – they are there to provide guidance.

Pay attention to any text highlighted in bold. It is highlighted to alert you to important information, so be sure to read it and take note!

The marks for each question are shown on the right-hand side of the page. Make sure that you note how many marks a question is worth as this will give you an idea of how long to spend on that question.

This is the resource – be careful, it may not be exactly like resources that you have seen before.

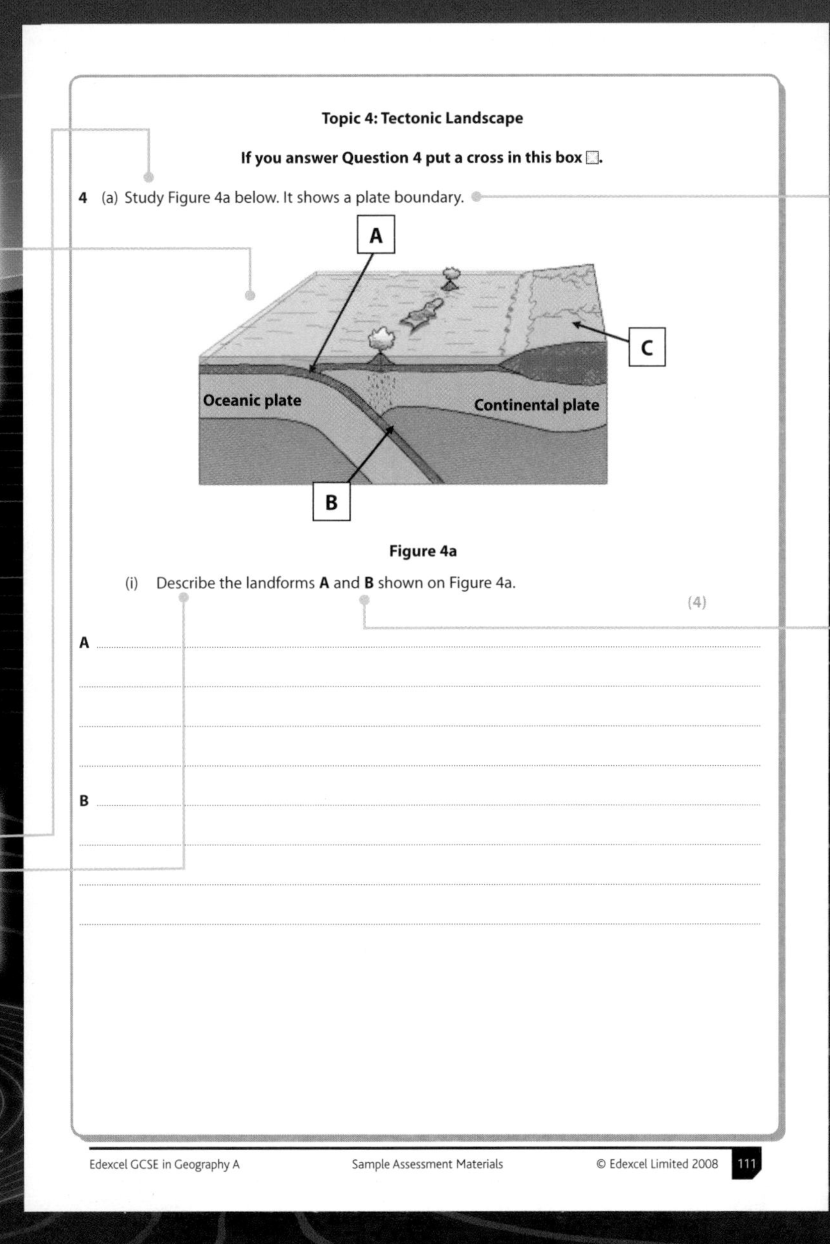

This is the 'stem' of a question – it often includes important information that you need to think about in your answers.

Words in bold are highlighted to catch your attention so be sure to note them.

These are the command words that tell you what to do.

Zone Out

Well done, you have finished your exam. So, what now? This section provides answers to the most common questions students have about what happens after they complete their exams.

About your grades

Whether you've done better than, worse than or just as you expected, your grades are the final measure of your performance on your course and in the exams.

When will my results be published?

Results for summer examinations are issued in August, with GCE first and GCSE second. January exam results are issued in March.

Can I get my results online?

Visit www.resultsplusdirect.co.uk, where you will find detailed student results information including the 'Edexcel Gradeometer' which demonstrates how close you were to the nearest grade boundary. Students can only gain their results online if their centre gives them permission to do so.

I haven't done as well as I expected. What can I do now?

First of all, talk to your subject teacher. After all the teaching that you have had, tests and internal examinations, he/she is the person who best knows what grade you are capable of achieving. Take your results slip to your subject teacher, and go through the information on it in detail. If you both think that there is something wrong with the result, the school or college can apply to see your completed examination paper and then, if necessary, ask for a re-mark immediately. The original mark can be confirmed or lowered, as well as raised, as a result of a re-mark.

If I am not happy with my grade, can I re-sit a unit?

Yes, you are able to re-sit each unit once before claiming certification for the qualification. The best available result for each contributing unit will count towards your final grade.

What can I do with a GCSE in Geography?

Geography is well known as a subject that links to all other subjects of the curriculum, so a GCSE in Geography is a stepping stone to a whole range of opportunities. A good grade will help you to move on to AS, Applied A Level or BTEC course. You may want to continue your study of Geography or take a course such as a BTEC National in Travel and Tourism which has a more work-related approach.

The skills that you develop can lead you to employment opportunities in journalism, media, engineering, ICT, travel and tourism, environmental management, marketing, business management and teaching. Geographers are everywhere!

Glossary

A-C

Ablation: the loss of ice from a glacier due to melting.

Abrasion: a process of erosion involving the scraping of a rock surface by moving particles of rock as they are transported by water, wind or ice.

Accessible countryside: a rural area beyond the commuter belt, but within day-trip reach.

Acid rain: precipitation that is unusually acidic because it contains dissolved pollutants, such as sulphur and nitrogen oxides, usually as a result of human activity.

Adventure holiday: a type of holiday that involves activities such as bungee jumping and snorkelling.

Aerial photograph: a photograph taken from the air.

Afforestation: the process of planting an area with trees.

Ageing (greying) population: a population in which there is a high percentage of people aged 65 or over.

Agribusiness: any large company involved in food production, including farming, seed supply, agrochemicals, farm machinery, food processing, and food marketing and retailing.

Agriculture: farming, using the land to grow crops or rear animals.

Air pollution: the addition of harmful chemicals, particles or biological material to the atmosphere.

Altitude: height of a point, in relation (usually) to sea-level, e.g. 3,000 metres above sea-level.

Annotation: adding written information (notes) to maps, diagrams and photographs to help the reader to understand them better.

Appropriate technology: equipment that the local community is able to use relatively easily and without much cost. Also known as Intermediate technology.

Aquifers: water-bearing rocks.

Arête: a long, narrow and sharp ridge that separates two corries.

Aspect: the direction in which a slope faces.

Attrition: the gradual wearing down of rock materials as they are transported by water, wind or ice.

Automation: the use of machinery in manufacturing and data processing.

Avalanche: a sudden and rapid movement of a mass of snow or ice down a slope.

Backwash: water from a breaking wave which flows under gravity down a beach and returns to the sea.

Bar: a barrier of sand or pebbles deposited offshore.

Biodiversity: the number and variety of living species found in a specific area.

Biofuel: fuel derived from biological material, such as palm oil.

Birth rate: the number of births per 1000 people in a year.

Bostis: the name given to squatter settlements in Bangladesh.

Brownfield/brownfield site: a piece of land that has been used and abandoned, and is now awaiting some new use.

Butler model: a model showing the sequence of changes experienced by holiday resorts.

Capitalism: an economic system in which most of the land and wealth is owned and controlled by people and companies rather than the government.

Carbon dioxide: an important greenhouse gas (which is increasingly a product of human activity).

Carbon footprint: a measurement of all the greenhouse gases we individually produce, expressed as tonnes (or kg) of carbon dioxide equivalent.

C

Carrying capacity: the maximum number of people that can be supported by the resources and technology of a given area.

Channelisation: a hard engineering method that involves deepening and/or straightening the river.

Choropleth map: a type of map that uses a progression of colour shades, from darker to lighter, to represent the distribution of a particular variable.

Circulation: a shift in location that is only temporary, such as a foreign worker on a six-month contract.

Climate change: long-term changes in global atmospheric conditions.

Coastal flooding: when seawater covers areas that are usually dry.

Coastal management: the decision-making procedures governing the coastal area, which include decisions about coastal flooding and defence.

Communication networks: the passing of information and ideas through the media (radio, TV, phone, internet, etc.).

Communism: an economic system in which the land and wealth is owned and controlled by the government.

Commuter: a person who travels from their home area to their place of work, on a daily basis.

Commuter belt: a countryside area with settlements that are used as dormitories by urban-based workers and their families.

Confluence: the point where two streams or rivers meet.

Congestion charging: a system in which drivers entering certain areas of a city, usually the centre, have to pay a daily charge.

Conservative plate boundary: where two tectonic plates slide past each other.

Conserve: to keep something in its current state by protecting it from harm or decay.

Constructive wave: a wave that adds more material to a beach than it removes.

Consumer culture: a culture in which there is an emphasis on the purchasing of goods and services.

Consumer industries: those industries making finished products, such as clothing, electrical goods and furniture, which are sold in shops.

Consumer society: an affluent society in which high volumes of goods and services are purchased.

Contour: a line on a map connecting places at the same height above sea-level.

Convection currents: (in tectonics) circulating movements of magma in the mantle caused by heat from the core.

Convergent plate boundary: where two tectonic plates move towards each other.

Corporations: large business organisations that are managed by a board of directors on behalf of their shareholders.

Corrie: an armchair-shaped depression formed by a glacier in a mountainous area.

Corrie lake: a small, circular lake in the base of the corrie.

Corrosion: a form of weathering involving the breaking down of material through chemical reactions.

Counterurbanisation: the movement of people and employment from major cities to smaller settlements and rural areas located just beyond the city, or to more distant smaller cities and towns.

Country of origin: (in migration) where a migrant comes from – their source.

Crop yield: the amount of useable crop produced per unit area of land.

Cross-section: a diagram, produced from information on a map, to show the shape of a landscape or structure, as if it had been cut through.

D-E

Dam: a barrier that retains water in a lake or reservoir.

Data-handling software: computer programs that can calculate, manipulate and display large amounts of statistical and other information.

Death rate: the number of deaths per 1000 people in a year.

Deforestation: the chopping down and removal of trees to clear an area of forest.

Deindustrialisation: the decline in industrial activity in a region or an economy.

Demographic transition model: a model of population change over time, based on the variations in a country's birth rate and death rate, in a series of stages.

Density: the number of people (or things) per unit area.

Dependent population: the people (e.g. children and the elderly) who are not working and therefore depend on others.

Depletion: the reduction and exhaustion of resources as a result of their use.

Depopulation: a decline in the number of people living in an area, usually through out-migration or changes in birth rate and death rate.

Deposition: the dropping of material that was being carried by a moving force.

Deprivation: the damaging lack of material benefits that are considered to be basic necessities in a society – employment, housing, etc.

Destructive wave: a wave that removes more material from a beach than it adds.

Development: the process of change whereby countries become better organised and better off. The basic fuel of development is economic growth.

Dispersed settlement: where individual dwellings are spread out.

Disposable income: the amount of money which a person has available to spend on non-essential items, after they have paid for their food, clothing and household running costs.

Distribution: a description of where a particular feature or features – such as people – appear within an area.

Divergent plate boundary: where two tectonic plates move away from each other.

Diversification: spreading business risks by adding new activities and removing complete dependence on the one original activity.

Divided bar-charts: a bar chart in which the individual bars are divided into different sections to show more detailed information.

Domestic: relating to the home and home life.

Download: to take information directly from the internet.

Drainage basin: the area of land drained by a river and its tributaries.

Drumlin: a low, elongated hill formed by the deposition and shaping of rock debris by a glacier.

Earthquake: a violent shaking of the Earth's crust.

Eco-tourism: a form of tourism that tries to minimise the environmental impacts of the tourists, by using local providers and resources and by keeping profits within the local area.

Eco-towns: proposed new towns that are designed to be much more sustainable than traditional settlements.

Economic migrant: a person seeking work, usually in another country.

Economy: what a country or area 'does for a living' in order to create wealth – its production, exchange, distribution and consumption of goods and services.

E-G

Educational holiday: a type of holiday that involves learning something, such as wine tasting or learning to cook.

Emigrant: a person who moves out of an area or country.

Energy deficit: a situation when the use of energy exceeds the production of energy.

Energy efficiency: the effective use of energy, without unnecessary waste.

Energy surplus: a situation when the production of energy exceeds the use of energy.

Enhanced greenhouse effect: the increase in the greenhouse effect that is widely thought to be caused by human activity.

Epicentre: the point on the surface directly above the focus of an earthquake.

Erosion: the wearing away and removal of material by a moving force, such as a breaking wave.

Erratic: a large rock that has been transported by a glacier and deposited in an area of different rock type, making it appear out of place.

Ethnicity: the national or cultural characteristics that allow people to identify themselves as members of a particular social group.

Evaluation: a reflection upon a piece of work, outlining its strengths and weaknesses and suggesting possible improvements.

Fetch: the distance a wave has travelled toward the coastline.

Field sketch: a simple drawing of a scene that can be labelled and/or annotated to emphasise particular features.

Focus: the site of the movement inside the Earth's crust that results in an earthquake.

Fold mountain range: a line of mountains formed by the uplift and buckling of continental crust at a convergent plate boundary.

Flood plain: a wide, flat area of land either side of a river in its lower course.

Food miles: the distance a food is transported from the place of its production to where it reaches the consumer. The greater this distance, the more fuel that has been used.

Forced migration: the movement of people who have no choice but to leave an area, because of persecution, war, etc.

Fossil fuels: non-renewable resources that can be burned – such as coal, oil or natural gas – that have been formed in the Earth's crust.

Freeze–thaw: a weathering process which causes rock to break down by the repeated action of water freezing and expanding in cracks.

Garden grabbing: a practice where property developers buy up large detached houses, knock them down and then squeeze a small estate of new homes on to the same plot.

Gated community: an area of wealthy private housing with a secure perimeter and a controlled entrance for residents and visitors.

Genetic modification (GM): the manipulation of the genetic material of a plant or animal to produce desired traits, such as nutritional value or resistance to herbicides.

Geology: the study of the Earth – its composition (rocks), structure and dynamic processes.

Ghetto: a residential area that is mainly occupied by one ethnic group.

GIS (Geographical Information Systems): a system that captures, stores, analyses, manages and displays geographical information.

Glaciation: the impact of glaciers on the landscape.

Global: relating to the whole world.

Global shift: the movement of manufacturing from HICs to cheaper production locations in MICs and LICs.

Global (information) superhighway: the internet/worldwide web/cyberspace, allowing the fast transfer of information to virtually anywhere on Earth.

G-I

Global warming: a trend whereby global temperatures rise over time, much of which is now linked with the enhanced greenhouse effect.

Globalisation: the process, led by transnational companies, whereby the world's countries are all becoming part of one vast global economy.

Greenfield/greenfield site: a piece of land that has not been built on before, frequently found in rural areas, but is now being considered for development.

Greenhouse gases: those gases in the atmosphere that absorb outgoing radiation, hence increasing the temperature of the atmosphere.

Gross national income (GNI): the total value of goods and services produced by a country's economy, plus overseas earnings.

Ground condition: the nature of the soil and immediate sub-surface material, especially its water retention.

Hanging valley: a tributary valley left high above a main valley that has been deepened by glacial erosion.

Hard engineering: using solid structures to resist forces of erosion.

Heritage: relating to something inherited from the past that is valued or celebrated – e.g. old buildings, archaeological remains, old customs and traditions.

Honeypot: a place of special interest or appeal that attracts large numbers of visitors and tends to become overcrowded at peak times.

Host country: (in migration) the country where a migrant now lives.

Hot spot: (in tectonics) a place where very hot magma rises from within the mantle to the crust, but not necessarily at a plate boundary.

Household: one or more people living in the same dwelling.

Hydraulic action: a form of erosion involving moving water, e.g. breaking waves forcing air into rock cracks thus compressing it and cracking the rock as a result.

Hydro-electric power: the use of fast flowing water to turn turbines which produce electricity.

Hypothesis: a testable statement.

Ice age: a period in the history of the Earth when temperatures are low, large ice sheets develop and glaciers advance.

ICT (Information and communications technology): the use of electronic computers and computer software to store, protect, process, transmit and retrieve information.

Immigrant: a person who moves into an area or country.

Impermeable: does not let water pass through it, e.g. an impermeable rock.

Incineration: destruction by burning, e.g. of waste materials.

Industrialisation: the move from an economy dominated by the primary sector to one dominated by manufacturing (the secondary sector).

Infiltration: (in hydrology) the process whereby water on the ground surface enters the soil and moves downwards.

Informal economy: forms of employment that are not officially recognised, e.g. people working for themselves on the streets of LIC cities.

Interlocking spurs: the landform resulting from the winding path of a river in its upper course – the hillsides on either side of the river jut out like the teeth of a zip.

International cooperation: the process of intergovernmental collaboration to address the issues that affect them all, e.g. global warming.

International migration: migration between different countries.

Irrigation: addition of water to farmland by artificial means.

L-N

Label: to add descriptive text to a map, diagram, photograph or other resource.

Landfill: disposal of rubbish by burying it and covering it over with soil.

Leisure: use of free time for enjoyment.

Levee: the naturally raised bank that extends along a river bank, which may be artificially strengthened or heightened.

Life cycle: the process of change experienced during their lifetime. The idea is also applied to products and tourist resorts.

Life expectancy: the average number of years a person might be expected to live.

Linear settlement: where dwellings and buildings are arranged along a road, a river valley, a ridge or a stretch of coastline.

Live sustainably/simply: campaigns that stress the need to change our own habits to live in a manner that preserves our environment.

Local scale: relating to a small area, as opposed to a region or the whole country.

Location: where something or someone is to be found.

Location factor: something which affects where a particular activity is found.

Lodgement: the pressing of sub-glacial debris into the valley floor during the advance of a glacier.

Longshore drift: the movement of sand along a coast by waves.

Lower course: the final part of a river's course as it approaches the sea.

Magma: hot, molten rock in the Earth's mantle.

Mantle: the zone of molten rock below the crust.

Mass movement: the downslope movement of material due to gravity.

Mass tourism: the large-scale movement of tourists to popular destinations, most often promoted through package holidays.

Meander: a large bend formed in a river as it winds across the landscape.

Mechanisation: the replacement of human (or animal) labour with machines.

Medical tourism: seeking private medical treatment abroad.

Mercalli Scale: a scale from I to XII, used to indicate the impact and effects of an earthquake.

Metering: the use of meters allowing a service such as water supply or electricity to be charged according to how much is used.

Methane: an important greenhouse gas (which is increasingly a product of human activity)

Middle course: the central section of a river's course.

Migration: the long-term movement of people (involving a change of address).

Milankovitch mechanism: changes in the orbital geometry of the earth that cause climate change.

Model: a representation of the real world that simplifies complex patterns, trends or distributions.

Moraine: a mixture of different-sized rocks and sediments eventually left behind in the landscape when a glacier melts.

Mouth: the point where a river leaves its drainage basin as it flows into the sea.

Multiplier effect: the impact of spending in a community, f rom tourists for example, which is re-spent and recycled in that community.

National migration: migration between different areas of one country.

Natural decrease: the fall in population caused by deaths exceeding births.

N-R

Natural increase: the rise in population caused by births exceeding deaths.

Net migration (migration balance): the difference between migrant arrivals and departures. If arrivals exceed departures, the balance is positive. If departures exceed arrivals, the balance is negative.

Non-point-source pollution: contamination due to emissions from a range of locations.

Non-renewable fuels: combustible sources of energy – like coal, oil or natural gas – that cannot be 'remade', because it would take millions of years for them to form again.

Nucleated settlement: where dwellings and buildings are packed closely together – e.g. around a crossroads.

Oblique aerial photograph: aerial photograph taken from an angle (not straight down).

One-child policy: a government policy in China designed to limit the number of children per woman to one.

Orbital geometry: the way in which the Earth orbits around the sun and the variations in that orbit.

Organic farming: an environmentally friendly form of agriculture that relies on methods such as crop rotation, green manure, compost and biological pest control rather than chemical fertilisers, pesticides and GM organisms.

Package holiday: a holiday in which travel and accommodation are put together by a tour operator and sold as a relatively cheap package.

Park and ride: congestion-reducing scheme that provides parking on the edge of a city and transport to ferry people to and from the centre.

Pattern: a description of how things are distributed in an area.

Per capita (Latin for 'each head'): per person.

Pie charts: a way of presenting data that divides up a circle (the 'pie') into 'slices' according to their proportion of the total.

Player: a person or organisation involved in the planning, management or running of a particular activity.

Point-source pollution: contamination due to emissions from a single location.

Population density: the number of people per unit area (usually km^2).

Population distribution: where people are located within a given area.

Population growth rate: the increase in population over a year, expressed as a percentage.

Population pyramid: a diagram to show how a population is composed in terms of gender and age.

Preserve: to maintain (something) in its existing state.

Primary sector: the economic activities that involve the working of natural resources – agriculture, fishing, forestry, mining and quarrying.

Protect: to shield from danger or change, e.g. allowing the public to have only very restricted access to an area of high biodiversity.

Push–pull mechanism: the process encouraging migration, resulting from the combination of push factors (negative aspects of home area) and pull factors (attractive qualities of destination area).

Pyramidal peak: the jagged mountain top produced when three or four corries form on different sides of a mountain and erode backwards towards each other.

Pyroclastic flow: a high-velocity mix of high temperature gas and debris erupted from a volcano.

Quaternary sector: the economic activities that provide intellectual services – information gathering and processing, universities, and research and development.

Recreation: activity during leisure time, to help one feel better.

R-S

Recycling: processing waste materials so that they can be used again.

Redevelopment: the attempt to change an area by investing capital in its renewal.

Refugees: people who have been forced to flee an area – because of persecution or war, for example – and have had to take refuge elsewhere.

Regeneration: the investment of capital in reviving the economic, social and environmental conditions in a declining area.

Relief: the shape of the land, especially in terms of its altitude.

Remittances: (in migration) money sent home by a migrant to family members.

Remote countryside: an almost completely rural area that takes about a day to reach from a city.

Renewable energy: energy sources that are potentially infinite.

Renewable fuels: combustible sources of energy – like biofuels – that can be regrown or regenerated.

Renewal: see Regeneration.

Reservoir: an artificial lake created as part of a water supply system.

Resort: a settlement where tourism is the main function.

Resource extraction: the removal of the Earth's resources, usually by mining.

Retirement migration: a type of voluntary migration where people choose to move when they retire.

Ribbon lakes: long, narrow lakes found on the floor of U-shaped valleys.

Richter Scale: an open-ended scale indicating the strength of an earthquake, as measured by a seismograph.

River cliff: the steep bank formed by erosion of the outside of a meander bend.

Rock fall: the movement downslope under gravity of loose fragments of rock.

Rural turnaround: the revival of rural areas previously suffering from depopulation.

Rural–urban migration: the movement of people from the countryside into towns and cities.

Satellite image: a photograph taken from a satellite in space.

Scale: the relationship between dimensions on a map, diagram or model and those in the real world.

Seasonal variability: the tendency for something to change according to the time of year.

Secondary sector: the economic activities that involve making things, either by manufacturing (TV, car, etc.) or construction (a house, road, etc.). The sector also includes public utilities, such as producing electricity and gas.

Secondary sources: sources of information that already exist, such as published material, the internet and previously conducted research results.

Sector shifts: changes in the relative importance of the economic sectors (primary, secondary, tertiary and quaternary) over time.

Self-catering holiday: a holiday in which the accommodation costs do not include food.

Site: the ground occupied by a settlement.

Situation: the location of a settlement relative to its surroundings, such as to other settlements, rivers, relief features and transport lines.

Sketch map: a simple map that emphasises particular features.

Slip-off slope: the gently sloping bank on the inside of a meander bend formed by deposition.

S-V

Social change: shifts taking place in society, such as changes in lifestyles, size of families and behaviour.

Soft engineering: using environmentally friendly methods of construction and management to cope with the forces of erosion.

Solar output: the energy emitted by the sun.

Source: the starting point of a stream or river, often a spring or a lake.

Spit: an embankment of sand, extending a beach into the open water.

Stack: an isolated column of rock, standing just off the coast, that was once attached to the land.

Stump: a collapsed stack.

Subduction: the process by which one tectonic plate is dragged down beneath another by convection currents.

Suburban intensification: raising building densities by developing vacant plots and encouraging non-residential activities.

Suburban sprawl: straggling, low density and unplanned spread of housing.

Suburbanisation: the outward spread of urban areas, often at lower densities compared with the older parts of a town or city.

Sustainable development: development that meets the needs of the present without compromising [limiting] the ability of future generations to meet their own needs.

Swash: the forward movement of water up a beach after a wave has broken.

Telecentre: a public place where people can access computers and the internet to gather information and communicate with others.

Telecottaging: working from a home in the country, using computer communication.

Temporary migration: a short-term move that does not involve a change of permanent address.

Tertiary sector: the economic activities that provide various services – commercial (shops and banks), professional (solicitors and dentists), social (schools and hospitals), entertainment (restaurants and cinemas) and personal (hairdressers and fitness trainers).

Throw-away society: a wealthy society in which people tend to dispose of goods once they are finished with, rather than reusing or repairing them.

Topography: the variations in relief.

Tourism: leisure time activity involving at least one overnight stay away from home.

Transnational companies (TNCs): huge businesses that operate on a global scale.

Trend: general direction in which something is developing or changing.

Tributary: a stream or small river that joins a larger one.

Truncated spurs: steep cliff faces either side of a valley where a glacier has removed interlocking spurs, to form a wider-floored valley.

Upper course: the higher part of a river's course and its source area.

Urban fringe: countryside that is being quickly lost to urban growth.

Urbanisation: the development and growth of towns or cities that increases the percentage of the total population living in urban areas.

U-shaped valley: a valley that has been created by glacial erosion widening and deepening the original river valley.

Valley glacier: a long, narrow stream of moving ice that deepens and widens the pre-existing river valley down which it flows.

V-Z

Vegetation: plant matter.

Visa: endorsement on a passport giving permission to enter a country for a specified time.

Volcano: a place from which molten magma, gas and debris from the mantle can escape through a vent in the Earth's crust.

Voluntary migration: when a migrant chooses to move – to retire, for example.

Washlands: areas on the flood plain that are allowed to flood (a soft engineering method).

Waste: any substance or object that the holder discards intends to discard or is required to discard.

Water-borne diseases: diseases passed on by microorganisms being present in water.

Water consumption: the amount of water used by a person or group of people.

Water deficit: a situation in which the usable water supply does not satisfy the demand.

Water pollution: the presence of contaminating and sometimes dangerous materials within a water body.

Water surplus: a situation in which the usable water supply exceeds the demand.

Waterfall: a vertical or near-vertical fall in a river's course.

Watershed: the boundary of a drainage basin – dividing it from other drainage basins.

Weathering: the breakdown and decay of rock by natural processes, without the involvement of any moving forces.

Youthful population: a population in which there is a high percentage of people under the age of 16 (or sometimes 19).

Zero population growth: when natural change and migration change cancel each other out, and there is no change in the total population.

Index

A-D

A

B

C

D

D-G

G-M

M-R

R-W

W-Z

Acknowledgements continued from page 2.

We are grateful to the following for permission to reproduce copyright material:

Figures

Figure 1.4 adapted from www.oxfordcartographers.com, copyright by Aademische Verlagsanstalt FL 9490, Vaduz, Auelestr 56. English version by Oxford cartographers, Oxford, UK. www.oxfordcartographers.com, copyright © Oxford Cartographers; Figure 1.8 adapted from 'How contours show concave and convex slopes', copyright © Prewitt & Company; Figure 2.1 adapted from Geophysical Research Letters (Mann et al 1999), copyright © 1999 American Geophysical Union. Reproduced/modified by permission of American Geophysical Union; Figure 2.7 adapted from Map "Top 10 Wheat Importers and Exporters, 2007 (%)", copyright © Food and Agricultural Organization of the United Nations; Figure 2.9 adapted from http://www.bdix.net/sdnbd_org/world_env_day/2002/current_issues/sea_label_rise/middlesex-univ.htm, copyright © Sustainable Development Networking Programme (SDNP); Figure 2.10 adapted from "Population density in areas below 10 metres elevation in Bangladesh, and above 10 metres", http://earthobservatory.nasa.gov/Features/GlobalWarming/global_warming_update6.php, copyright © NASA's Earth Observatory; Figure 2.11 adapted from Sustainability & Cities, Island Press (Newman, P. and Kenworthy, K. 1999) copyright © 1999 by Peter Newman & Jeffery Kenworthy. Reproduced by permission of Island Press, Washington, D.C.; Figure 2.13 adapted from Beyond the Limits: Confronting Global Collapse, Envisioning a Sustainable Future (Meadows, et al 1992), copyright © Donella Meadows, with permission of Chelsea Green Publishing (www.chelseagreen.com); Figure 3.1 adapted from http://cgz.e2bn.net/e2bn/leas/c99/schools/cgz/accounts/staff/rchambers/GeoBytes%20GCSE%20Blog%20Resources/Images/Coasts/Destructive_Waves.jpg, Figure 3.2 adapted from http://cgz.e2bn.net/e2bn/leas/c99/schools/cgz/accounts/staff/rchambers/GeoBytes%20GCSE%20Blog%20Resources/Images/Coasts/Constructive_Waves.jpg, and Figure 3.4 adapted from http://cgz.e2bn.net/e2bn/leas/c99/schools/cgz/accounts/staff/rchambers/GeoBytes%20GCSE%20Blog%20Resources/Images/Coasts/Headlands_and_Bays.jpg, copyright © Rob Chambers; Figure 3.3 adapted from www.tulane.edu/~sanelson/geol111/slump.gif, copyright © Professor Stephen A. Nelson; Figure 3.5 adapted from "Landforms produced by the erosion of a headland", http://www.georesources.co.uk/sea6.gif, copyright © David Rayner (GeoResources); Figure 3.7 adapted from "Longshore Drift", http://geographyfieldwork.com/LongshoreDrift.htm, reproduced by permission of Barcelona Field Studies Centre http://geographyfieldwork.com; Figure 3.9 adapted from 'The bar and lagoon at Slapton in Devon', http://www.le.ac.uk/bl/gat/virtualfc/217/images/ley2.jpg, copyright © Dr E. Gaten, University of Leicester; Figure 3.12 adapted from 'Map of Swanage Bay and Durlston Bay in Dorset', Ordnance Survey, copyright © www.collinsbartholomew.com Ltd, reproduced with kind permission of HarperCollins Publishers; Figure adapted 4.5 from http://cgz.e2bn.net/e2bn/leas/c99/schools/cgz/accounts/staff/rchambers/GeoBytes%20GCSE%20Blog%20Resources/Images/Rivers/ox-bow_lake.gif, copyright © Rob Chambers; Figure 4.10 adapted from http://www.swenvo.org.uk/environment/images/Flooding_properties_1996_2006.gif, copyright © Environment Agency; Figure 4.11 adapted from Tomorrow's Geography John Murray (Harcourt, M., Warren, S., and Warn, S 2001), reproduced by permission of John Murray (Publishers) Ltd; Figure 4.13 adapted from Edexcel GCSE Geography, paper 1312/4H (Figure 1a) June 2006 copyright © Edexcel Limited; Figure 5.10 adapted from www.geographyhigh.connectfree.co.uk, by Principle Knox originally from Physical Geography in Diagrams, Longman (Ron B Bunnett 1977) copyright © Pearson Education Ltd; Figure 5.13 adapted from The Nature of the Environment, 4th ed. (Goudie, A, S. 2001), copyright © Blackwell Publishing; Figure 5.17 adapted from figure 1 Annual (1 October–30 September) avalanche fatalities in Russia as a whole and in the northern Caucasus, for the period 1996–2007 in "Assessment and mapping of snow avalanche risk in Russia", Annals of Glaciology 49 2008 (Yuri Seliverstov, Tatiana Glazovskaya, Alexander Shnyparkov, Yana Vilchek, Ksenia Sergeeva, Alexei Martynov) Research Laboratory of Snow Avalanches and Debris Flows, Faculty of Geography, M.V. Lomonosov Moscow State University; Figure 6.2 adapted from Smithsonian Institution, http://earth.rice.edu/MTPE/geo/geosphere/hot/volcanoes/volcanoes_map.gif, copyright © Smithsonian Institution, Global Volcanism Program; Figure 6.3 adapted from http://www.physicalgeography.net/fundamentals/images/tectconvection.gif, Figure 6.5 adapted from Figure 10i-2 "Creation of oceanic crust on the ocean floor", Figure 6.6 adapted from Figure 10i-6 "Collision of a oceanic plate with a continental plate", and Figure 6.7 adapted from Figure 10i-5: "Collision of two oceanic plates" http://www.physicalgeography.net/fundamentals/10i.html, US Geological Survey; Figure 6.9 adapted from Explorevolcanoes, http://explorevolcanoes.com; Figure 6.10 adapted from National Earthquake Information Center (NEIC), http://earthquake.usgs.gov/eqcenter/recenteqsus/, U.S. Geological Survey; Figure 6.11 adapted from Diagram showing the focus and epicentre of an earthquake, www.usgs.gov, US Geological Survey; Figure 6.13 adapted from Long Valley Web Team, http://lvo.wr.usgs.gov/zones/TephraFall.gif, USGS Long Valley Volcano Observatory; Figure 7.5 adapted from INCPEN, "Towards greener households", June 2001 http://www.wasteonline.org.uk/resources/InformationSheets/Packaging.htm, copyright © INCPEN; Figure 7.10 adapted from "Why do we need to reduce the amount of